E. Fitzer · L. M. Manocha
Carbon Reinforcements
and Carbon/Carbon Composites

Springer-Verlag Berlin Heidelberg GmbH

E. Fitzer · L. M. Manocha

Carbon Reinforcements and Carbon/Carbon Composites

With 236 Figures

Springer

Professor Dr. E. Fitzer
Institut für Chemische Technik der Universität Karlsruhe
Kaiserstraße 12
76131 Karlsruhe
Germany

Professor Dr. Lalit M. Manocha
Insitute of Materials Science
Sadar Patel University
Vallabh Vidyanagar 338120
Gujarat
India

ISBN 978-3-642-63707-0

Library of Congress Cataloging-in-Publication Data
Fitzer, Erich, 1921 – Carbon reinforcements and carbon/carbon composites /
E. Fitzer, L.M. Manocha . p. cm.
Includes bibliographical references and index.
ISBN 978-3-642-63707-0 ISBN 978-3-642-58745-0 (eBook)
DOI 10.1007/978-3-642-58745-0
1. Carbon composites. 2. Carbon. I. Manocha, L.M. (Lalit M.), 1951 – . II. Title.
TA 418.9.C6F567 1988 620. 1'93--dc21

Originally published by Springer-Verlag Heidelberg New York in 1998
Softcover reprint of the hardcover 1st edition 1998

Production Editor: Christiane Messerschmidt, Rheinau
Typesetting: Fotosatz-Service Köhler OHG, Würzburg
Cover design: Design & Production, Heidelberg

SPIN: 10082193 2/3020 – 5 4 3 2 1 0 – Printed on acid-free paper

Preface

Advanced composite materials have been a major research topic for the past four decades. Conventional materials including glass, ceramics, and metals have been investigated and refined for composites applications. As a reinforcement and a matrix for these materials, carbon has proven to be the most successful. The atomic and chemical configuration of carbon give it the flexibility of producing materials in bulk form with a wide variety of properties. Coated, uncoated, or surface-modified carbon fibers are used to reinforce polymers, metals, and ceramics. Carbon/Carbon composites have been the most effective materials for extreme temperature conditions. Applications of carbon/carbon composites have ranged from brake linings to missile nose cones. Worldwide interest in carbon/carbon composites has advanced basic carbon science and identified future research needs. Finally, it has also generated substantial literature in the form of research papers in scientific and technical journals, conference proceedings, and technical reports, etc. Many of these deal with highly specialized facets of carbon/carbon research; some are even classified.

To encourage new applications of carbon/carbon composites and facilitate further research, this information needs to be made available in comprehensive form. Therefore this book is intended to provide the broad scientific community with information about the present state of carbon fibers and carbon/carbon composites research, as well as projections for other "high tech" materials and their applications.

The first part of this book reviews recent developments in carbon fibers. Processing, structure, and properties of different types of carbon fibers derived from PAN, pitch, and hydrocarbon vapors are described. Surface modification of carbon fibers through chemicals and coating techniques are described in the context of fiber/matrix bonding in composites. The role of reinforcement fibers in determining the unique properties of composite materials is detailed.

The second part of the book is devoted to those composites which have the same thermal stability as carbon fibers, the carbon-fiber-reinforced carbon and graphite or so called carbon/carbon composites. Various aspects of carbon/carbon technology from fiber architecture to processing of unidirectional and multidirectional composites, microstructure, properties, and applications of the products are discussed. Finally, the principal drawback to the use of carbon/carbon composites is addressed: although these composites exhibit the highest thermal stability with the best mechanical properties, their long-term applications are limited to non oxidizing environments. Coatings which inhibit oxidation and which make the use of carbon/carbon composites practical for a wide range of application are therefore reviewed.

The project of writing the book was initiated by Professor Erich Fitzer and if was planned that we write it together. Unfortunately Professor Fitzer suddenly became ill and was unable to continue. In order to disseminate the vast scientific knowlegde he had already invested and the high value of information on material development generated worldwide, it was left to me to complete the project as a tribute to Professor Fitzer. Many friends and admirers of Professor Fitzer provided all the support I needed to complete the book.

I am especially indebted to Professor Dan Edie, Clemson University, USA for reviewing and supplementing the first part of the book on carbon fibers. In addition, I gratefully acknowledge his suggestions and material for the second part on carbon/carbon composites. It has been with his kind cooperation that the book was able to be prepared in the present form.

Furthermore I would like to thank Professor Robert A. Meyer, University of California, USA, Dr. Maurice A. Wright, Materials Technology Center, Southern Illinois University, USA, Professor Peter Thrower, Penn State University, USA, Dr. P. Ehrburger, CNRS, Mulhouse, France, Dr. Karl-Heinz Köchling, Institut für Chemische Technik, Karlsruhe, Germany and Professor Eiichi Yasuda, Tokyo Institute of Technology, Japan for their support in various forms.

I am grateful to Professor Fitzer family's for negotiating with Springer-Verlag and for providing technical information from Professor Fitzer's library. Thanks are due to Mr. S.M. Dhawan, the Scientist-in-charge of the Library at the National Physical Laboratory, New Delhi for encouragement and providing literature needed for this venture.

I am also grateful to my family for their encouragement and support and especially to my colleague and wife, Dr. Satish Manocha who, in addition, contributed to the technical part of the book as well as to the Institute management while I was busy in writing the book.

June 1997 L. M. Manocha

Contents

Reinforcement

1 Carbon Fibers

1.1 Introduction

Today, carbon fibers are, by far, the main high strength, high modulus reinforcing material used in the fabrication of high performance composites. The need for these carbon fiber reinforced structural materials originated shortly after World War II, as military aircraft manufacturers in the USA found that existing materials were limiting the performance of their designs. Initially, boron fibers were developed to reinforce plastics, yielding the first "advanced structural composites" [1, 2]. Even though these fibers were extremely strong, stiff, and relatively light compared to existing metals, they had numerous drawbacks: (i) their large diameter, 100–150 μm, made the fibers difficult to handle; (ii) since the loop strength of the fibers was quite low, they could be used only as monofilaments; (iii) the outer shell of the fiber was boron and the core was a different material, creating thermal mismatch problems; (iv) most importantly, these boron fibers were extremely expensive.

Since the early 1950s many varieties of reinforcing fibers have been developed. Glass fibers are often considered the baseline for comparing reinforcing fibers for polymer composites [3, 4]. Although the strength of early varieties of glass fiber was much lower than that of many other reinforcing fibers, they could, however, be manufactured using relatively simple technology and, thus, sold at a much lower price. During the past 20 years, improved quality glass has been developed, allowing the strength of glass fibers to be increased to the extent that they are a viable reinforcing material for composites. The use of these low-cost, relatively high-strength fibers has permitted the evolution of a spectrum of techniques for fabricating composite materials with a variety of contours and shapes [5]. Nevertheless, often the application temperature or other mechanical deficiencies, such as their low specific stiffness, limits the use of even these newer varieties of glass fiber. Since the brittleness of glass fibers prevents their use

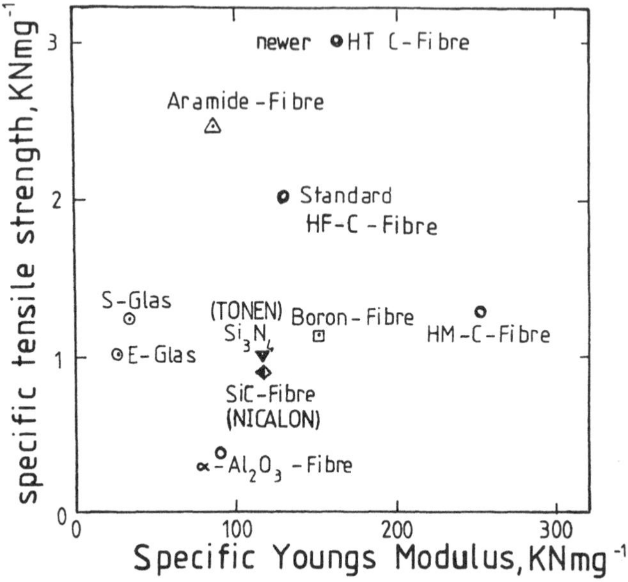

Fig. 1.1. Specific mechanical properties of reinforcing fibers at room temperature

in certain applications, organic fibers have been developed where high strength and impact resistance are the prime requirements. However, because of their unique combination of properties, as well as their potential for additional improvement, today carbon fibers are the overwhelming choice for aircraft and aerospace composite applications.

The strength and stiffness of carbon fibers is superior to that of all other reinforcing fibers. The specific properties (tensile strength and moduli divided by density) of various fibers are compared in Fig. 1.1 [6]. Note specifically the newer high-tenacity carbon fibers, commonly designated as HT, which are slightly more stiff than standard type A carbon fibers and exhibit specific tensile strengths which are significantly higher than all other reinforcing fibers. These newer HT carbon fibers, also termed IM (intermediate modulus) carbon fibers have been commercially available for several years. Another variety, HM (high-modulus) carbon fibers, surpasses all other reinforcing fibers in terms of specific stiffness. Figure 1.1 also includes the more exotic types of fibers, such as SiC, Si_3N_4, Al_2O_3 and boron since they are competitive reinforcements for metallic-matrix and ceramic-matrix composites.

For fiber-reinforced carbon matrix composites, the stability and the retention of properties at high temperatures are primary requirements for

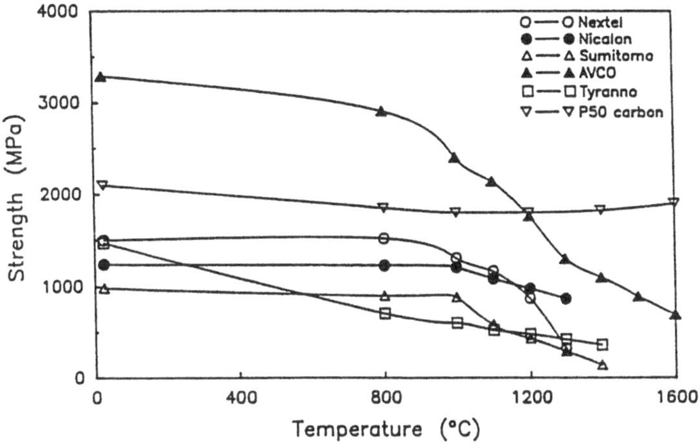

Fig. 1.2. The effect of temperature on the tensile strength of various commercial reinforcing fibers [7]

reinforcing fibers. Here again, as is evident in Fig. 1.2, carbon fibers out do all other high performance fibers [7]. Thus, carbon fibers can be regarded as the fiber of choice for reinforcing all classes of composite materials (polymer-matrix, metal-matrix and ceramic-matrix).

1.2 Key to the Strength and Stiffness of Carbon Fibers

The unique properties of all carbon fibers can be directly attributed to the highly anisotropic nature of the graphite crystal. The crystallographic structure of graphite is shown in Fig. 1.3 [6]. The graphite crystal is composed of stacks of multiple sheet-like layers of carbon atoms (graphene layers) which are separated by a distance of 0.335 nm. In the plane of the sheets, the carbon atoms are linked by covalent bonds with a strength of 400 kJ/mol which are created by the sp^2 hybridization of the electron orbitals. These high strength bonds make the graphite structure extremely strong and stiff in the crystallographic a direction. In fact, when loaded in this crystallographic direction, the theoretical tensile modulus of graphite is 1060 GPa, and tensile strengths of 106 GPa should be possible. On the other hand, in the direction normal to these basal planes, the crystallographic c direction, these carbon layers are held by weak van der Waals bonds, explaining the low tensile modulus of graphite (36.5 GPa) when graphite is loaded in this direction [8].

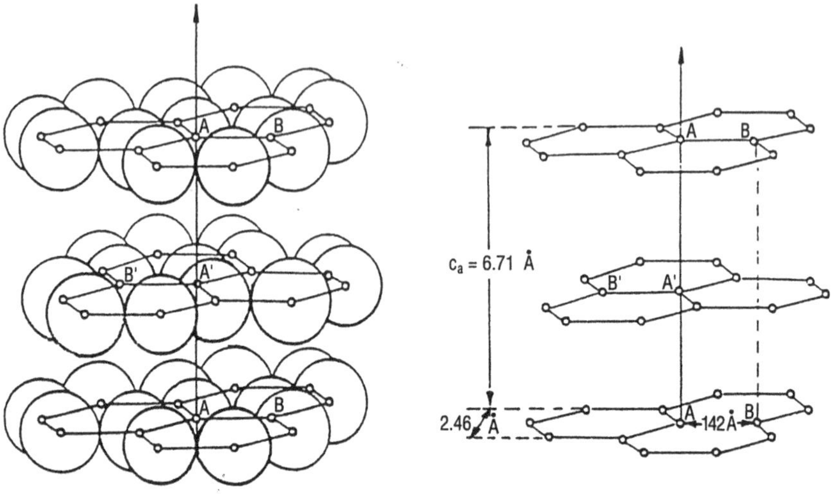

a b

Fig. 1.3 a, b. The anisotropic structure of graphite. **a** The densely packed graphite layer structure. **b** Hexagonal lattice structure of graphite

Although it is the high elastic constant in direction parallel to the graphene layers of the graphite crystal which have been the driving force for developing carbon fibers, the mechanical properties of the fibers are often limited by the low transverse properties of the graphite structure. Nevertheless, the theoretical properties of graphite provide an irresistible target for continuing efforts to improve the properties of carbon fibers.

If it was possible to make a hypothetical carbon fiber from infinitely large polyaromatic layers of elemental carbon arranged in the ideal graphite structure, it would fail in compression because of the weak bond between the graphene layers. This can be quantitatively described in terms of the difference in the elastic constants of the graphite single crystal. In particular, the value of C_{11} for graphite crystal is high compared with that of C_{33}, a direct result of the high bond strength in the crystallographic a direction relative to the low bond strength in the crystallographic c direction. The low shear strength between the graphene layers in structurally perfect graphite, as described by elastic constant C_{44}, would be expected to deteriorate the compressive properties of this hypothetical carbon fiber even more. As Fig. 1.4 illustrates, the low shear resistance of crystalline graphite allows it to be easily bent, in spite of its high strength in the direction parallel to the surface [9].

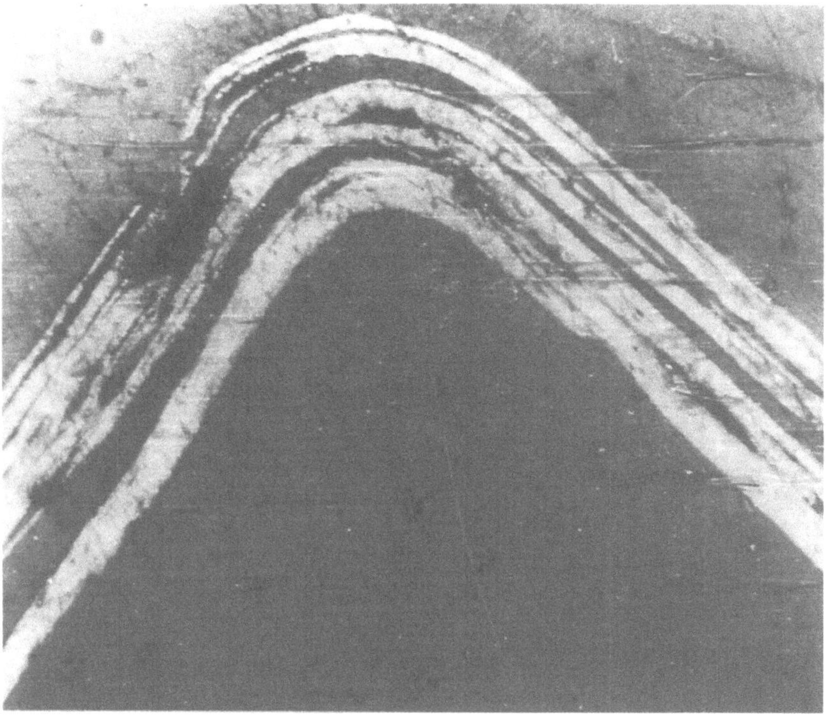

Fig. 1.4. Bent single crystal of graphite [6]

Considerable carbon fiber research has been directed toward finding ways to overcome this natural shear sensitivity of the graphite structure. Perhaps the most promising approach involves inhibiting shear failure by modifying the molecular structure of the carbon fiber. Instead of infinitely large crystallographic graphitic layers, researchers have found that small-sized polyaromatic layers, containing disclinations and deviating from perfect preferred orientation, lead to improved tensile and compressive properties in carbon fibers. One such structure, termed turbostratic, is shown in Fig. 1.5 [10]. Since the molecular orientation of fibers with a turbostratic structure is still more or less parallel to the fiber axis, the axial properties of the fiber are still a reflection of the covalent bonding parallel to the basal planes within the fiber. Therefore, properties which directly reflect molecular orientation, such as modulus and thermal conductivity, will tend to increase as the degree of alignment of the basal planes parallel to the fiber axis increases. However, both the tensile and the compressive strengths of the fiber will tend to decrease as the size of the graphitic layers increases. The implications of this are discussed in subsequent sections.

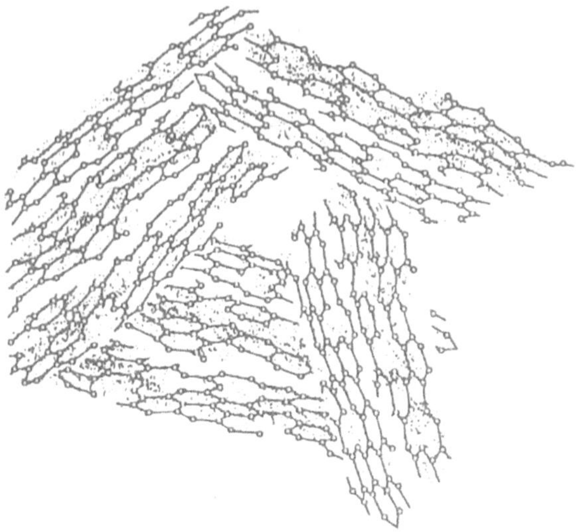

Fig. 1.5. The turbostratic structure of graphite

1.3
Short History of Carbon Fiber Development

Carbon fibers were first produced by Edison in the late 19th century [6]. After initial trials using bamboo fibers, Edison found that regenerated cellulose (rayon) could be converted to carbon filaments for use in incandescent lamps. In the early 1960s the Union Carbide Corporation (UCC) used the same precursor, rayon, to produce the first commercial carbon fiber. Initially, two varieties of this first carbon fiber were marketed by Union Carbide, termed VYB and WYB. Since the VYB variety was heat treated at only 1000 °C, it contained a small quantity of non-carbon elements. Structural studies showed that these VYB carbon fibers exhibited the fibrillar structure of the rayon precursor and were porous. In addition, X-ray analysis showed that their structure was that of isotropic and nongraphitizing carbon [11]. The WYB variety was heat treated at 2500 °C and, thus, consisted of pure carbon. However, like all carbon fibers formed from polymer fiber precursors, their structure was nongraphitizing. Nevertheless, the WYB carbon fiber was described as a "graphite fiber" when it was introduced to the market. This early, incorrect use of the term graphite to describe a fiber that had been heat treated at a high temperature to convert it to nearly pure carbon was unfortunate and continues to cause confusion today.

1.3.1 Early High Performance Carbon Fibers

The story of high performance carbon fibers began in the mid 1960s when the UCC demonstrated that isotropic carbon fibers could be converted to a high modulus product. This was accomplished by stretching the isotropic fiber by as much as 100% of its original length during a high temperature heat treatment. Stretching at an elevated temperature resulted in plastic deformation, in turn orienting the basal planes in the direction of the applied stress, parallel to the fiber axis. This increased the strength and stiffness of the carbon fiber considerably [12, 13]. Using this hot-stretching technique, tensile moduli as high as 500 GPa were obtained. Initially, three commercial types of these new high performance carbon fibers were marketed by UCC – Thornel 25, 50 and 75 – where the number indicated the fiber modulus in Msi.

Another technical and commercial breakthrough for high performance carbon fibers occurred in the mid 1960s – the development of carbon fibers produced from polyacrylonitrile (PAN) precursor fibers. This process proved to be more economic due to the lower cost of the PAN precursor fiber and the simpler process needed to convert PAN fiber to carbon fiber. The use of PAN as a precursor for carbon fibers was first reported by Shindo [14], a Japanese researcher. However, the carbon fibers produced in Shindo's initial trials exhibited only moderate tensile strengths and moduli (550-700 MPa and 170 GPa, respectively), partly because of inadequate molecular orientation within the fibers. Independent of this Japanese research effort, Phillips, Watt, and Johnson [15] of the Royal Aircraft Establishment in Farnborough, UK, developed the first successful commercial process for converting PAN precursor fibers to carbon fibers. In the British process, the change in length of the fibers was controlled during their thermal treatment, greatly improving the mechanical properties of the final carbon fiber. A continuous version of this same process was later commercialized by Morganite, Ltd., and Courtaulds, Ltd. As a result, in 1966 the first handleable, high performance carbon fibers were introduced to the market. These British PAN-based carbon fibers were designated as type I (similar to the current HM classification) and type II (similar to the HT classification). Since 1966, numerous patents have been granted for producing carbon fibers from a PAN precursor. Basically all of these processes involve first the stretching of the PAN precursor fiber to orient the polymer molecules parallel to the fiber axis and then stabilizing this oriented structure under tension in an oxidizing environment. Finally, after stabilization the fibers are heat treated at high temperatures, ranging from 1000 °C to

3000 °C, in an inert environment. As shown in Fig. 1.6, over the last 25 years intensive research and development efforts have continuously improved the mechanical properties of PAN-based carbon fibers. However, since PAN-based carbon fibers are nongraphitizing, researchers realized that the potential for improving certain properties, such as tensile modulus and thermal conductivity, was limited. Therefore, interest grew in developing a truly graphitic fiber. By the 1960s it was realized that synthetic carbons formed from liquid crystal pitch materials could develop a true graphite structure. It was this discovery that led Otani, in 1965 to begin investigating pitch as a precursor material for carbon fibers [16]. Since then, research on both forming the pitch precursor and converting the precursor to carbon fibers has been continuous [17–19]. In fact today more research groups are working on pitch-based carbon fibers than on PAN-based carbon fibers. As a result, a commercial carbon fiber formed from liquid crystalline pitches is now available with an ultra-high modulus, 850 GPa (very close to the theoretical modulus). Pitch-based carbon fibers are discussed in detail later in this chapter.

Until the 1970s, the use of carbon fibers was limited to applications, such as aerospace and sporting goods, where the volume was limited and cost was not an issue. However, the production capacity, the market volume, and the product cost for materials are intimately related. Therefore, it was only after producers anticipated broader applications for carbon fibers in non-aerospace fields, such as machinery and automobiles, that they increased capacity sharply (Fig. 1.7a), a decision which partly accounted for the simultaneous decrease in the price of carbon fibers (Fig. 1.7b) [20]. However, this decrease in price was also the result of the introduction of different varieties of carbon fibers specifically for these new applications. While the higher cost, small carbon fiber tows (3k, 6k, and 12k) are produced specifically for aerospace and other high-tech applications, the larger, lower cost tows (40k and greater) are made for non-aerospace markets. Although the price of carbon fibers has decreased significantly over the last 20 years, intensive research and development worldwide should continue to reduce their cost even further. Continued cost reduction is critical if carbon fibers, particularly pitch-based carbon fibers, are to be used in high volume, low cost applications.

1.3.2 Japanese Dominance of Carbon Fiber Market

Today, the world carbon fiber market is dominated by Japanese companies. While one might interpret this as a direct result of the first Japanese

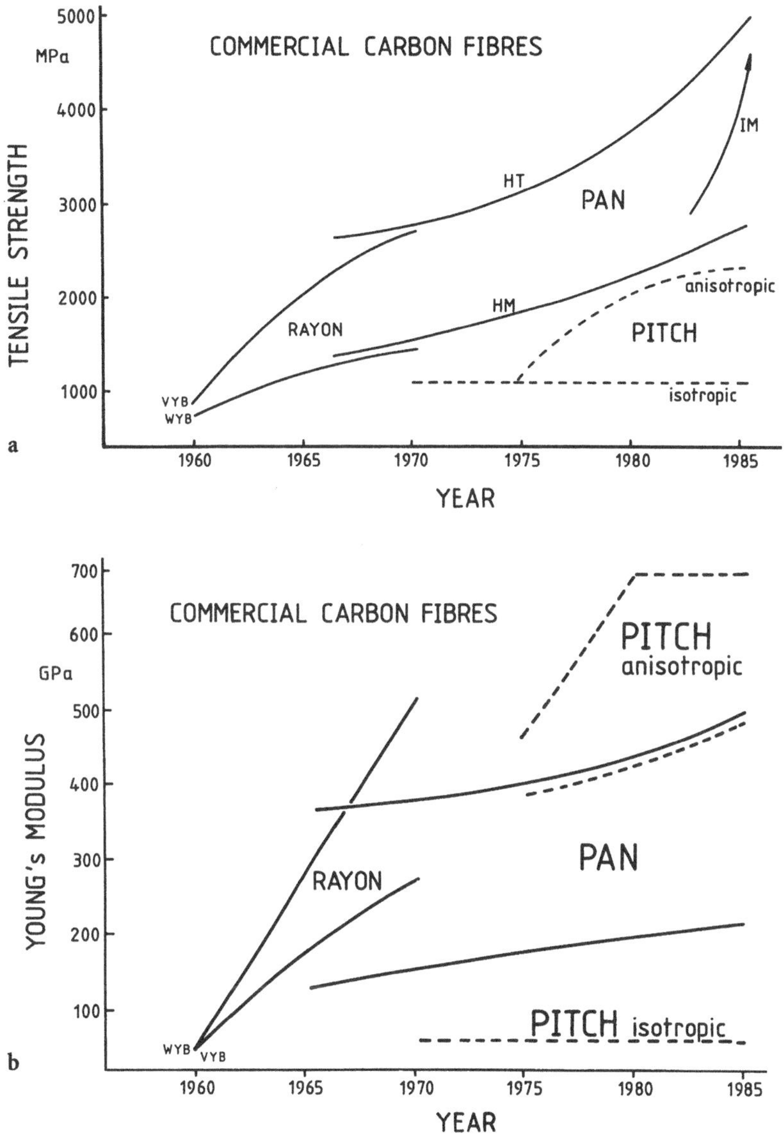

Fig. 1.6 a, b. Improvements in the properties of carbon fibers over the past 20 years. **a** Increase in the tensile strength of carbon fibers. **b** Increase in the modulus of carbon fibers

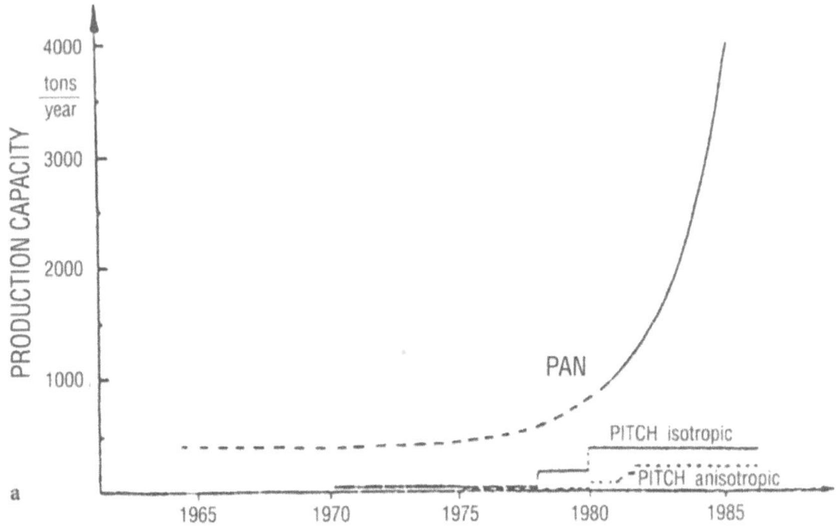

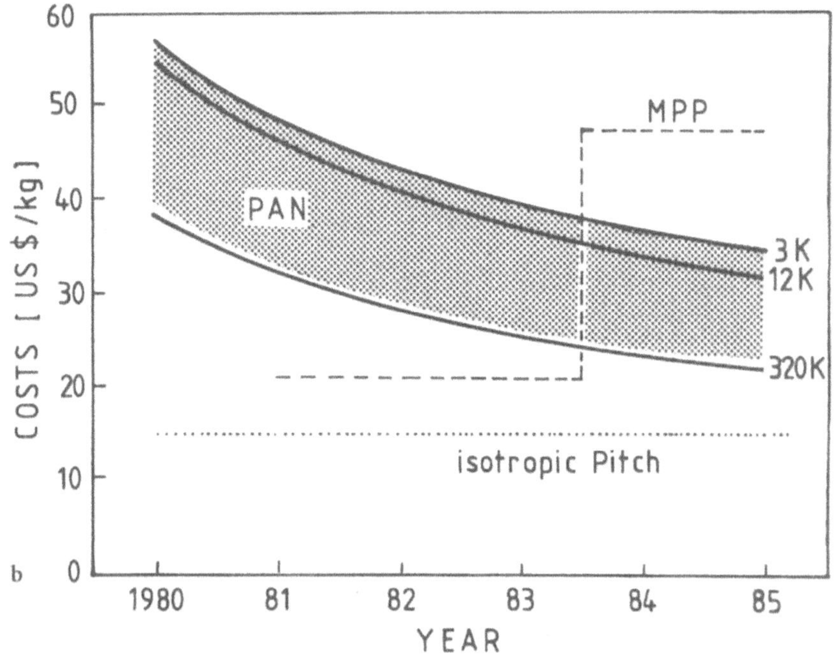

Fig. 1.7 a, b. Production and price of carbon fibers over the past 20 years. **a** World production capacity of carbon fibers. **b** Price evolution for carbon fibers based on polyacrylonitrile (PAN), mesophase pitch (MPP) and isotropic pitch

patent [21], issued in 1960 (Fig. 1.8), the technical reason for this domination can be explained by the PAN textile fiber processes used in various countries. In the 1960s, Japanese industry was using primarily a wet-spinning process with inorganic solvents to produce PAN textile fibers. By contrast, in the United States and Germany either the original dry-spinning process or the newer wet-spinning process, utilizing organic solvents, were used in the production of PAN textile fibers. Research has shown that PAN fibers which are wet spun from inorganic solvents are, in many respects, superior precursors for the production of carbon fibers. In fact, the ideal PAN precursor is actually a copolymer. The reason for this is that, as shown in Fig. 1.9, the exotherms developed during oxidative stabilization of some PAN copolymers are more controllable than those of PAN homopolymers [22]. In these early years, no industry developed a special precursor specifically for the production of carbon fibers. Instead, developing a carbon fiber product depended on the availability of a suitable precursor. Thus, shielded by their patent umbrella, several innovative Japanese companies with access to the best wet-spun PAN copolymers, began producing carbon fibers as soon as the initial market in the United States became obvious.

Industry in the UK found itself in a similar, fortunate situation. The British company Courtaulds also wet spun PAN fibers from inorganic solvents. Their PAN fiber was a copolymer, consisting of mostly acrylonitrile and a small amount of itaconic acid. This PAN copolymer had a major disadvantage as a textile fiber – it yellowed when heated slightly in air. However, this deficiency actually proved to be a distinct advantage when it was used as a precursor fiber for carbon fiber. The yellowing in air was an indication of the copolymer's tendency to catalyze the cyclization of PAN, a characteristic which made the copolymer an excellent carbon fiber precursor. Courtaulds was, until recently, the only company which freely marketed a suitable PAN precursor fiber. In fact, researchers as well as many producers of carbon fibers, who were not licensees of the Japanese patents depended on this raw material from Courtaulds.

Another key patent issued to British researchers (Fig. 1.10) demonstrated that, during stabilization, preoriented PAN must be held at a constant length [23]. Unfortunately, the priority of the Japanese PAN patent and prior art detailed in a patent on inflammable PAN fibers weakened the protection offered by this key patent. In 1972 one of the leading developers of carbon fibers in the UK, Rolls Royce, encountered severe problems with the turbine blades of a jet engine that they were supplying to Lockheed, contributing to the eventual bankruptcy of the company. Combined with other

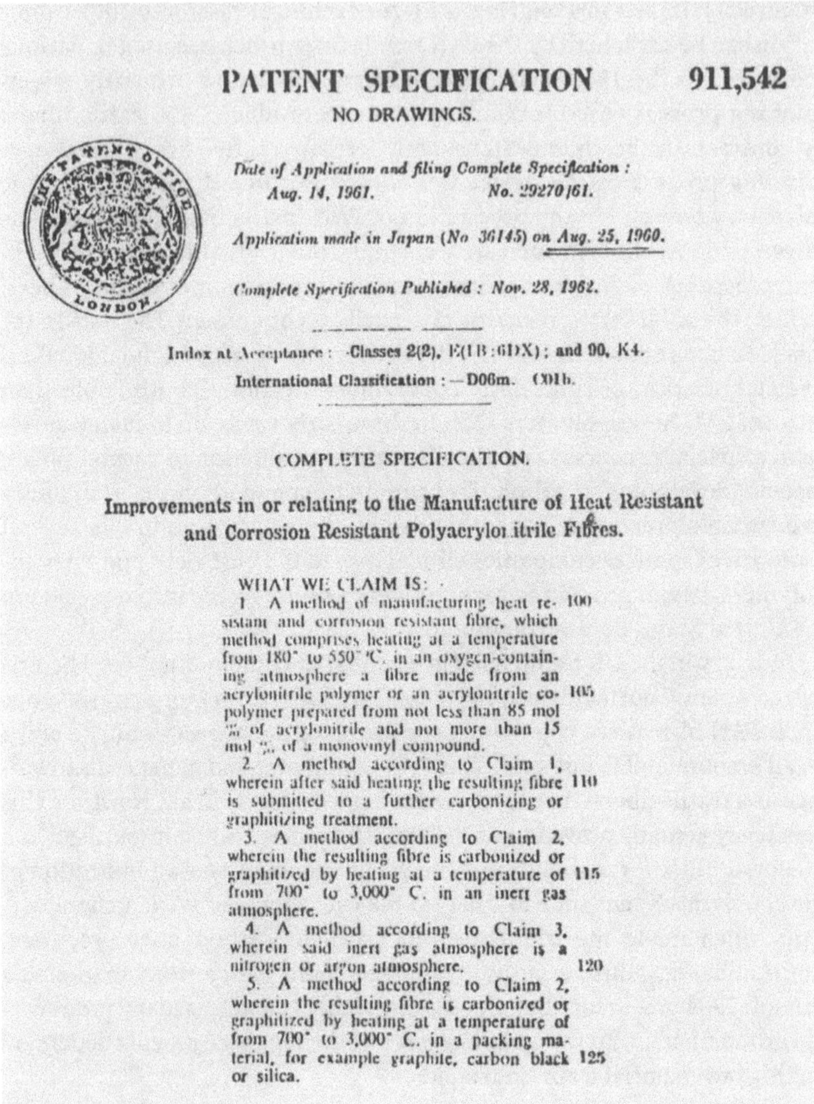

PATENT SPECIFICATION 911,542

NO DRAWINGS.

Date of Application and filing Complete Specification: *Aug. 14, 1961.* *No. 29270/61.*

Application made in Japan (No 36145) on Aug. 25, 1960.

Complete Specification Published: Nov. 28, 1962.

Index at Acceptance: -Classes 2(2), E(1B:6DX); and 90, K4.

International Classification: —D06m. C01b.

COMPLETE SPECIFICATION.

Improvements in or relating to the Manufacture of Heat Resistant and Corrosion Resistant Polyacrylonitrile Fibres.

WHAT WE CLAIM IS:

1. A method of manufacturing heat resistant and corrosion resistant fibre, which method comprises heating at a temperature from 180° to 550° °C. in an oxygen-containing atmosphere a fibre made from an acrylonitrile polymer or an acrylonitrile copolymer prepared from not less than 85 mol % of acrylonitrile and not more than 15 mol %, of a monovinyl compound.

2. A method according to Claim 1, wherein after said heating the resulting fibre is submitted to a further carbonizing or graphitizing treatment.

3. A method according to Claim 2, wherein the resulting fibre is carbonized or graphitized by heating at a temperature of from 700° to 3,000° C. in an inert gas atmosphere.

4. A method according to Claim 3, wherein said inert gas atmosphere is a nitrogen or argon atmosphere.

5. A method according to Claim 2, wherein the resulting fibre is carbonized or graphitized by heating at a temperature of from 700° to 3,000° C. in a packing material, for example graphite, carbon black or silica.

Fig. 1.8. Fundamental Japanese patent covering the production of carbon fibers from PAN

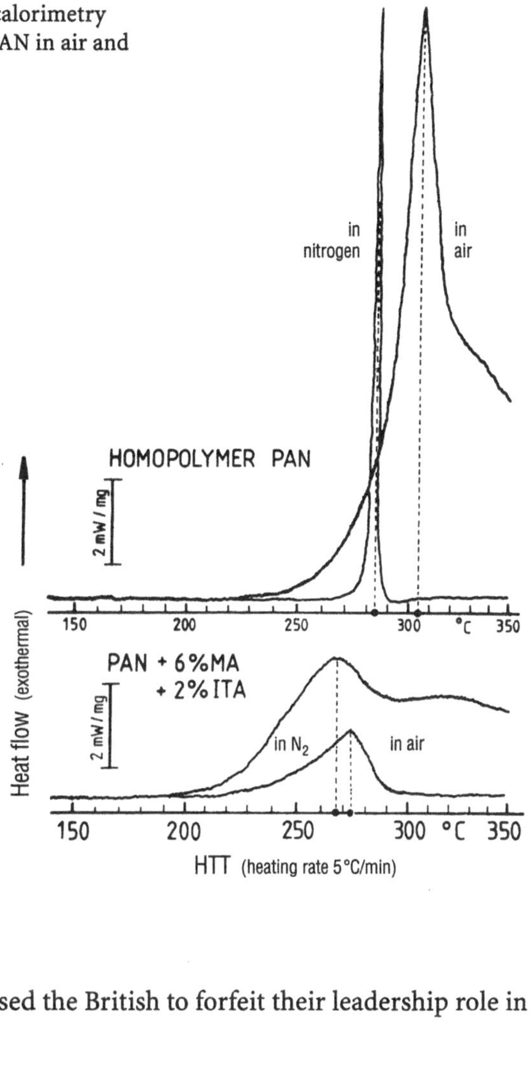

Fig. 1.9. Differential scanning calorimetry analysis of the stabilization of PAN in air and in nitrogen [20]

problems, these events caused the British to forfeit their leadership role in carbon fiber development.

1.4 Carbon Fiber Process

Although many different precursor materials can be used to produce carbon fibers, only three – rayon, PAN, and pitch – are considered suitable for commercial processes. A schematic comparison of these three commercial routes for producing carbon fibers is shown in Fig. 1.11 [20]. Precursor yield (kilogram of carbon fiber per kilogram of precursor fiber), a vital

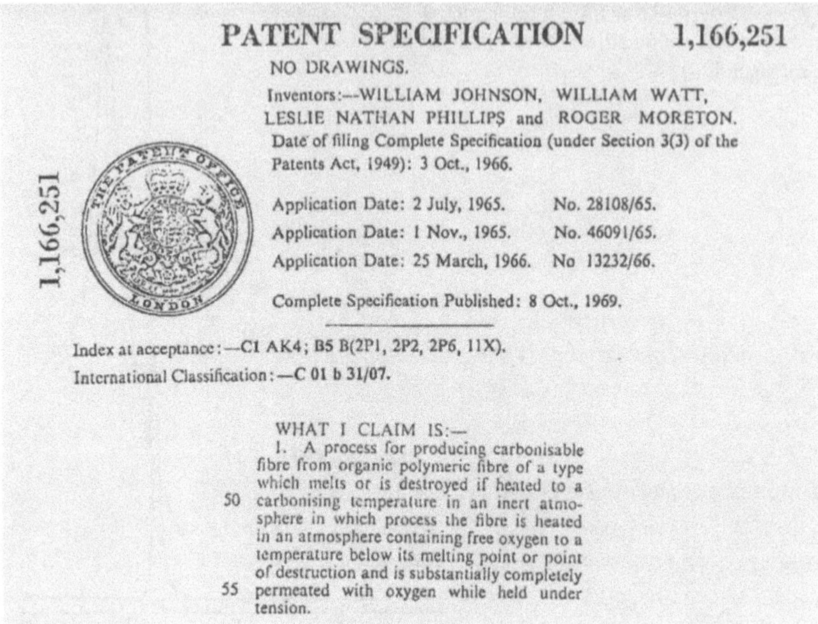

PATENT SPECIFICATION 1,166,251

NO DRAWINGS.

Inventors:—WILLIAM JOHNSON, WILLIAM WATT, LESLIE NATHAN PHILLIPS and ROGER MORETON.

Date of filing Complete Specification (under Section 3(3) of the Patents Act, 1949): 3 Oct., 1966.

1,166,251

Application Date: 2 July, 1965. No. 28108/65.

Application Date: 1 Nov., 1965. No. 46091/65.

Application Date: 25 March, 1966. No 13232/66.

Complete Specification Published: 8 Oct., 1969.

Index at acceptance:—C1 AK4; B5 B(2P1, 2P2, 2P6, 11X).

International Classification:—C 01 b 31/07.

WHAT I CLAIM IS:—

1. A process for producing carbonisable fibre from organic polymeric fibre of a type which melts or is destroyed if heated to a carbonising temperature in an inert atmosphere in which process the fibre is heated in an atmosphere containing free oxygen to a temperature below its melting point or point of destruction and is substantially completely permeated with oxygen while held under tension.

Fig. 1.10. British patent covering stabilization under stress to produce carbon fibers with improved mechanical properties

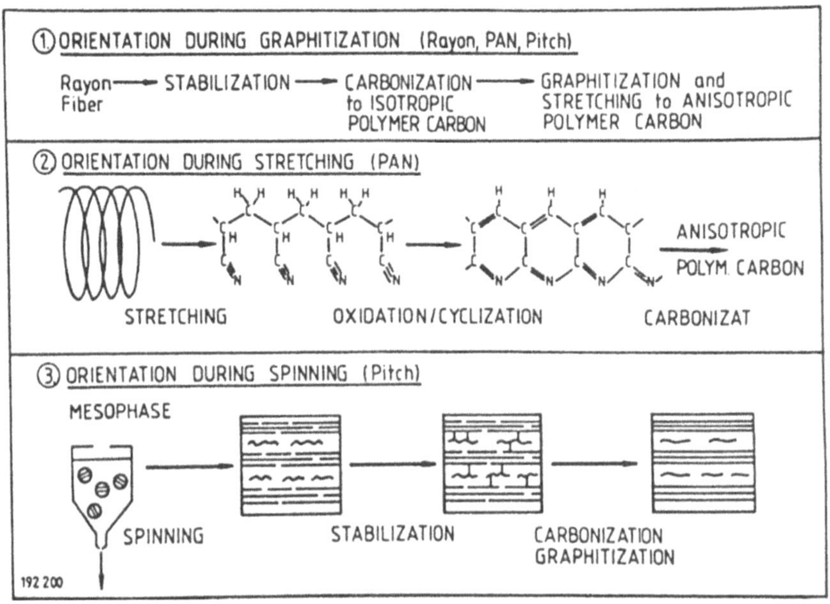

Fig. 1.11. Three techniques for producing anisotropic carbon fibers

Table 1.1. Carbon yield for various carbon fiber precursors

Precursor	Process yield[a]
Rayon	20 – 30
PAN	45 – 50
Pitch	75 – 90

[a] Kilogram of carbon fiber per kilogram of precursor fiber.

issue in both the processing and economics of a commercial carbon fiber, for these three precursors is compared in Table 1.1. Although rayon was the precursor used to produce carbon fiber for the first fiber-reinforced composites, the carbon fibers used in the vast majority of current advanced composites are made using PAN and pitch precursors. Since there are many detailed reviews on the production of carbon fibers from all three precursors, this chapter does not cover the details of carbon fiber processing. However, some recent results, relevant to the subject of this book, are discussed in this section.

1.5 PAN-Based Carbon Fibers

Of the three precursors, PAN has proven to be the most suitable for developing the carbon structure needed to produce high strength fibers. Therefore, the PAN process is nowadays the standard commercial method used for the manufacture of the overwhelming majority of carbon fibers. Briefly, the process begins with the PAN precursor fiber, a linear polymer containing about 68% carbon. Subsequently a controlled transformation to carbon fiber is effected through a series of heat treatment steps in different environments at progressively increasing temperatures. The technology has been slightly modified and optimized by each producer; however, the basic process involves three steps: preoxidation or stabilization (heating in air at temperatures ranging from 200 °C to 230 °C); carbonization (heat treatment in an inert atmosphere at temperatures ranging from 1000 °C to 1500 °C); and a final high temperature thermal treatment (at approximately 2000 °C–3000 °C in an inert atmosphere). Fig. 1.12 shows the chemical structure of the fibers after the various process steps [24, 25]. As Fig. 1.13 indicates, while different transport problems are encountered in each stage of the heat treatment, heat and mass transport pose the key problems during the critical stabilization step.

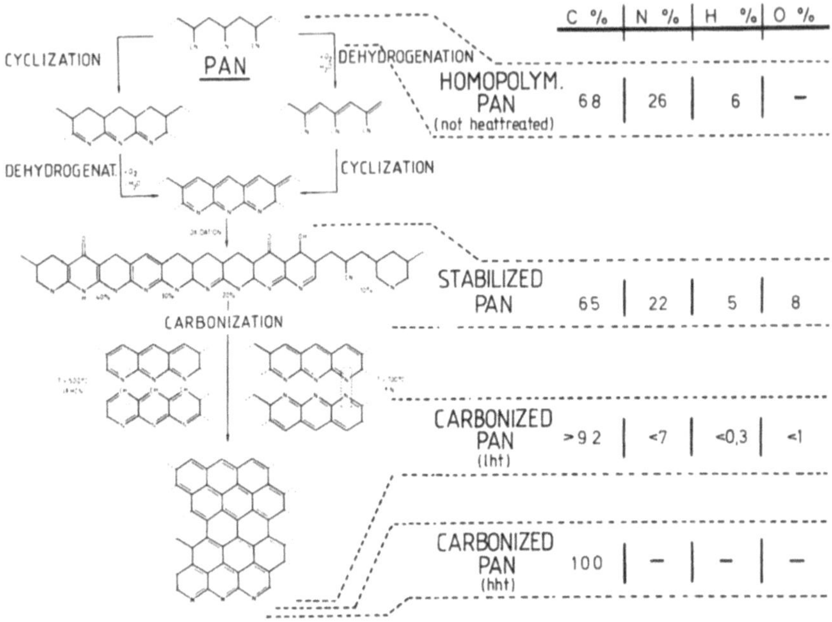

Fig. 1.12. The mechanism of formation of graphitic layers within carbon fibers made from PAN precursor [24] and elemental analysis of the PAN precursor fiber, stabilized PAN, carbon fiber heat treated at low temperature (lht), and carbon fiber heat treated at high-temperature (hht) [25]

The stabilization step converts the precursor from a linear polymer to a highly condensed, thermally stable structure [24–27]. From a process engineering viewpoint, the principal feature of the overall reaction is that the cyclization of the nitrile groups is highly exothermic (Fig. 1.9). The exact temperature at which cyclization occurs depends primarily on the comonomer content of the PAN precursor, the stretch applied to the precursor fiber before and during stabilization, and the surrounding atmosphere during stabilization [28–31]. If the PAN precursor fiber is heated too rapidly to the cyclization temperature, cyclization is spontaneous, generating a large amount of heat. On a large scale, this rapid release of heat can cause loss of molecular orientation within the precursor fiber and even melting of the fiber itself [32, 33]. Obviously, this will have a detrimental effect on the properties, particularly the strength and modulus, of the final carbon fiber. Thus, to ensure optimum properties, the cyclization is usually conducted so as to promote a slow release of heat, minimizing both the loss of orientation and the fragmentation of the polymer molecules. Because of the critical importance of the cyclization step in the PAN carbon fiber pro-

Stabilization	Carbonization
• Mass transport - Diffusion of Oxygen into the fiber - Volatile products such as CO_2, HCN and H_2O	• Mass transport - Volatile products such as HCN, H_2O, NH_3, CO_2, CH_2, and N_2
• Heat transport - Main problem of overheating the fiber	• Heat transport - Heat transport facilitated because of thermal conductivity of the fiber
• Shrinkage - Must be inhibited	• Shrinkage - Mainly in the cross-section

Fig. 1.13. Transport problems and shrinkage phenomena during fabrication of carbon fibers

cess and the obvious advantages offered by a more controlled reaction and a decreased reaction time, the kinetics and mechanism for the conversion of PAN to a cyclized ladder structure have been thoroughly investigated [30, 31]. To make matters more complex, oxidation, the incorporation of oxygen-containing groups into the polymer chain, occurs concurrently with the cyclization reaction (Fig. 1.14). These oxygen-containing groups promote cross-linking reactions in the early stages of carbonization. While the cyclization reaction is highly temperature dependent, the oxidation process is diffusion controlled. Hence, the rate of oxidation depends on the diameter and chemical composition of the precursor fibers, as well as the temperature and concentration of oxygen in the reacting environment. Therefore, in the typical commercial process a series of hot air circulating ovens are employed to stabilize the PAN precursor fibers in steps, starting at low temperatures and finishing at high temperatures. This approach allows the producer to improve process economics by providing a uniform and high quality final product, while reducing the total reaction time [34–36].

After stabilization, the chemical composition of the PAN precursor fiber can have a dramatic influence on the mechanical properties of the final carbon fiber. The oxygen content required for optimum final fiber properties varies for different PAN copolymers (Fig. 1.15). However, in general, during stabilization the oxygen content of the precursor fiber increases from 8% to 11%. Simultaneously, the evolution of CO_2, generated by decarboxylation reactions, and HCN, evolved as uncyclized nitrile groups react, causes the carbon content of the PAN fiber to decrease slightly from about

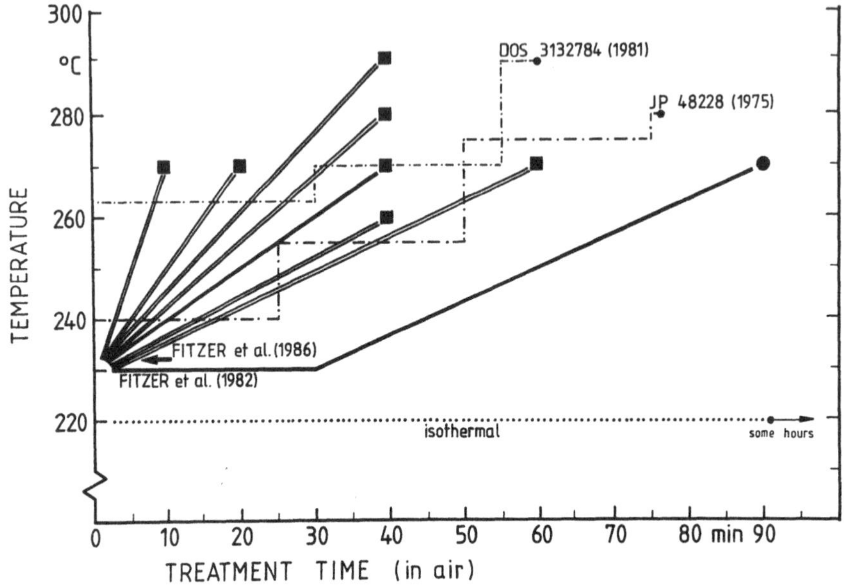

Fig. 1.14. Temperature profiles employed to study the effect of linear heating during stabilization of PAN in air [34], and proposals found in the patent literature [35, 36]

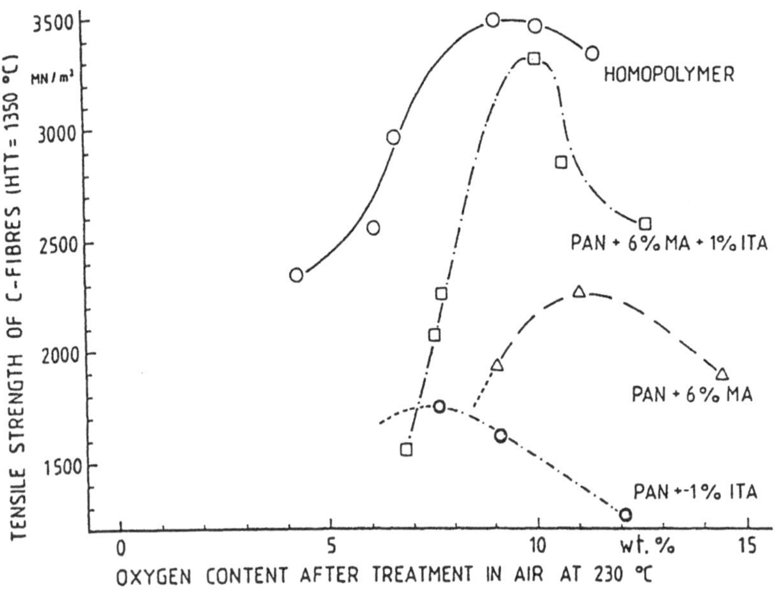

Fig. 1.15. Influence of oxygen content after stabilization in air at 230 °C on the mechanical properties of carbon fibers after final heat treatment at 1350 °C. All oxidations were performed after a 6 hour pretreatment at 230 °C in nitrogen

68% to 65% during stabilization. The evolution of HCN, as well as dehydrogenation reactions involving H_2O, also cause the hydrogen content of the fiber to decrease. Therefore, during the stabilization process the fiber suffers a net weight loss ranging from 5% to 8%.

As previously mentioned, the preferred orientation needed for the final high performance carbon fiber is generated by stretching the precursor fiber prior to stabilization while it is still a thermoplastic polymer. However, the chemical reactions which take place during the low temperature stabilization of PAN fibers can disrupt the orientation of the molecular chains and result in fiber shrinkage, significantly influencing the modulus of the final carbon fibers [37]. Therefore, mechanical means must be employed to control the natural shrinkage which occurs during the stabilization process, i.e. as the thermoplastic polymer is transformed into a thermally-stable ladder polymer. The amount of shrinkage or elongation permitted depends on the type of precursor fiber and its prior process history. While certain precursors must be allowed to shrink to some extent during stabilization, others actually require elongation during stabilization to achieve optimum final properties.

Once the PAN fibers are stabilized, they are subjected to pyrolysis (carbonization). This high temperature treatment eliminates most of the non-carbon elements from the fiber, essentially converting it to a carbon fiber. Generally, carbonization is conducted in an inert atmosphere, usually ultra-pure nitrogen, at temperatures ranging from 1000 °C to 1500 °C. During this stage a thermally stable pyridinic structure forms and then subsequently collapses into a turbostratic, stacked ring structure. The development of mechanical properties in PAN-based carbon fibers is directly related to the perfection and orientation of this turbostratic structure. Various mechanisms have been proposed for the reactions which take place during the carbonization stage [25, 34, 38]. In the temperature range of 300 °C to 600 °C, most of the chemical reactions involve intermolecular cross-linking, reorganization and coalescence of cyclized sections. These reactions lead to the evolution of water, ammonia, hydrogen cyanide, carbon dioxide, carbon monoxide, and methane, together with a lower amount of low molecular weight nitriles. This initial low temperature carbonization stage is extremely important since it is during this stage that the foundation of the carbon building block is laid. Thus, even during the early stage of carbonization both temperature and rate of heating is extremely important. In the temperature range of 600 °C to 1000 °C, intermolecular cross-linking takes place, causing hydrogen, hydrogen cyanide, and water to be evolved. At this stage, the carbon content of the fiber increases to about 92 %, the nitrogen content decreases to about 7 %, and the hydrogen content is less than 0.3 %. After carbonization at temperatures of between 1200 °C and 1400 °C, the PAN-based carbon fiber achieves a tensile strength of about 3 GPa and a modulus of approximately 250 GPa. Fig. 1.16 clearly shows the strong influence that the heating rate and final temperature, both during stabilization and carbonization, exert on the ultimate mechanical properties of the carbon fiber [22].

As Fig. 1.17 illustrates, at temperatures below about 1400 °C, increasing the carbonization temperature improves physical properties such as density, tensile strength, tensile modulus, and thermal and electrical conductivity, reflecting an increase in both the preferred orientation and the degree of crystallinity [39, 40]. Higher heat treatment temperatures in argon will continue to improve the preferred orientation and thus the modulus. However, the effect of heat treatment temperature on the tensile strength of the final carbon fiber is more complex. While increasing the heat treatment temperature results in an increased final fiber strength below 1400 °C, above this temperature the strength begins to decrease, reaching a minimum at temperatures of 1800 °C. This reduction in strength is related to the loss of nitrogen and, thus, the formation of flaws. The nitrogen content of the fiber decreases with heat treatment temperatures approaching zero at

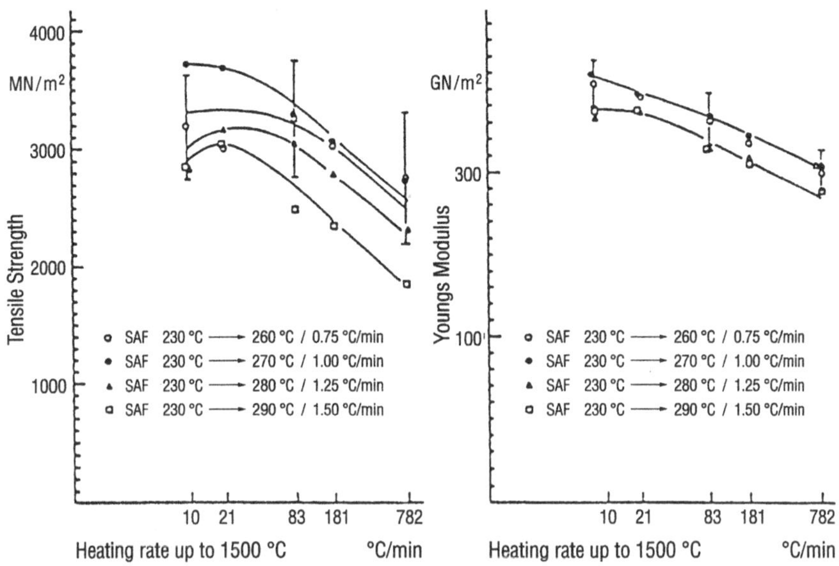

Fig. 1.16. Influence of rate of heating to 1500 °C during carbonization on the final mechanical properties of PAN-based carbon fibers [22]

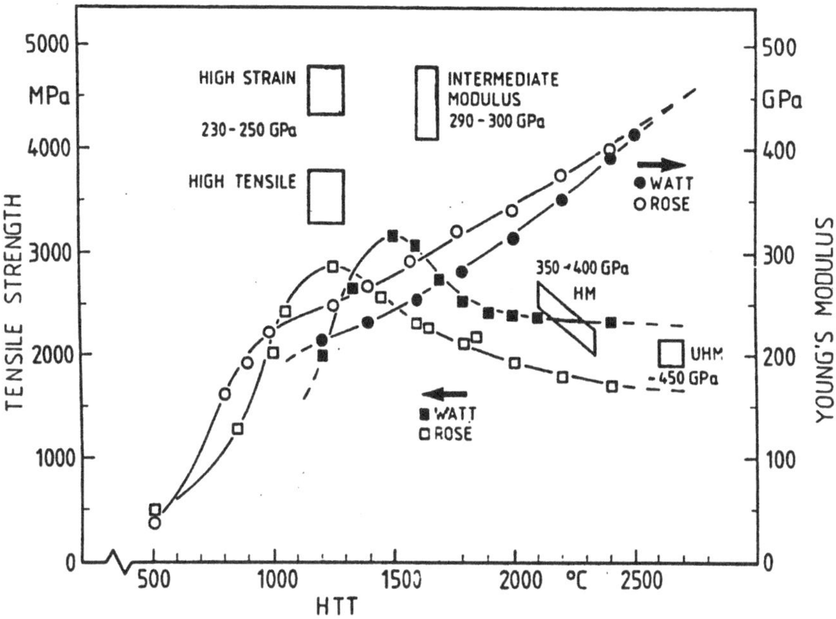

Fig. 1.17. Influence of final heat treatment temperature (HTT) on the mechanical properties of carbon fibers [40, 41]

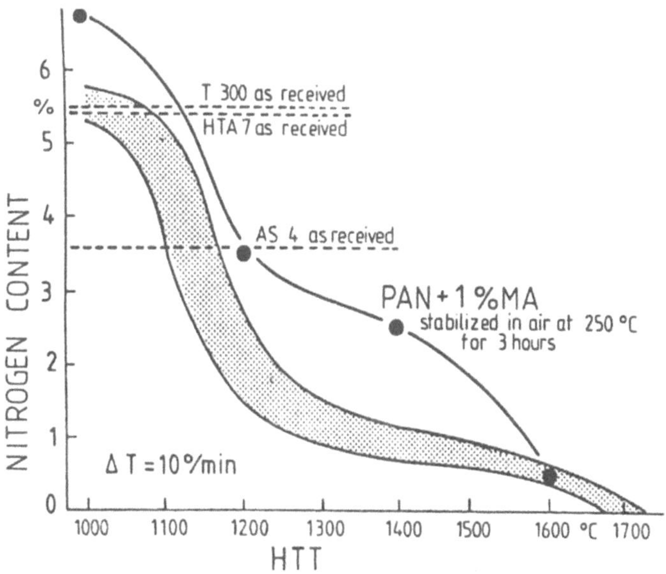

Fig. 1.18. Decrease in the nitrogen content during carbonization treatment for a stabilized PAN fiber and post-treated commercial carbon fiber (T300S from Torayca) [41]

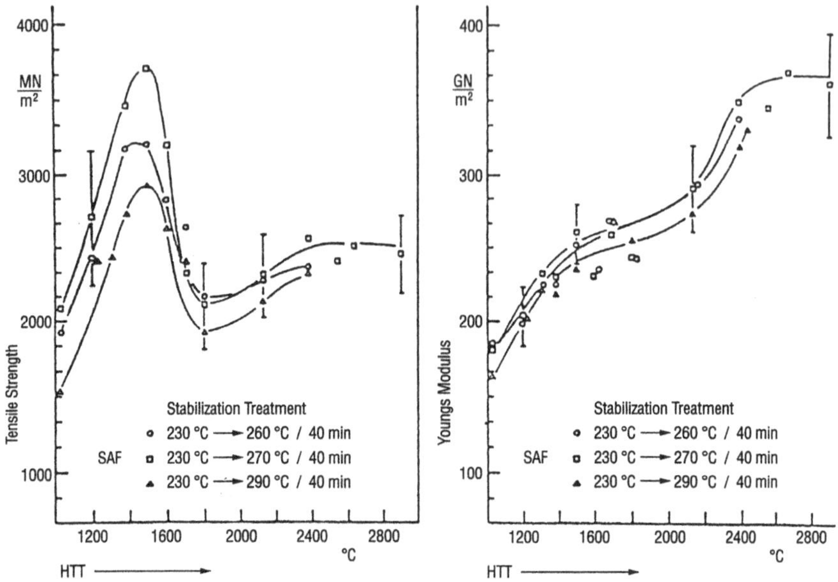

Fig. 1.19. Influence of final heat treatment temperature on the mechanical properties of fibers stabilized at various conditions [22]

1800 °C (Fig. 1.18) [41]. This final removal of nitrogen permits the structural rearrangement of lattice defects, leading to another increase in fiber strength at temperatures above 1800 °C – the second maximum seen in Fig. 1.19. This second maximum may be one reason for the improved properties of the new generation of intermediate (IM type) high tenacity carbon fibers.

Recent studies [22] have shown that the final carbonization temperature necessary to develop maximum tensile strength depends on the precursor employed [22]. Nevertheless, for any particular precursor (as Fig. 1.19 shows), the maximum attainable strength depends on the process conditions employed during stabilization and carbonization.

1.6 Pitch-Based Carbon Fibers

Two types of pitch-based carbon fibers are commercially available: isotropic carbon fiber, produced from an isotropic pitch precursor, and anisotropic carbon fiber, produced from an anisotropic, mesophase precursor. The development of pitch-based carbon fibers was, to a large extent, made

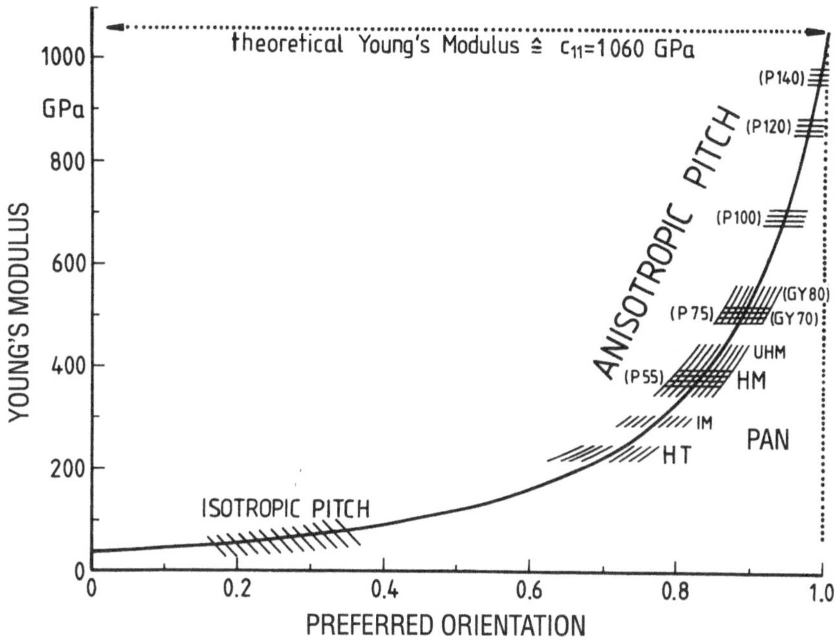

Fig. 1.20. The effect of preferred orientation on the Young's modulus of carbon fibers [34]

possible by two events which occurred in the 1960s: the pioneering investigations of Otani and his coworkers on isotropic pitches and fibers, and the remarkable discovery of mesophase by Brooks and Taylor [42]. Otani developed a procedure for producing carbon fibers from an isotropic pitch precursor. In his process, which has been commercialized by the Kureha Company of Japan, the isotropic precursor is melt-spun into fiber form. These isotropic precursor fibers are then oxidized to render them infusible and subsequently carbonized. While the final isotropic carbon fibers have very low strengths and moduli, and, thus, are of no value for advanced composites, they can be used in insulation felts and as fillers for concrete.

For the carbon fiber to develop a high modulus, it is critical that the preferred orientation of the polyaromatic layers is parallel to the fiber axis. Fig. 1.20 shows this well-documented dependence of Young's modulus on the degree of preferred orientation. This high degree of preferred orientation can be achieved in rayon-based and isotropic pitch-based carbon fibers only through an extremely difficult and costly high-temperature stretching process. In the PAN-based carbon fiber process, preferred molecular orientation parallel to the fiber axis is developed by first stretching the molecules of the polymeric precursor fiber and then preventing recoil during subsequent heat treatment. However, the necessary preferred orientation also can be developed within the carbon fibers during melt spinning. In particular, when liquid crystalline mesophase pitches are melt-spun, extremely high degrees of preferred orientation can be achieved. Flow through the extrusion die causes the mesophase molecules to align, and final drawdown further enhances the alignment.

1.6.1 Mesophase Pitches

Pitch can be produced either by the distillation of coal tar or from the high molecular weight residues produced during the distillation of crude oil. All pitches consist of a wide range of polyaromatic and cyclic aliphatic hydrocarbons, containing linear aliphatic side chains. All of the compounds in pitch have not yet been identified. As Fig. 1.21 illustrates, the molecular weights of the polyaromatic compounds vary from 200 to 1000 [43]. Because of this varation in molecular composition, evaporation and melting occurs over a wide range of temperatures (Fig. 1.22) [44]. A mesophase can be formed by the thermal or catalytic polymerization of a suitable petroleum or coal tar pitch. In fact, the reaction occurs by a classic free radical

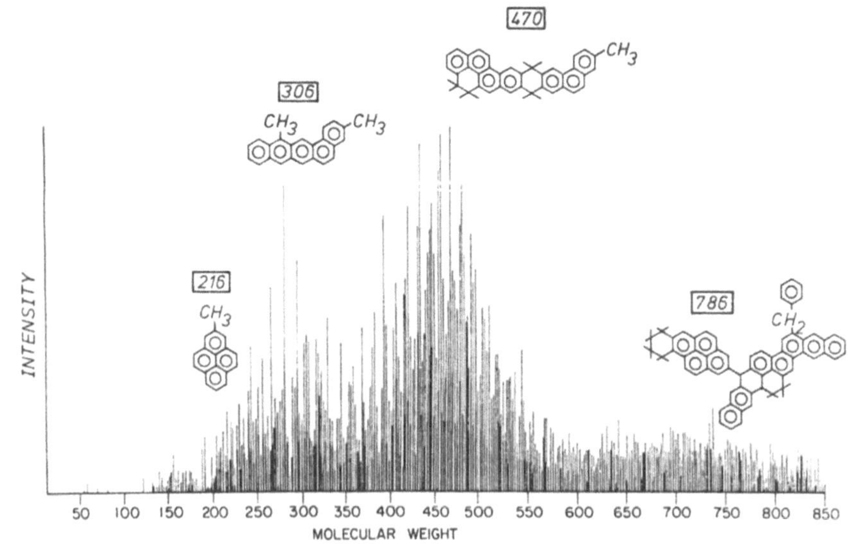

Fig. 1.21. Molecular weight distribution for a typical petroleum pitch [43]

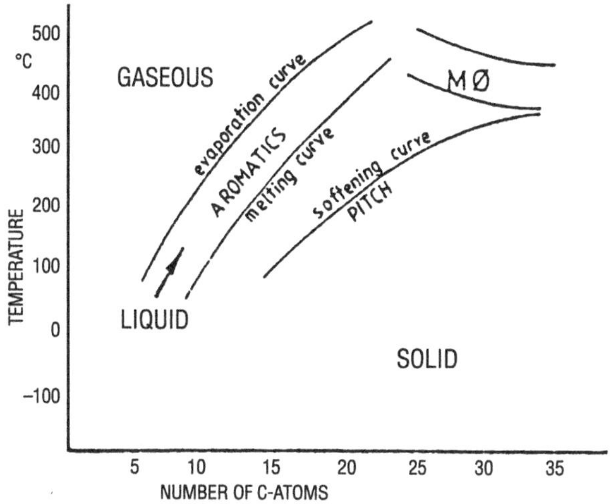

Fig. 1.22. Melting/evaporation temperatures for pure aromatics as a function of the number of carbon atoms [44]

mechanism. Thus, when a highly aromatic pitch is heated to temperatures of 400 °C–450 °C for extended periods, an optically anisotropic material, a mesophase, is formed. The formation reaction depends on the temperature and the carbon-to-hydrogen ratio of the aromatic during this heat treatment. Figure 1.23 details the chemical structures of some typical polynuclear aromatic molecules in mesophase pitch [45, 46]. Initially, small spheres of mesophase are formed, with the polyaromatic layers arranged as shown in Fig. 1.24. As they grow larger, the spheres coalesce to form larger spheres. Insoluble impurities in the molten pitch will inhibit coalescence and, therefore, hinder the formation of extended areas of anisotropy. The coalescence behavior of this anisotropic mesophase is similar to that of nematic liquid crystals formed from rod-like or disc-like molecules. A multiphase diagram for describing the formation of this liquid crystalline phase has been detailed by Diefendorf [47]. Phenomenologically, the steps in the formation and growth of liquid crystals are similar to those of crystallization, beginning with nucleation and continuing through the complete conversion to mesophase.

To lower their softening temperature and allow them to readily orient under applied stress, the mesophase pitches used to produce carbon fibers normally are a mixture of high molecular weight molecules containing a small number of side groups. Both of these characteristics are vital if the material is to be melt-spun. Mesophase composed of high molecular weight compounds with no side groups, or containing components with high specific softening points, often decompose before flow is possible.

	Coal tar Pitch	Petroleum Pitch	Soluble in
A	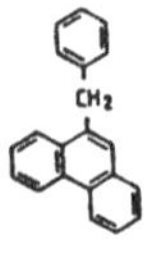 MW = 268 C/H = 1,31 Harom/Haliph = 7 Carom/Caliph = 20	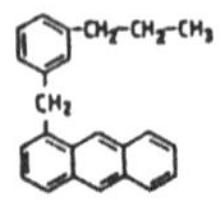 MW = 306 C/H = 1,08 Harom/Haliph = 0,73 Carom/Caliph = 25	Cyclohexane / Acetone 9:1 δ = 8,37
B	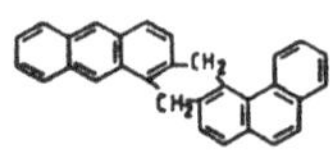 MW = 380 C/H = 1,5 Harom/Haliph = 4 Carom/Caliph = 14	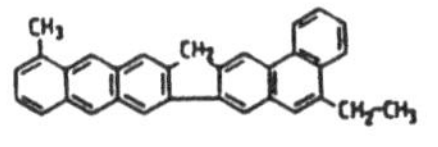 MW = 408 C/H = 1,33 Harom/Haliph = 1,4 Carom/Caliph = 7	Cyclohexane / Acetone 1:1 δ = 9,05
C	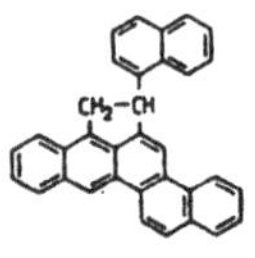 MW = 429 C/H = 1,6 Harom/Haliph = 6,3 Carom/Caliph = 16	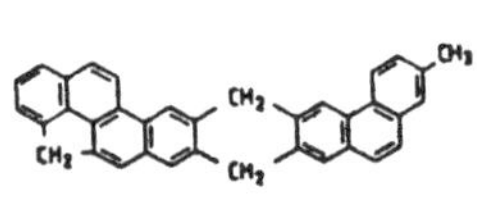 MW = 456 C/H = 1,5 Harom/Haliph = 1,66 Carom/Caliph = 8	Acetone δ = 9,9
D	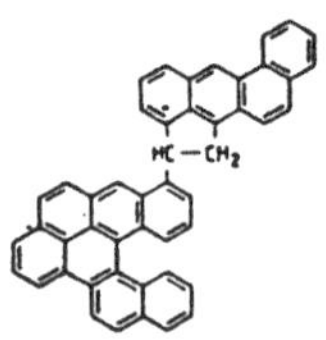 MW = 602 C/H = 1,48 Harom/Halips = 7,6 Carom/Caliph = 23	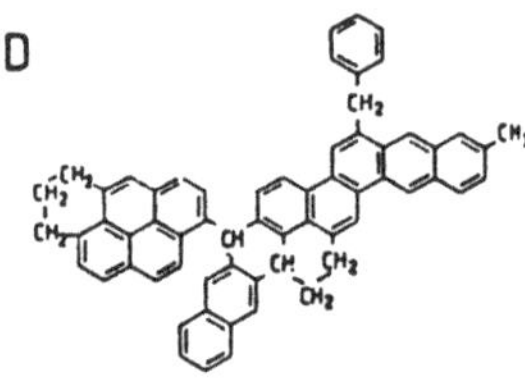 MW = 812 C/H = 1,45 Harom/Haliph = 1,59 Carom/Caliph = 6	Toluene δ = 8,9
E	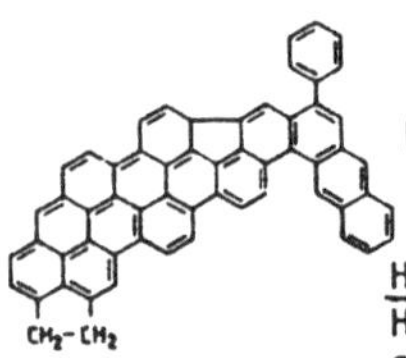 MW = 737 C/H = 2,03 Harom/Haliph = 6,3 Carom/Caliph = 29	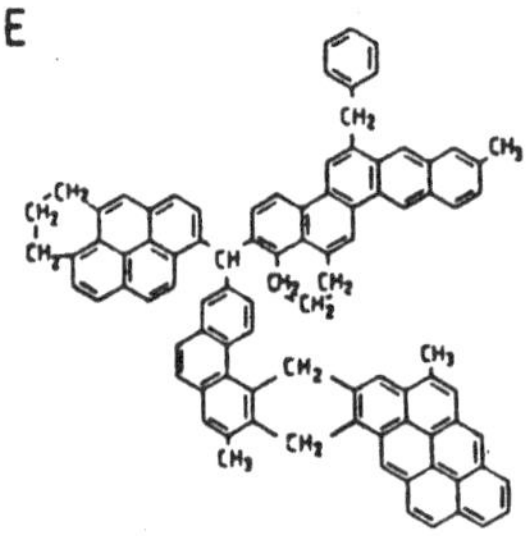 MW = 1203 C/H = 151 Harom/Haliph = 1,10 Carom/Caliph = 5,7	THF δ = 9,1
F	C/H = 2,14	C/H = 1,84	Quinoline δ = 10,8
G	C/H = 2,90	C/H = 1,95	

Fig. 1.23. Model molecules for various pitch fractions as derived from analytical data [45, 46]

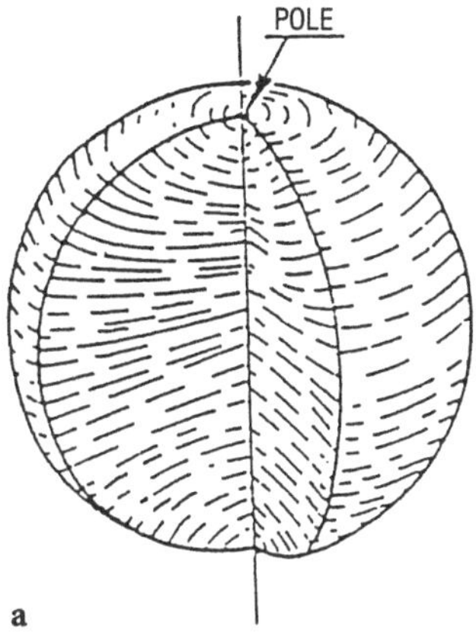

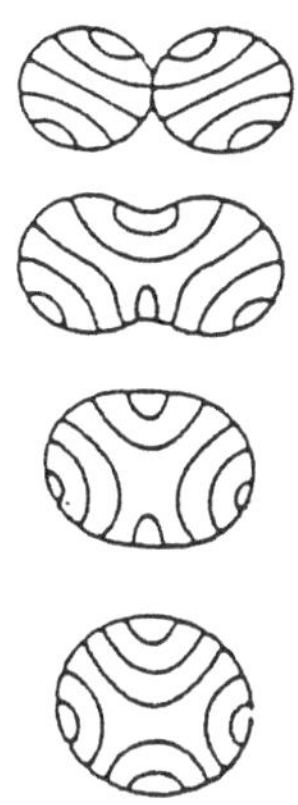

Fig. 1.24 a, b. Mesophase spherule. **a** Schematic [10]. **b** Model of mesophase formation during heat treatment of a pitch at 400 °C

In addition to the production of mesophase from petroleum and coal pitches, significant efforts are currently underway to produce synthetic mesophases from pure aromatic hydrocarbons [48]. Such an approach may provide a mesophase with improved uniformity and purity, all at high process yields. By using the appropriate material to catalyze the condensation of aromatic hydrocarbons, the yield of mesophase has been increased to as high as 90% [49].

1.6.2 Melt Spinning Mesophase Pitch Precursor Fibers

The first step in the production of all carbon fibers is to spin the precursor polymer or pitch into fibers. Since mesophase pitch is a thermoplastic material which softens and flows prior to decomposing, it can be melt-spun into fiber form. This ability to be melt-spun is a significant advantage.

Since the viscosity of the mesophase pitch determines the extrusion temperature, the drawdown rate and even, to a large extent, the tension on the fiber during drawdown, it is extremely important in melt-spinning. The flow behavior of most mesophase pitches is shear thinning at low rates of shear, but becomes relatively Newtonian at high shear rates [50]. Perhaps the most unusual flow characteristic of mesophase pitches is that their viscosities are extremely temperature dependent. Figure 1.25 shows the apparent viscosity of different fractions of a coal tar pitch, Ashland 240 –

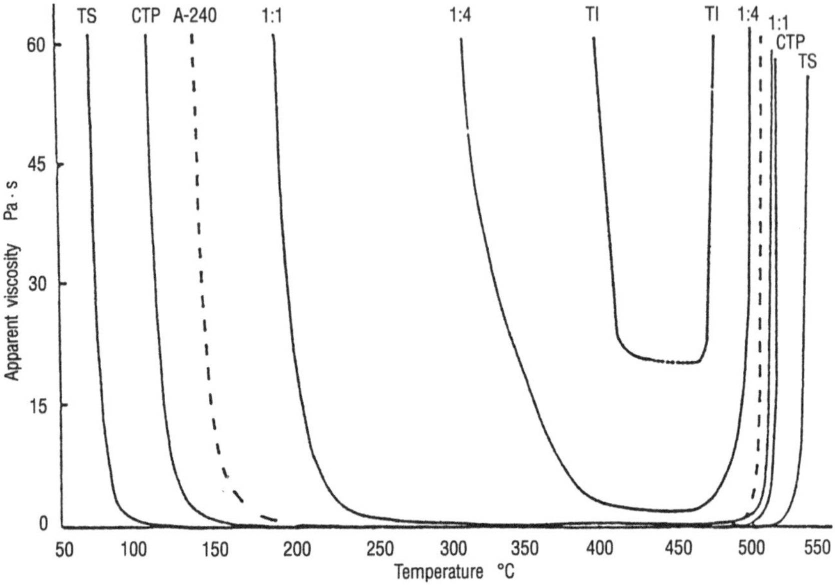

Fig. 1.25. Apparent viscosity of different fractions of coal tar pitch (CTP) and petroleum pitch (A 240) as a function of heat treatment temperature [50]

toluene soluble and insoluble fractions, and mixtures of toluene soluble and insoluble fractions – as a function of temperature. In all cases, the drop in viscosity is caused by softening. The increase in viscosity at temperatures above 400 °C is caused by the increase in the average molecular weight as the pitch is thermally polymerized, forming particles of mesophase. Then, as mesophase spheres coalesce to form bulk mesophase, the viscosity normally decreases.

Figure 1.26 shows the softening behavior of pitch which contains both mesophase and isotropic fractions. Note the range of viscosities recorded as the pitch mixture begins to flow. It is well documented that complex interactions occur when two-phase pitches, such as these, are processed. Since the two phases are somewhat soluble in one another, large fluctuations in viscosity occur during flow. This, combined with the natural temperature sensitivity of the viscosity, makes the melt-spinning of pitches containing less than 100% mesophase extremely difficult.

Figure 1.27 shows a schematic diagram of the melt-spinning process; in fact, the process consists of melt-extrusion, followed by fiber drawing [51]. The extruder contains a screw which forces the pitch through three zones: (i) a feed zone, (ii) a melting zone, (iii) and a pumping zone. In the first zone

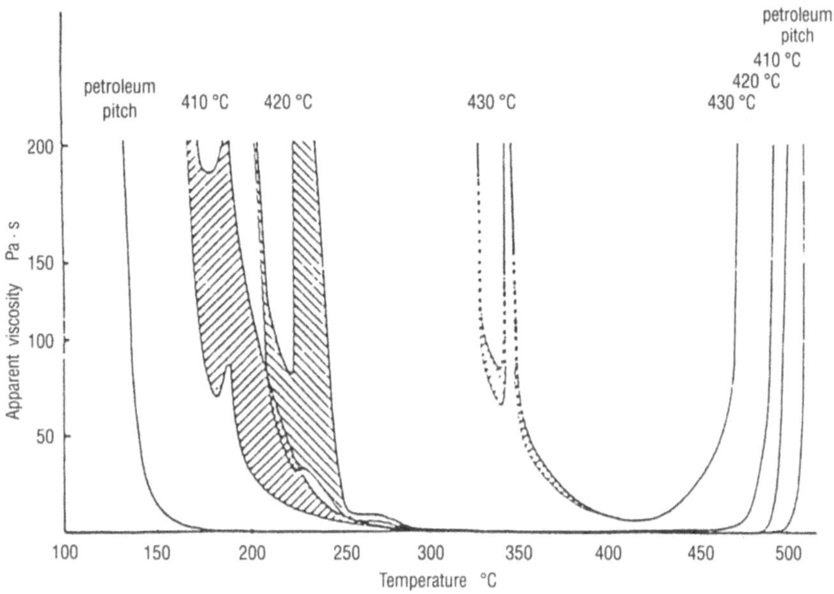

Fig. 1.26. Softening and solidification behavior of mesophase containing pitch [50]

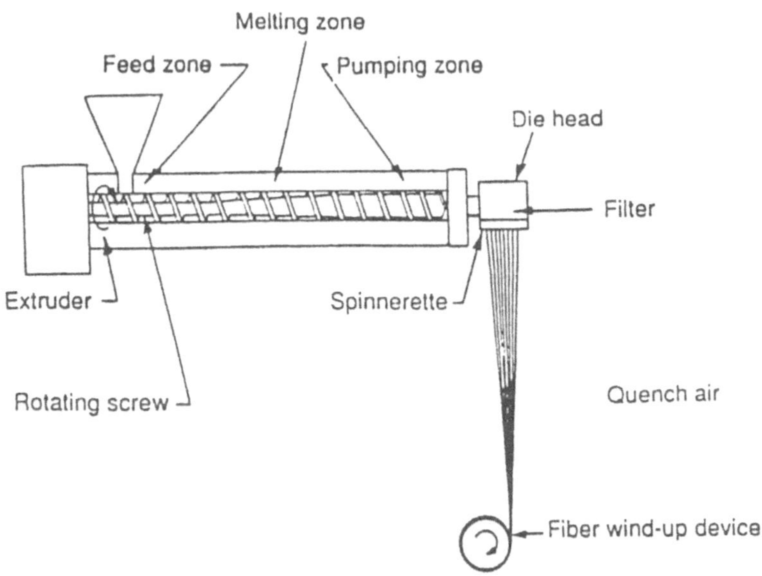

Fig. 1.27. Schematic diagram of a melt-spinning process

the solid pitch is merely metered into the screw. In the second zone the pitch is heated to well above its softening temperature. Then, in the third zone the pitch is compressed and fed to the die head. Often the die head contains a secondary gear pump to ensure even flow and a filter to remove any solid impurities from the mesophase. Finally, the molten mesophase is extruded through a multihole die, called a spinnerette. As the molten mesophase exits the spinnerette capillaries, it cools, forming fibers which are drawn to their final diameter by the take-up winder. Since the mesophase precursor is a lyotropic liquid crystal, shear forces tend to orient the molecules during flow through the small capillaries in the spinnerette. It is here that the transverse structure of the fiber (radial, radial-folded, random, etc.) is first formed. Then, during drawdown, elongational stress further orients the molecular structure parallel to the fiber axis (Fig. 1.28), resulting in a highly oriented precursor fiber.

The severe temperature dependency of mesophase, combined with the brittle nature of the as-spun fiber, makes the choice and precise control of

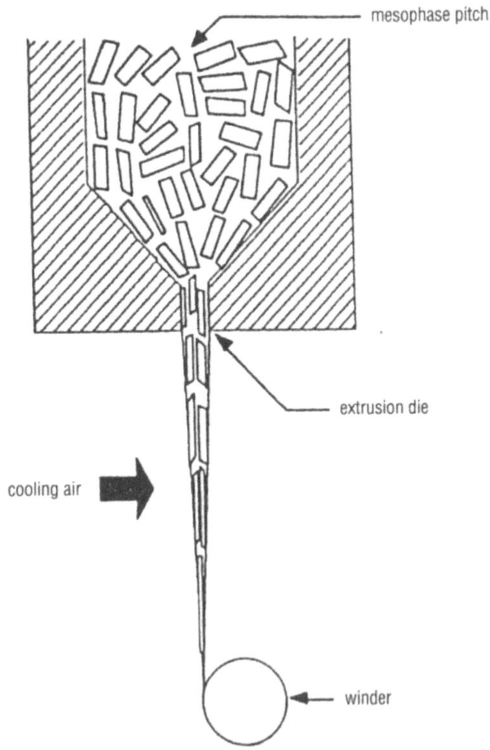

Fig. 1.28. The development of orientation during the extrusion of mesophase

melt-spinning parameters extremely critical. In fact, process conversion and product uniformity depend on the proper control of temperatures and pressures at the various stages of spinning as much as on the mesophase pitch itself. In a recent analysis of the melt-spinning of mesophase pitch, Edie and Dunham [52[showed that only a ± 3.5 °C variation in temperature across the spinnerette can cause a ± 15% variation in the diameter of the as-spun filaments. They demonstrated that small changes in conditions can have a tremendous effect on the tensile stress in the filaments during melt-spinning, which in turn influences spinnability. They found that the most effective way to reduce this threadline stress is to increase mesophase flow rate and/or decrease winding speed. However, the drawdown ratio is critical for controlling the degree of induced molecular orientation in the fiber. Fibers melt-spun at a small drawdown ratio exhibit a relatively low degree of preferred orientation. Additionally, the molecular orientation is non-uniform, decreasing rapidly toward the core of the fiber. By comparison, fibers melt spun at a high drawdown ratio possess a much higher, and more uniform, degree of preferred orientation. Therefore, for any given mesophase the drawdown conditions represent a compromise between maximizing the degree and uniformity of preferred orientation in the as-spun fiber and minimizing the stress in the fiber during melt-spinning.

1.6.3 Thermosetting Mesophase Pitch Fibers

In order to prevent the molecular orientation from relaxing and to avoid additional problems such as fiber–to–fiber sticking, pitch precursor fibers must be thermoset prior to final carbonization. Thermosetting can be accomplished either by heating the pitch precursor fiber to just below its softening temperature in an oxygen containing atmosphere or by immersing the precursor fiber in a strong oxidizing liquid. During this stabilization step oxygen diffuses into the precursor fiber and reacts, creating carbonyl and phenoxy groups which cross-link the large plate-like molecules. As these groups form, the oxygen content and the mass of the fiber increases, and these increases are often monitored to determine the degree of stabilization. Stevens and Diefendorf [53] have reported that mesophase fibers are adequately stabilized when the increase in mass is approximately 6%. However, the exact value, like the time and temperature required for stabilization, depends on the molecular structure of the pitch precursor. Similar to PAN precursor fibers, the stabilization of pitch precursor fibers is diffusion-controlled. Since higher stabilization temperatures increase the reaction rate, they decrease the total time required for this process step. However, high stabilization temperatures also result in loss of molecular orientation and, hence, disimprove the mechanical properties of the final carbon fibers. Therefore, the optimum stabilization temperature for any mesophase fiber is actually a compromise between minimizing the time required to stabilize the precursor fiber and optimizing the final properties of the carbon fiber. Typically, heating in air to a temperature of between 250 °C and 300 °C for a period of up to 2 h will adequately stabilize mesophase precursor fibers. The extremely low strength of the as-spun fiber (in the order of 30 MPa) makes handling another important factor which influences the commercial production of pitch-based carbon fiber [49, 54]. Because of this, the fibers often are stabilized batch-wise, rather than continuously. Prior to carbonization, the fiber is too weak to carry its own weight over an extended distance. Thus, in a continuous stabilization the precursor fiber must be supported as it passes through the oven. Supporting the fiber during stabilization and, at the same time, avoiding surface damage poses a challenging design problem.

1.6.4 Carbonization and Graphitization

After stabilization the final mechanical properties of the mesophase fibers are developed in a two-step final heat treatment. In the first step, often termed pre-carbonization, the fiber is heated to approximately 1000 °C in the presence of an inert gas. Since gases such as CH_4 and H_2 are evolved during this initial step, heating to this temperature must be carried out in a controlled manner. In the second step, the fiber is heated to a temperature ranging from 1500 °C to 3000 °C, depending on the final properties required. Unlike PAN-based fibers, pitch-based precursor fibers must be heated to temperatures of approximately 1800 °C to produce high strength carbon fibers (HT type). For producing high modulus carbon fibers, heat treatment to temperatures of more than 2500 °C is often necessary, although this depends greatly on the mesophase precursor employed.

1.7 Vapor-Grown Carbon Fibers

Another variety of reinforcement with significant commercial potential is vapor-grown carbon fibers. This short-fiber product is produced from the gas phase by decomposing a hydrocarbon gas on a heated substrate in the presence of a transition metal. Because a transition metal is employed, these filaments are sometimes referred to as catalytic chemical vapor deposition (CCVD) carbon filaments. In this process a submicron activated catalyst, normally a transition metal (Fe, Ni, or Co), serves to nucleate the filament. Once nucleated, the filament grows away from the hot substrate, and normally the catalyst particle remains at the tip of the filament throughout the growth process [55]. Several different types of CCVD filaments can be grown, depending on the temperature and type of hydrocarbon employed during the vapor deposition process [56]. Low temperature deposition (> 900 °C) normally results in vermicular filaments, while high temperatures (1500 °C to 2500 °C) favor the growth of long, straight filaments. Characteristically, vermicular filaments have a low density and a nonuniform ring-like structure, consisting of an outer sheath which is resistant to oxidation and a core region which can be easily oxidized. By comparison, filaments produced at high temperatures have hollow cores and smaller diameters. This is important since in the early stages of formation, the growth rate of the filament is inversely proportional to its diameter.

The final CCVD product consists primarily of discontinuous fibers with diameters ranging from 1 to 100 μm and lengths ranging from a few millimeters to a few hundred millimeters. When low temperature is employed, the as-deposited filaments exhibit poor interlayer correlation, and the graphene layer planes are relatively straight over only short distances. However, vapor-grown filaments are graphitizable. When the filaments are heated to 3000 °C, massive structural reorganization occurs as crystallite size increases (Fig. 1.29). At sufficiently high temperatures, this can even result in facetting on the external surface of the filament [57].

Since vapor-grown fibers have been produced in only pilot-scale quantities, process conditions have been neither optimized nor standardized. While the structure, and thus the mechanical properties, of vapor-grown fibers is independent of the gas source employed (i.e., benzene or methane), it is extremely dependent on other process parameters, such as the temperature of growth, type and distribution of catalyst, position of the substrate, etc. An example of this sensitivity is the filament diameter which, even when the gas and temperature of growth are held constant, can vary from 3 to 130 μm. Like all carbon fibers, the strength of vapor-grown fibers is a function of their diameter. Therefore, it is not surprising that the reported mechanical properties of vapor-grown fibers vary widely.

Tibbetts and Beetz [58] investigated the structure and properties of vapor-grown fibers and observed that vapor-grown fibers exhibit a nonlinear stress strain curve. While the strain increases with increasing load, the

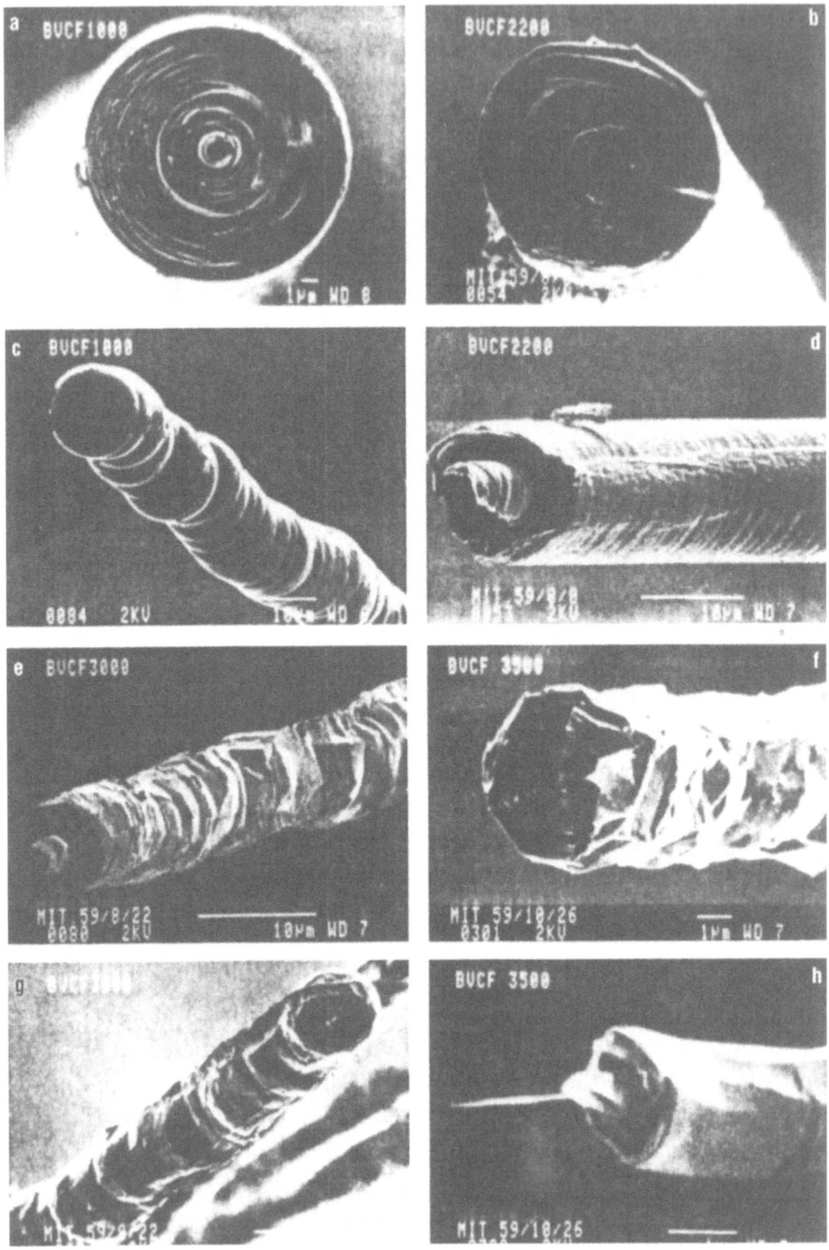

Fig. 1.29a–h. SEM photographs showing the morphology change on the surface and the cross section of benzene-derived filaments at different heat treatment temperatures. **a, c** asdeposited (1000 °C); **b, d** 2200 °C; **e, g** 3000 °C; **f, h** 3500 °C

strain per unit load decreases. The result is that Young's modulus of the vapor-grown fiber increases with strain, making the ultimate strain-to-failure of vapor-grown fiber (1 % to 1.3 %) lower than that of polymer-derived carbon fiber. As previously mentioned, these fibers possess a ring-like core-sheath structure. The bonding between the rings of the structure is relatively weak and responsible for the unusual tensile failure mechanism of these filaments. As the filament is loaded in tension, cracks are initiated at the surface and propagate inward. Since the bonding between the ring-like layers of the filament is poor, the crack is often deflected after penetrating only a short distance into the fiber and then propagates around the circumference of the layers. This transfers the load to the more aligned core layers. Continued loading causes stress concentrations to develop in these core layers, eventually initiating another circumferential crack. After a series of such events, fiber failure will eventually occur. Since the failure events are sequential along a progression of circumferential rings within the filament, the resulting failure surface appears as a series of thin cylinders of graphite, telescoping to increasing length. This unusual fracture mode, termed "sword and sheath", results in a noncatastrophic failure. The strong diameter dependence of the tensile strength and Young's modulus of vapor-grown filaments, illustrated in Fig. 1.30 [58], is most likely the result of this ring-like failure mode.

CCVD filaments, like pitch-based carbon fibers, can be classified as graphitizable carbons. Therefore, when heat treated to temperatures of approximately 2800 °C, vapor-grown filaments develop a well-oriented graphitic structure, resulting in an increase in Young's modulus but a decrease in tensile strength. Once graphitized at these temperatures, vapor-grown filaments no longer exhibit a nonlinear stress/strain curve. Thus, strain-stiffening appears to be limited to as-grown CCVD filaments.

Since no stabilization is required, CCVD filaments can be produced in a one-step process. This simplicity, combined with the low-cost feed and isothermal growth conditions, makes the process economically quite attractive. Because of this, and in spite of the fact that their diameters and mechanical properties vary widely, CCVD filaments may become an alternative to fiberglass for use in short-fiber reinforced composites. In addition, since the vapor-grown filaments do not exhibit catastrophic failure, they may prove to be an excellent reinforcement for ceramic-matrix composites. General Motors, Inc., in the United States and a group at Shinshu University in Japan are currently refining the production of CCVD filaments and evaluating potential applications in advanced materials. While early CCVD filaments were produced by a batch process,

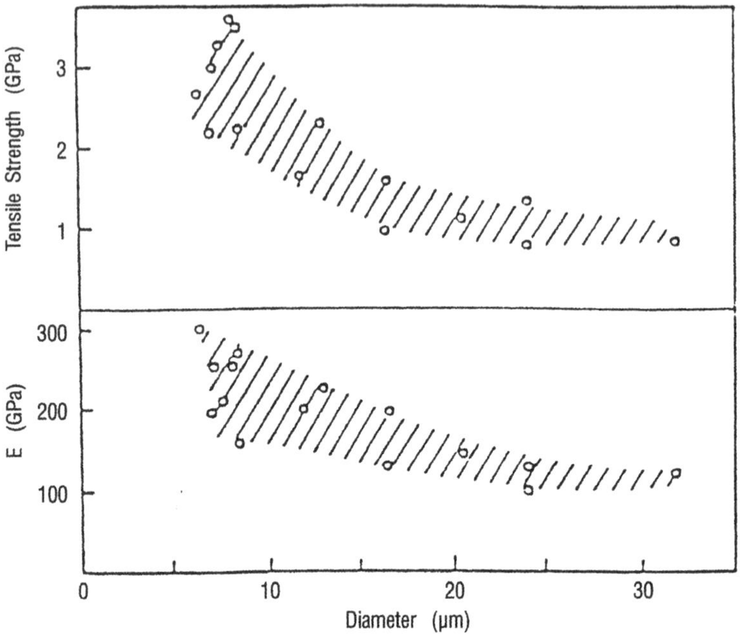

Fig. 1.30. Tensile strength and Young's modulus (E) of fibers of different diameter grown from the vapor phase [58]

Koyama and Endo recently patented a method for the production of continuous vapor-grown fibers [59].

1.8
Current Status of Carbon Fibers

As previously mentioned, two varieties of PAN-based carbon fibers were developed during their early years of commercialization: high strength (HT or type II) fibers and high modulus (HM or type I) fibers. As these new advanced fibers were used to improve the performance of materials, applications were encountered where further increases in fiber strength and stiffness were required. With the improvement of fiber properties, designers continuously increased performance requirements, forcing carbon fiber producers to further improve fiber mechanical properties. This fostered a continuous development race. For instance, when varieties of PAN-based carbon fiber with increased stiffness had been developed, engineers found that their designs were limited by the low strain-to-failure ratio of these fibers. Thus, improved composite toughness became the next

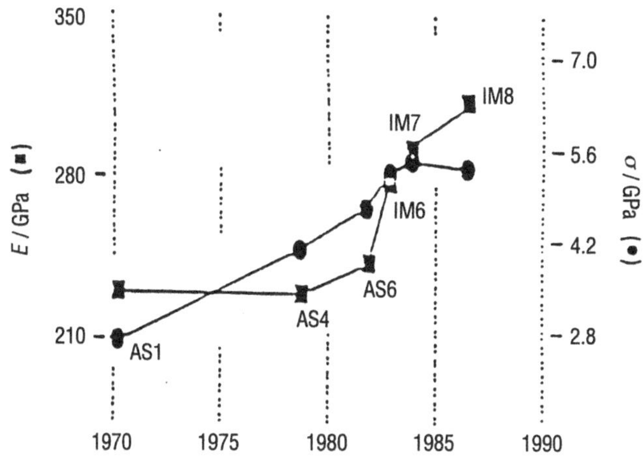

Fig. 1.31. Improvements in the tensile strength, σ, and Young's modulus, E, for Hercules carbon fibers (Hercules 1990 release)

critical design issue. To solve this problem, PAN-based carbon fiber producers developed a high strain-to-failure-fiber by increasing the fiber strength without sacrificing its modulus.

Today, improvements in the PAN precursor fiber, a better understanding of the structure-property relationship for carbon fibers, and advances in carbon fiber technology have resulted in significant increases in the range of properties available in commercial carbon fibers. Figure 1.31 illustrates this rapid improvement in properties for a typical commercial PAN-based carbon fiber. The mechanical properties of a variety of currently available carbon fibers are compiled in Table 1.2. Note in particular the properties of the newer generation intermediate modulus (IM) carbon fibers. This variety of PAN-based carbon fibers, which has higher specific tensile strengths than any other variety of carbon fiber and higher moduli than the standard HT variety, has been commercially available for a number of years. Nearly all carbon fiber manufacturers have focused their development efforts on this IM variety of fibers.

One of the most important steps in improving the properties of the IM variety was reducing the final diameters of the carbon fibers. Since carbon fibers fail at critical flaws, reducing the fiber diameter reduces the probability of encountering a critical flaw in a given test length. Thus, as shown in Fig. 1.32, the strength of a carbon filament increases as the diameter decreases. Two different approaches can be used to produce fine diameter fibers: (i) as suggested by Pepper and Harris, the PAN precursor fiber can

Table 1.2. Mechanical properties of some commercially available PAN-based carbon fibers

	Tensile strength [MPa]	Tensile modules [GPa]
1. Standard HT-Fibers		
T-300 (Toray)	2800	230
AS-4 (Hercules)	3700	250
2. IM-Fibers "New Generation"		
T-800 (Toray)	4900	290
T-1000 (Toray)	7050	290
Celion G 40-700 (BASF)	5000	300
3. HM-Fibers		
M-40 (Toray)	2740	390
M-50 (Toray)	2450	490
Celion G 50-300 (BASF)	2500	360
Celion GY-80 (BASF)	1860	570

be drawn prior to stabilization and carbonization, and (ii) the PAN precursor fiber can be elongated during stabilization. Both approaches reduce surface flaws and other intrinsic flaws in the fiber. Also, both approaches produce a fiber diameter which is so small that skin-core effects caused by differential stabilization are eliminated. The improved uniformity and reduction in flaws both contribute to the improvement in tensile strength observed for IM carbon fibers. In addition, the extra elongation imparted to the fibers results in greater orientation of the stacked condensed ring structure parallel to the fiber axis, accounting for the higher tensile moduli of IM carbon fibers.

Today, the PAN-based carbon fiber industry is relatively mature, and the processes used by the various manufacturers have been optimized to produce a uniform product with consistent mechanical properties. However, the production of mesophase pitch-based carbon fiber has not yet reached this level of maturity. Because of this, the mechanical properties quoted by producers are often difficult to confirm. While companies such as Amoco Performance Products, Inc., produce a relatively standardized product, the property data supplied by some producers is based on laboratory-produced materials. With this in mind, the reported properties of various pitch-based carbon fibers are compared in Fig. 1.33.

In general, the densities of mesophase pitch-based carbon fibers are from 10% to 15% higher than those of PAN-based carbon fibers. Also, the

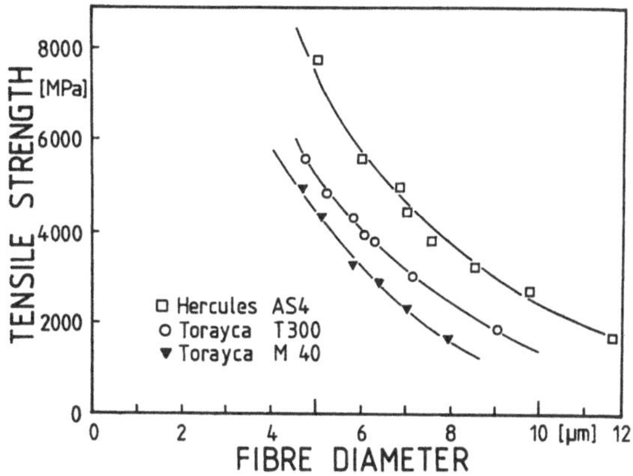

Fig. 1.32. Influence of fiber diameter on the tensile strength of carbon fibers

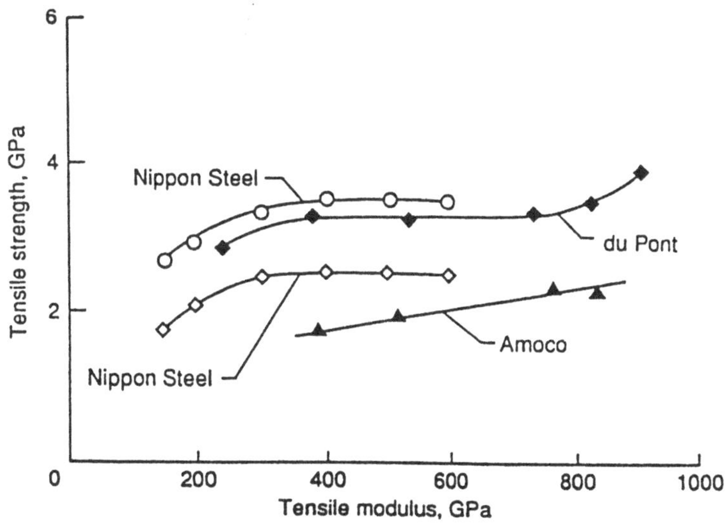

Fig. 1.33. Mechanical properties of various pitch-based fibers [60]

high degree of orientation created as the liquid mesophase is melt spun and the graphitic structure of the final fiber result in thermal and electrical conductivities of mesophase pitch-based carbon fibers which are 1.5 to 2.5 times higher than those of PAN-based carbon fibers. For the same reasons, significantly higher levels of tensile modulus can be achieved with pitch-based carbon fibers. The properties of those pitch-based fibers which are currently commercially available are listed in Table 1.3. The high modulus combined with the excellent thermal properties of these fibers, make pitch-

Table 1.3. Mechanical properties of commercially available pitch-based carbon fibers

Fiber	Tensile strenght [GPa]	Tensile modulus [GPa]	Strain %
Amoco			
Thornel P-25	1.4	160	0.9
Thornel P-55	1.9	380	0.6
Thornel P-75	2.1	520	0.4
Thornel P-100	2.2	724	0.3
Thornel P-120	2.4	830	0.3
Carbonic			
HM 50	2.8	490	0.7
HM 60	3.0	590	0.5
HM 80	3.5	790	0.4

based carbon fibers an attractive reinforcement for advanced ceramic composites – in particular, carbon/carbon.

Figure 1.34 graphically illustrates the present properties of carbon fibers and indicates the additional improvement in properties which are required for the most critical applications [6]. However, every increase in properties comes at a price, and for high volume applications, such as automobiles, reduction in price is also a requirement. Indeed, the use of carbon fiber reinforced composite materials in such applications could significantly reduce energy consumption. Nearly 10 years ago an enormous effort was undertaken to market high performance fibers for a broad range of applications. Unfortunately, the high cost of carbon fibers made their use practical in only a small number of these applications. As Fig. 1.7 shows, the present prices for standard qualities of carbon fiber are as low as 30–40 dollars per kilogram. It is only during the past few years that the price of carbon fibers has begun to decrease significantly. This decrease in price is primarily due to three reasons: (i) the market has grown to the point where economy of scale is now a factor, (ii) new producers have entered the market, in turn increasing competition, and (iii) a significant international market in civilian applications has developed. Nevertheless, the prices for high modulus and special grade carbon fibers remain exorbitant and must be reduced significantly if advanced composites are to be cost competitive with conventional structural materials.

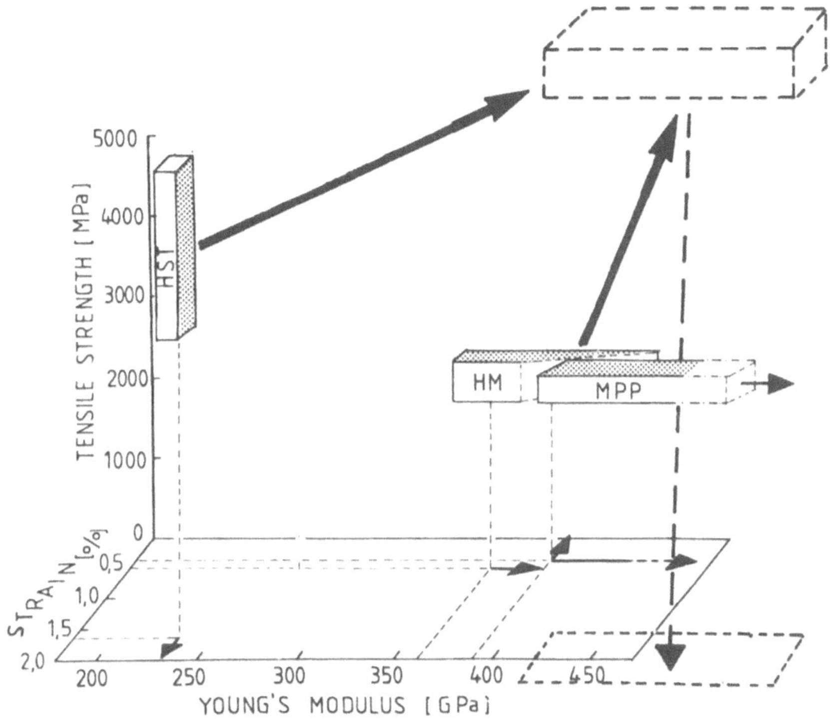

Fig. 1.34. Three-dimensional representation of the current mechanical properties of commercially available carbon fibers [20]

1.9 Structure of Carbon Fibers

Since X-ray and electron diffraction, and scanning and transmission electron microscopy are, perhaps, the most effective methods for investigating the structure of carbon fibers, they have been used by numerous authors to develop and test models for carbon fibers [61–64]. Wide-angle X-ray diffraction is the simplest and quickest technique for the structural characterization of any material. Characteristically, the X-ray diffraction pattern for high-strength PAN-based carbon fibers contains a broad 002 reflection on the equator and a diffuse 100 reflection on the meridian. This two-dimensional graphite-like structure of PAN-based carbon fibers is termed turbostratic. Even though the 002 and 100 reflections become quite sharp when the PAN-based carbon fiber is heat treated to more than 2600 °C, no reflection characteristic of three-dimensional order appears. This differentiates

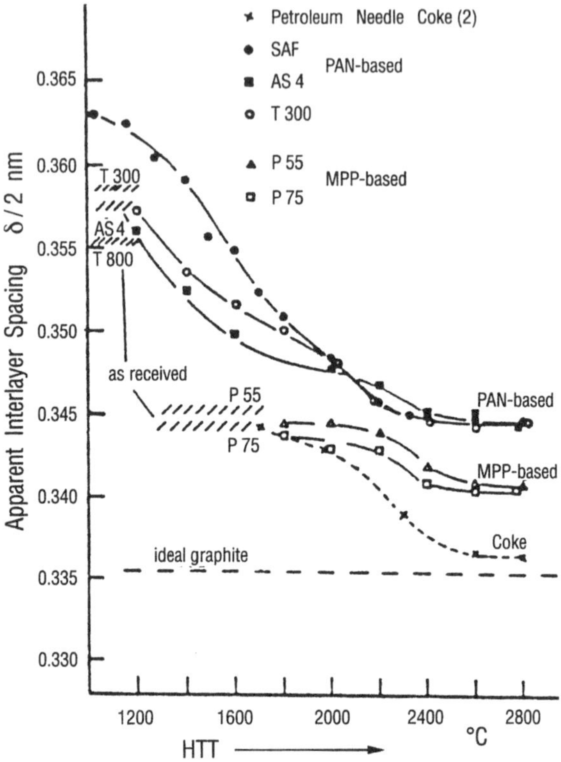

Fig. 1.35. The apparent interlayer spacing, d002, for PAN-based and mesophase pitch-based (MPP-based) carbon fibers compared with that of petroleum needle coke [6]

PAN-based carbon fibers from mesophase pitch-based carbon fibers, which exhibit definite three-dimensional ordering. The apparent interlayer spacings for a variety of carbon fibers, as determined by X-ray diffraction measurements, are plotted in Fig. 1.35. As the figure indicates, the interlayer distance for all commercial HT-type carbon fibers ranges from 3.5 to 3.6 Å. Even when PAN-based carbon fibers are heat treated to temperatures of 3000 °C (HM fibers), the interlayer spacing is still greater than 3.44 Å. By comparison, mesophase pitch-based carbon fibers exhibit three-dimensional ordering interlayer spacings of less than 3.45 Å and, therefore, can be truly called graphite fibers.

Even though X-ray analysis is an important tool for studying the structure of carbon fibers, it has significant limitations. Since it provides only average values, X-ray analysis alone cannot describe the complex crystalline arrangement within the carbon fiber. This need for structural information throughout the fiber makes transmission electron microscopy (TEM) an important tool for analyzing the complex microstructure of car-

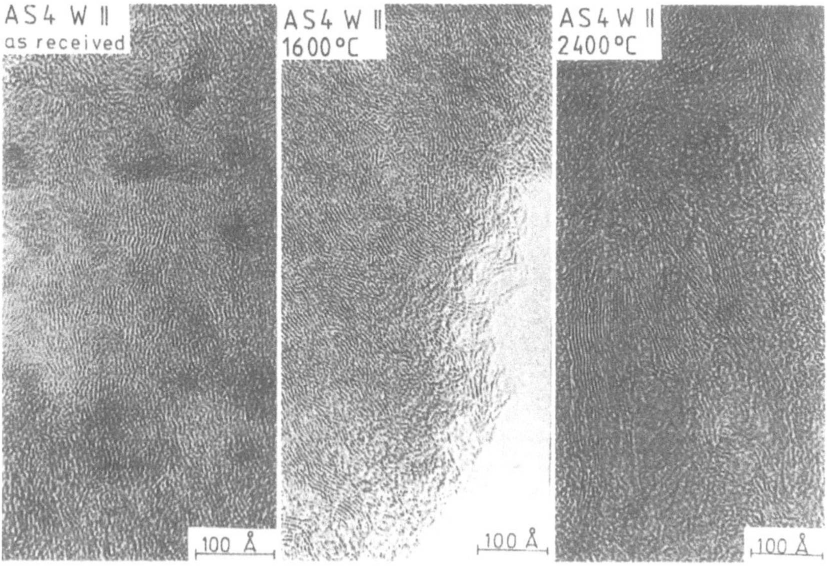

Fig. 1.36. Bright-field lattice-fringe image (longitudinal section) for PAN-based HT-type carbon fiber (Hercules, AS 4 W) [65]

bon fibers. Figure 1.36 shows bright-field images of a commercial HT-type fiber (Hercules AS 4 W). These TEM images show that the microstructure of this PAN-based carbon fiber contains only small regions where the polycarbon layers are arranged parallel to one another. In addition, a wide variety of other irregularities are apparent [65]. Bright-field images of mesophase pitch-based fibers, on the other hand, exhibit a much more developed microstructure and, like graphite, contain many extended areas of parallel layers (Fig. 1.37).

Based on x-ray diffraction results and TEM analyses, numerous models have been proposed for the microstructure of carbon fibers. In the first model, Perret and Ruland [66] proposed that the basic structural unit of PAN-based carbon fiber consists of ribbon-shaped monotonic layers of sp^2-type carbon oriented parallel to the fiber axis and forming microfibrils which undulate over a range of several thousand angstroms (Fig. 1.38). The major deficiency of this early model was that it failed to explain the lateral cohesion within PAN-based carbon fibers. Subsequently, Diefendorf and Tokarsky [67] proposed a basket weave model (Fig. 1.39a). Like the original model of Perret and Ruland, they proposed that undulating ribbon-shaped layers of sp^2-type carbon formed the fundamental building block for PAN-based carbon fibers. They also proposed that these ribbons thicken and the

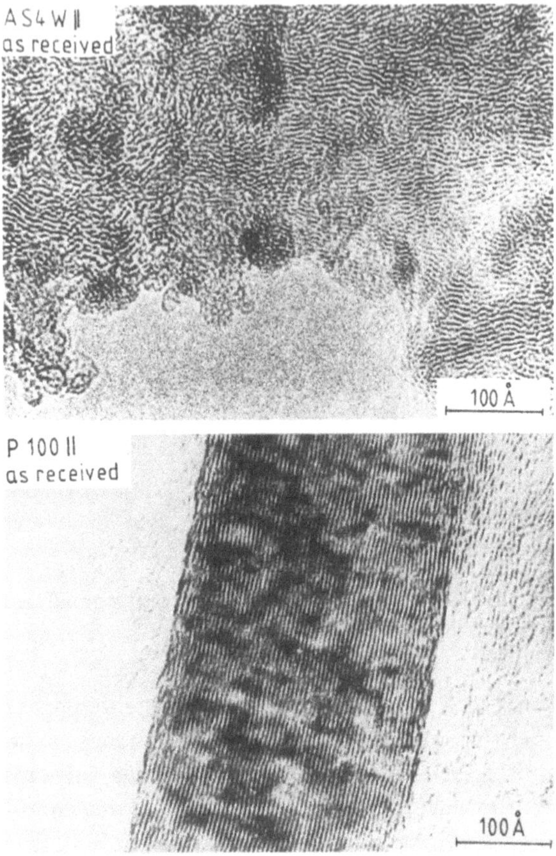

Fig. 1.37. Bright-field lattice-fringe image (longitudinal section) for mesophase pitch-based HM-type carbon fiber (UCC, P 55)

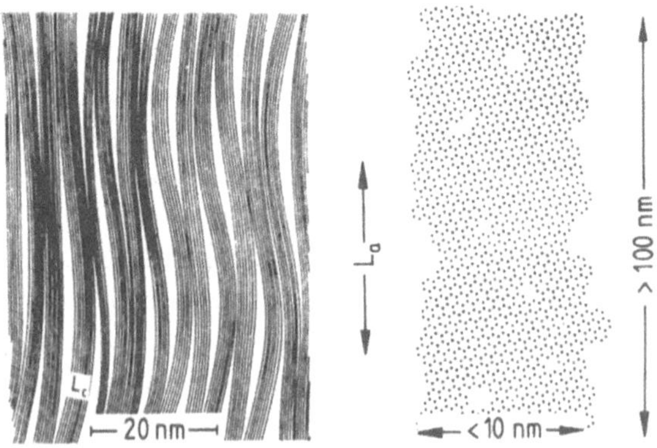

Fig. 1.38. Two-dimensional Ruland model based on stretch-graphitized rayon-based fibers; L_c, stack height; L_a, stack width [66]

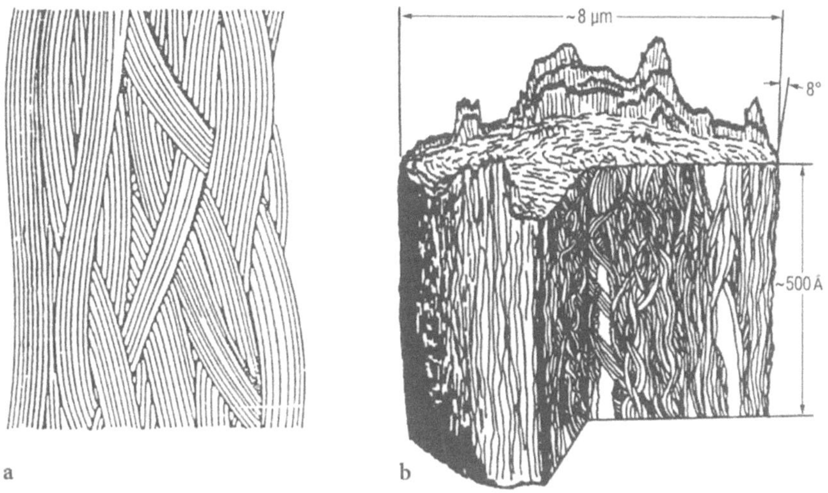

Fig. 1.39a, b. Diefendorf and Tokarsky model for the structure of PAN-based carbon fibers [67]. **a** Basket weave model. **b** Three-dimensional model of Fortafil 5-Y carbon fiber

amplitude of the undulation increases as the modulus of the fiber increases. However, to account for the lateral cohesion of carbon fibers, they proposed that the ribbons form a basket weave structure. Diefendorf and Tokarsky also presented the three-dimensional model shown in Fig. 1.39b in which the ribbons are randomly, radially oriented in the core but arranged in an onion-skin orientation at the fiber surface. This would imply that the axial orientation of the ribbons is stronger, and therefore the modulus is higher, at the surface than in the core of PAN-based carbon fibers. This may account for the change in surface energies when carbon fibers are processed at different temperatures.

Johnson [68] accounted for the voids found in PAN-based carbon fibers by proposing that misoriented crystallites interlink with other oriented and misoriented crystallites forming enclosed, sharp-edged voids, as shown in Fig. 1.40a. However, Johnson also observed distinct differences between the lattice fringes at the skin and in the core of high modulus fibers. Johnson found that the layer planes are oriented more or less parallel to the fiber axis in the core of the fiber, with intermittent voids formed by the complex crystalline interlocking. By comparison, he found that the layer planes in the skin region are essentially parallel to the surface, but in addition to the type of complex crystalline interlocking seen in the core, many fold through angles of up to 180° in a “hairpin” fashion. Based on these results,

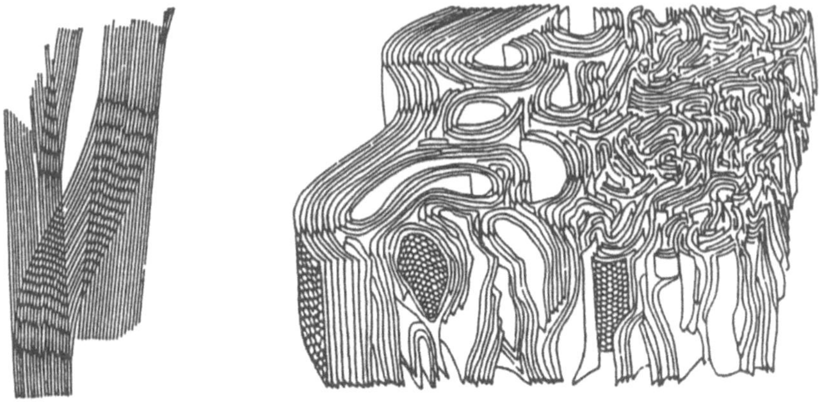

a b

Fig. 1.40 a, b. Johnson's model for the structure of HM PAN-based carbon fibers [68]. **a** Two dimensional representation of longitudinal structure showing interlinking of aromatic layers. **b** Three-dimensional model

Johnson developed the three-dimensional model for the microstructure of high modulus PAN-based carbon fibers shown in Fig. 1.40 b.

After analyzing several types of PAN-based carbon fibers using dark-field and high resolution TEM, Oberlin and Guigon proposed other lamellar models. In their models the lamellae consist of folded and crumpled sheets of layer planes which, although entangled, are not considered to be interlinked. Instead, the sheets are linked to each other at their boundaries. Two different models have been proposed: one for high-strength fibers, and another for high-modulus-fibers. In the high-strength fiber model, lamellae are formed by sheets of basic structural units associated edge–to–edge in a zigzag form, with twisting and tilting at the boundaries. When these large wrinkled sheets touch each other, either transversely or longitudinally, they become bonded together. In the model for high-modulus fibers, the lamellae also can be pictured as crumpled sheets of paper (Fig. 1.41 b). However, in this model the bent layers are more perfect and consist of larger basic structural units (BSU), formed through coalescence of the small BSU seen in high-strength fibers, with elongated and entangled pores. In high-modulus fibers there are fewer boundaries between the BSUs; thus the borders approach the classical definition of grain boundaries.

Because the layer plane structure in pitch-based carbon fibers is highly graphitic, optical microscopy and scanning electron microscopy (SEM) have been extensively used to study their macrostructure. As previously

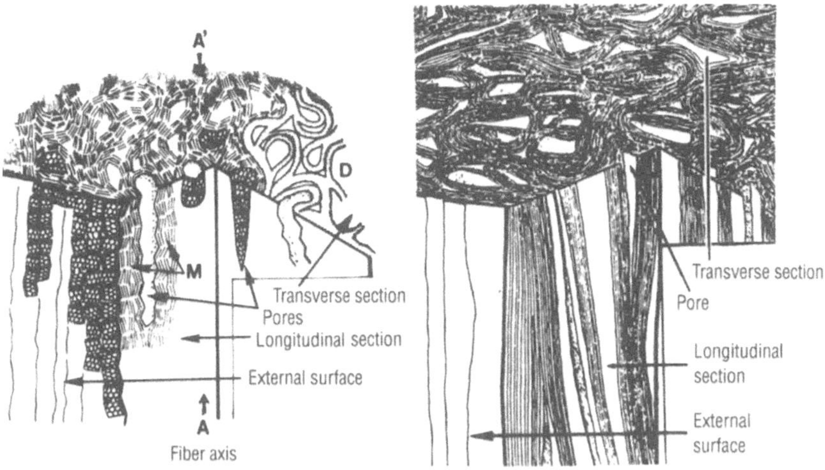

Fig. 1.41. Structural models for PAN-based HT and HM carbon fibers based on transmission electron microscopy studies [69]

mentioned, pitch-based carbon fibers, unlike PAN-based carbon fibers, are melt-spun from a liquid crystalline precursor. Hence, a variety of microstructures can be produced merely by modifying the spinning conditions. While numerous macrostructures have been produced, most commercial pitch-based carbon fibers exhibit either the radial, random, flat-layer or radial-folded morphologies (Fig. 1.42). In a recent study, Endo [70] compared both the macrostructures and the microstructures of Thornel and Carbonic high modulus pitch-based fibers, and PAN-based high modulus fibers. SEM micrographs showed that Thornel P-120 fibers exhibit the characteristic bent flat-layer macrostructure, while the macrostructure of

Fig. 1.42. Microstructures of pitch-based carbon fibers [60]

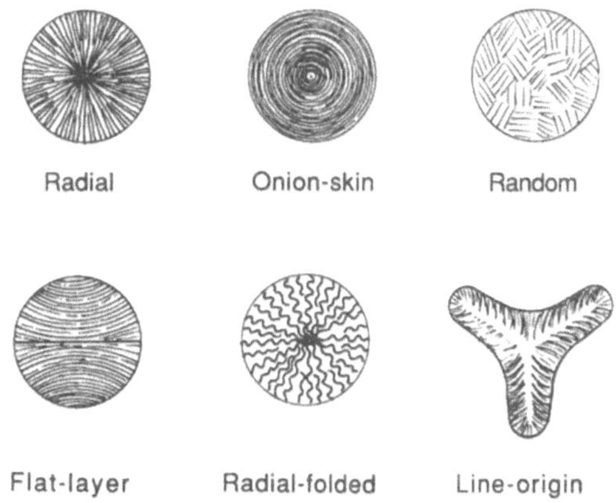

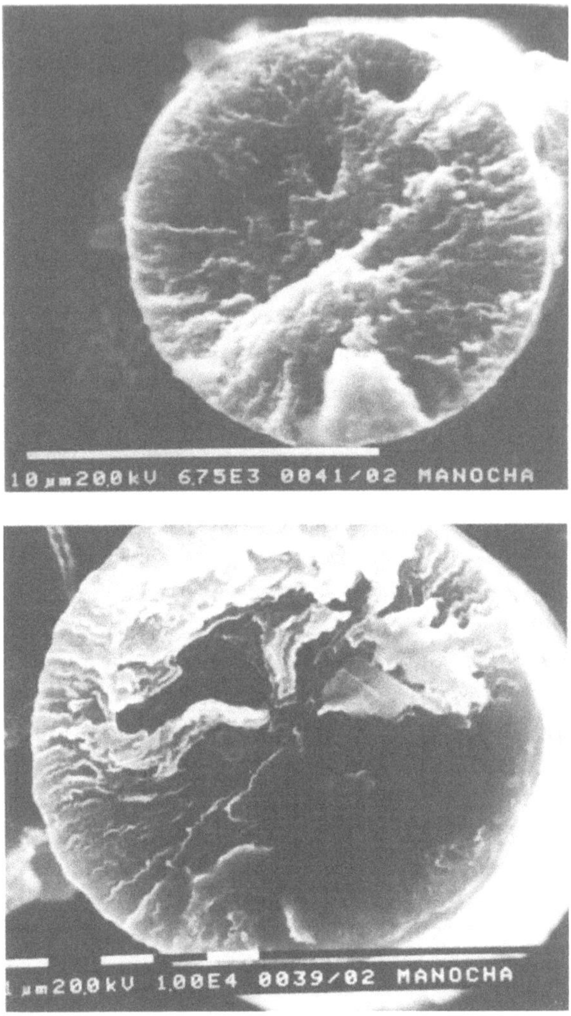

Fig. 1.43 a, b. Scanning electron micrographs showing the morphology of pitch-based carbon fibers. **a** Thornel P-120. **b** Carbonic HM 60

Carbonic HM 60 fibers is radial-folded (Fig. 1.43). Also, the Thornel fibers have a high degree of preferred orientation and possess three-dimensional graphitic structure. By comparison, the preferred orientation of the Carbonic HM 60 pitch-based carbon fibers was slightly less. The structure of the Carbonic fibers consisted of turbostratic layer planes of carbon which were oriented within 15° of the fiber axis. In addition, the degree of graphitization exhibit by the Carbonic fibers was intermediate between that of the Thornel and the Torayaca fibers. Figures 1.44 shows Endo's schematic representation of the microstructure of these Thornel and the Carbonic

Fig. 1.44a, b. Structure models for mesophase pitch-based carbon fibers [70]. a Thornel P-120. b Carbonic HM60

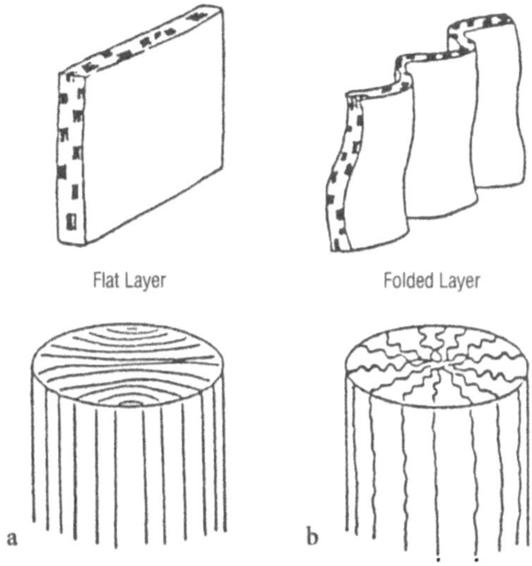

pitch-based carbon fibers. Endo concluded that pitch-based fibers with distorted textures, such as Carbonic, are more likely to exhibit balanced properties than pitch-based fibers with flat-layered textures. In other words, mesophase pitch-based carbon fibers with distorted transverse textures and, thus, less highly graphitized structures, are likely to exhibit increased tensile strengths.

1.10 Fracture Behavior of Carbon Fibers

The direct relation between preferred orientation and Young's modulus is well established for carbon fibers. For both PAN-based and pitch-based carbon fibers, their tensile moduli increase as the orientation of the graphite-like layer planes within the fibers becomes more parallel to the fiber axis. Therefore, improving the orientation of the graphitic structure, through techniques such as stress graphitization or boron doping, has always been the primary objective in the production of a carbon fiber with a high level of stiffness. However, the development of tensile strength in carbon fibers is not nearly as straightforward or well understood. Numerous investigations [41, 71] have shown that carbon fibers, like all brittle materials, are flaw-sensitive. Brittle materials fail at randomly distributed critical flaws, and as the length of the fiber increases, the probability of encountering a critical flaw increases. Thus, the measured tensile strength of a carbon fiber will depend upon the gauge length subjected to testing. Based on this principle, Chawatiak et al. [72] and Beetz [73] estimated the limiting value of tensile strength for PAN-based and pitch-based carbon fibers to be 7 GPa and 3.2 GPa, respectively. However, many new varieties of PAN-based and pitch-based carbon fibers possess tensile strengths that are significantly higher than these estimated limits. This implies that, in addition to flaw sensitivity, other factors such as the structure of the carbon fiber itself can limit its strength. Recently, Johnson [68] presented an excellent explanation of the structure-property relationship for carbon fibers, based on a mechanism originally proposed by Reynolds and Sharp [74]. Johnson noted that since even in highly oriented carbon fibers all of the crystallites are not perfectly aligned, the locking of some misoriented crystallites into the fiber structure is inevitable (Fig. 1.45a). When tensile stress is applied to carbon fibers, these misoriented crystallites will tend to align, until their movement is restricted by a disclination in the structure. If the stress is sufficient, the basal plane of this misoriented crystallite will rupture (perpendicular to the fiber axis), relieving the stress within the fiber (Fig. 1.45b). If the crystallite size in the directions of propagation,

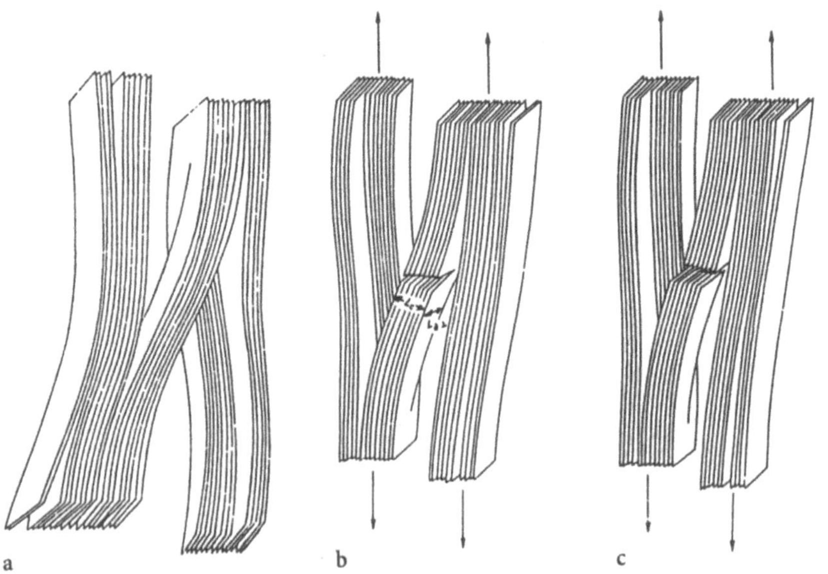

Fig. 1.45a–c. Model for tensile failure of carbon fiber [68]. **a** A misoriented crystallite linking two crystallites parallel to the fiber axis. **b** Basal plane rupture under applied stress. **c** Complete failure of the misoriented crystallite. L_a, stack width; L_c, stack height

either L_a or L_c, is greater than the critical flaw size, C, the crack will continue to propagate, causing the fiber to fail catastrophically. This critical crack size can be determined by the Griffiths relationship for brittle solids,

$$\sigma = \sqrt{\frac{2\,E\,\nu_a}{\pi\,C}}$$

where, σ = tensile strength,
E = tensile modulus,
ν_a = surface energy.

Even if the ruptured crystallite is smaller than the critical flaw size, catastrophic failure can still occur if the structure of the crystallite initiating failure is sufficiently continuous with the neighboring crystallites to permit the crack to continue propagating. In fact, as Edie and Stoner explain [60], crack propagation between crystallites would appear to be the dominant failure mechanism for most carbon fibers. If we assume a surface energy of 4.23 J/m^2 (the commonly accepted value), the above equation predicts the critical flaw size for typical type I, PAN-based carbon fibers with a tensile strength of 2.8 GPa and a modulus of 400 GPa to be 136 nm. Assuming the same surface energy for type II carbon fibers with a strength of 5.7 GPa and a modulus of 300 GPa, the critical flaw size should be 25 nm. By comparison, the measured values of L_a and L_c, for either type I or type II fibers, are much smaller than these predicted critical flaw sizes. Since the crystallite dimensions of these fibers are considerably smaller than the critical flaw size, it would appear unlikely that the failure of a single crystallite would result in catastrophic failure of the fiber. This would indicate that most PAN-based carbon fiber failures result from crack propagation outside the initiating crystallite into the neighboring crystallites.

As previously mentioned, the microstructure of PAN-based carbon fibers lacks three-dimensional order and consists of folded and crumpled sheets of layer planes with relatively large interlayer spacing. Because of this, the crystallites in PAN-based carbon fibers are much smaller than the critical flaw size. This not only accounts for the characteristic high strength of PAN-based carbon fibers but also explains their high failure strain. When a tensile load is applied to the fiber, these small individual crystallites are able to fail and relieve the applied stress without causing a catastrophic failure.

The same fracture criteria would also appear to hold true for pitch-based fibers. Using the data from Endo's study of Thornel and Carbonic pitch-based fibers [70], the above equation predicts that the critical length is approximately 130 nm. While this is smaller than that estimated for the Torayca PAN-based fiber studied by Endo (200 nm), it is still ten times larger than the average crystallite thickness. Endo noted that the Thornel fibers exhibited a flat-layer transverse texture and contained relatively large regions of graphite-like crystallinity with three-dimensional ordering. Unlike the smaller, more disrupted structure found in PAN-based fibers, this extended structure creates few natural crack absorbing or crack stopping mechanisms, making crack propagation more likely. The resulting flaw sensitivity makes carbon fibers with this type of structure poorly suited for advanced ceramic composites.

While this extended structure was typical for most early pitch-based carbon fibers, newer types exhibit a variety of structures. For example, Carbonic pitch-based fibers contain comparatively small crystallite units arranged in a turbostratic structure and still exhibit excellent preferred orientation along the fiber axis. The transverse texture of the Carbonic fiber can be best described as folded-layered, a more disrupted structure which might be expected to inhibit crack propagation. Therefore, it is not surprising that these fibers exhibit higher tensile strengths and failure strains than Thornel pitch-based carbon fibers. As with PAN-based fibers, these improvements can be explained by the cumulative effect of numerous non-catastrophic failures in misoriented crystallites, which reduce flaw sensitivity.

It is interesting to note that while the graphitic structure, the tensile strength, and the failure strain of Carbonic pitch-based fibers are similar to those of PAN-based carbon fibers, their tensile modulus is considerably higher. In fact, their tensile modulus and degree of preferred orientation are nearly equal to those of Thornel pitch-based carbon fibers. This demonstrates that a high degree of graphitization is not a prerequisite for high-modulus fibers. Instead, a high degree of preferred orientation is the primary requirement. Thus, it is quite possible to produce pitch-based carbon fibers which exhibit high strengths, high moduli and high failure strains.

While the desire for increased tensile properties has driven most recent carbon fiber development efforts, in many structural applications the compressive properties of the reinforced composite are more critical. Like most other properties, compressive strength of the composite is directly proportional to the compressive strength of the reinforcing fiber. Here again, the properties of PAN-based and pitch-based carbon fibers differ. As Fig. 1.46 shows, round PAN-based carbon fibers exhibit higher compressive

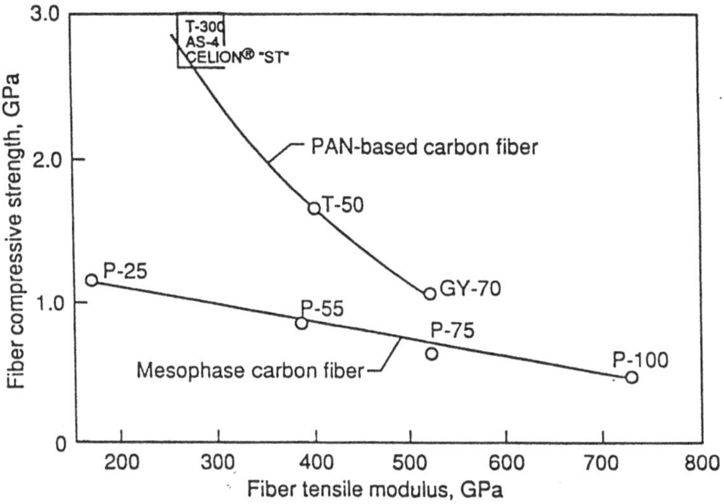

Fig. 1.46. Compressive strength as a function of tensile modulus for carbon fibers [75]

strengths than round pitch-based carbon fibers with similar tensile moduli [75]. Based on experimental observations, Johnson concluded that fiber microstructure has a major influence on both the compressive properties and the compressive failure mode of carbon fibers. While PAN-based fibers and low-modulus pitch-based carbon fibers failed by buckling, high modulus pitch-based fibers failed in a simple shear mode. Thus, it appears that less extended structures not only reduce flaw sensitivity of carbon fibers but also inhibit compressive shear failure. Figure 1.47 com-

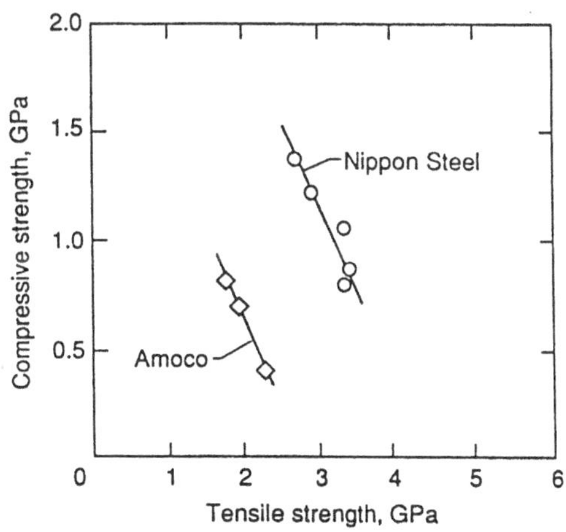

Fig. 1.47. Compressive strength as a function of tensile strength for two different pitch-based carbon fibers [60]

pares the compressive strengths of two commercial pitch-based carbon fibers – Amoco fibers which have a flat-layered structure and Nippon Steel fibers which have a more random structure. When equivalent tensile strengths are compared, the fibers with the random structure exhibit higher compressive properties than those with a flat-layered structure (further evidence that disrupted structures inhibit compressive shear failure).

1.11 Emerging Carbon Fibers

1.11.1 PAN-Based Carbon Fibers

A number of PAN-based carbon fibers with improved properties, particularly higher tensile strength and higher modulus, are currently under development. Several new fibers which have already been commercially announced by the manufacturer were described in the previous sections. Continuing development will certainly lead to further improvements; in particular, one might anticipate that intermediate modulus carbon fibers with strain to failures exceeding 2% will be produced. While these high strain to failure fibers will make new designs possible, they will, in turn, require the development of new matrix systems.

In addition to this evolutionary development of improved fiber properties, entirely new PAN-based carbon fiber processes are being considered. Producers in Europe and in the United States have explored the use of utilizing melt-spun technology to form the PAN precursor fibers. This route eliminates many of the environmental problems associated with solvent processes. In addition, the fiber coagulation step is eliminated, simplifying the spinning process. In this process the acrylonitrile is polymerized in an aqueous suspension, purified, and fed to an extruder. After it is melted in the extruder, the plasticized polymer is extruded into fiber form. Besides eliminating solvent recovery and fiber coagulation, the process offers additional advantages. In solution spinning large amounts of residual solvent are removed during the coagulation step, leading to a skin-core variation in the fiber morphology. Since the melt-spinning technique eliminates the need for this solvent removal step, the cross-sectional morphology of fibers produced by this process should be more uniform. The melt-spun technology also allows fibers with a variety of cross-sectional shapes to be produced, merely by changing the shape of the extrusion die. BASF has used melt-spun technology to produce pilot-scale quantities of noncircular PAN-based carbon fibers with rectangular and tetralobal cross sections (Fig. 1.48). These noncircular carbon fibers could offer several advantages. Since PAN-based carbon fibers apparently fail in buckling when they are subjected to a compressive load, the higher moment of inertia of these noncircular fibers should result in an increase in compressive strength. Certain fiber shapes, such as rectangular, yield higher packing densities in composites than round fibers. Also, noncircular shapes make it possible to increase the interfacial surface area between fibers and matrix. This in turn could in-

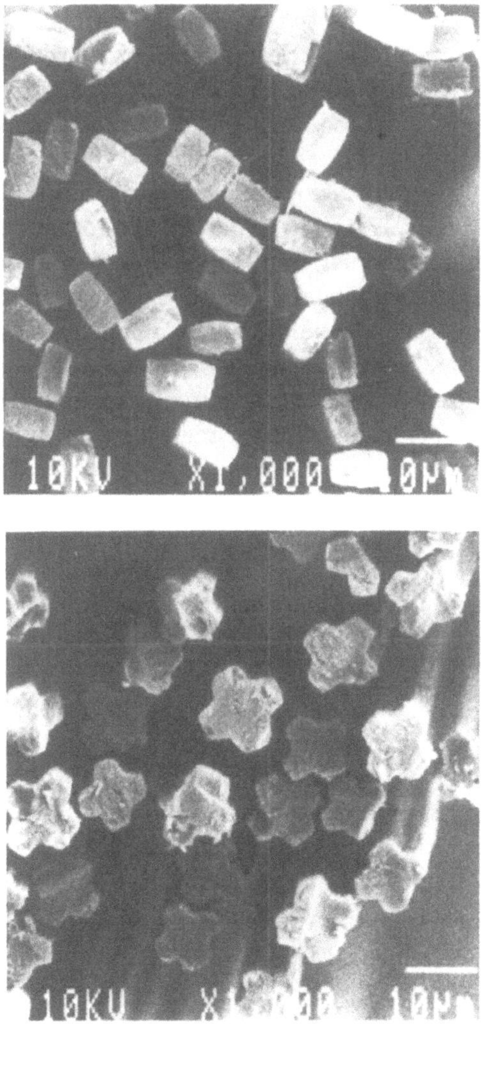

a

b

Fig. 1.48 a, b. Noncircular PAN-based carbon fibers produced with melt-spun technology [76]. **a** Rectangular fiber, **b** Tetralobal fiber

crease the interlaminar shear strength of the composite, a primary consideration for ceramic matrix and carbon matrix composites.

1.11.2 Pitch-Based Carbon Fibers

Although commercial pitch-based carbon fibers exhibit extremely high moduli, their tensile and compressive properties are much less than those of PAN-based carbon fibers. These deficiencies are not necessarily inherent to pitch-based fibers, but rather they are the direct result of the technology currently employed by commercial manufacturers. Precursor impurities, even submicron in size, can dramatically reduce the tensile strength of pitch-based carbon fibers. Unfortunately, the current heat-soaking processes used to produce the mesophase precursor in many pitch-based carbon fiber processes tend to concentrate the impurities found in raw pitch feed. Since mesophase is more dense than isotropic pitch, during the heat-soaking process it settles to the bottom of the reactor along with many of the solid impurities found in the feed. Even though the mesophase product is subsequently filtered, the removal of submicron impurities from the molten mesophase is difficult, if not impossible. Newer methods for producing the mesophase precursor appear to overcome this difficulty. One technique employs supercritical extraction to separate the mesophase from the raw pitch. In this method the raw pitch is completely dissolved at elevated temperatures and pressures in a solvent such as toluene. The solution is then separated by passing it through a series of stages with controlled continuously decreasing pressures. The solid impurities, along with any extremely high molecular weight material, are removed at the first stage of the separation process. Another process, currently being developed by Mitsubishi Chemical, Inc., begins with a pure chemical feed to produce a relatively pure mesophase precursor. In this method a strong acid catalyst, a mixture of HF and BF_3, is used to convert a naphthalene or anthracene feed to a mesophase product. Both of these newer methods offer two distinct advantages over current heat-soaking processes. The processes are continuous, and the mesophase product is more uniform than that produced by the heat-soaking technique. Recent studies by Mochida et al. [77] and Edie et al. [78] indicate that both supercritical extracted mesophase and Mitsubishi mesophase are more spinnable than heat-soaked mesophase. In addition, fibers formed from these newer precursors exhibit tensile strengths which are 50% greater than those of current commercial pitch-based carbon fibers.

As mentioned earlier, the strength of pitch-based carbon fibers is also influenced by their structure. Pitch-based carbon fibers with radial and flat-layer structures are more flaw sensitive than those with more disrupted transverse structures. Also, fibers with radial or flat-layer structures tend to split open at high heat treatment temperatures. Current research indicates that the properties of pitch-based carbon fibers can be improved dramatically if novel techniques, such as modifying the fiber structure and/or shape are employed. Disrupting the flow of mesophase during extrusion can create random or radial-folded transverse textures in the final carbon fiber. As previously mentioned, fibers with these more disrupted transverse textures appear to be less flaw sensitive and to exhibit higher strengths, both tensile and compressive.

Another approach to modifying the balance of properties involves modifying the cross-sectional shape of the fiber. In fact, since pitch-based fibers are produced by melt-spinning, a variety of shapes can be produced by merely changing the shape of the extrusion capillary. Additionally, varying the shape of the capillary also changes the shear stress profile that the precursor is subjected to during extrusion. Since the precursor is a liquid crystal, the transverse orientation of the as-spun fiber is a direct result of this imposed stress. Thus, varying the shape of the extrusion capillary not only modifies the fiber shape, but also changes the transverse structure of the final carbon fiber. Non-circular pitch-based carbon fibers were first produced by Edie et al. [79]. In this early study, a solvent extracted mesophase pitch was extruded through trilobal and round capillaries to form trilobal and round fibers of equal cross-sectional area. After heat treatment, the tensile strengths of the trilobal fibers were found to be as much as 27 % greater than those of the round fibers. Edie et al. postulated that this increase in strength was most likely the result of the unusual transverse texture of the trilobal fiber. When fibers of equal cross-sectional area are compared, stabilization tests show that non-circular fibers oxidize more rapidly than solid round fibers. This might be expected since the limbs of trilobal fibers are thinner than the diameter of the round fiber. In a later study, Edie et al. [80] extruded a heat-soaked mesophase into trilobal and c-shape fibers, using the spinnerette capillaries shown in Fig. 1.49 a and 1.49 b, and determined the optimum processing parameters as well as the process conversion for each fiber shape. In this investigation processing and properties of larger noncircular fibers were compared with smaller round fibers with equivalent limb thickness and diameters, respectively. An SEM micrograph of a typical trilobal fiber produced by Edie et al. [80] is shown in Fig. 1.50 a. This comparison was selected to ensure that diffusion effects would be

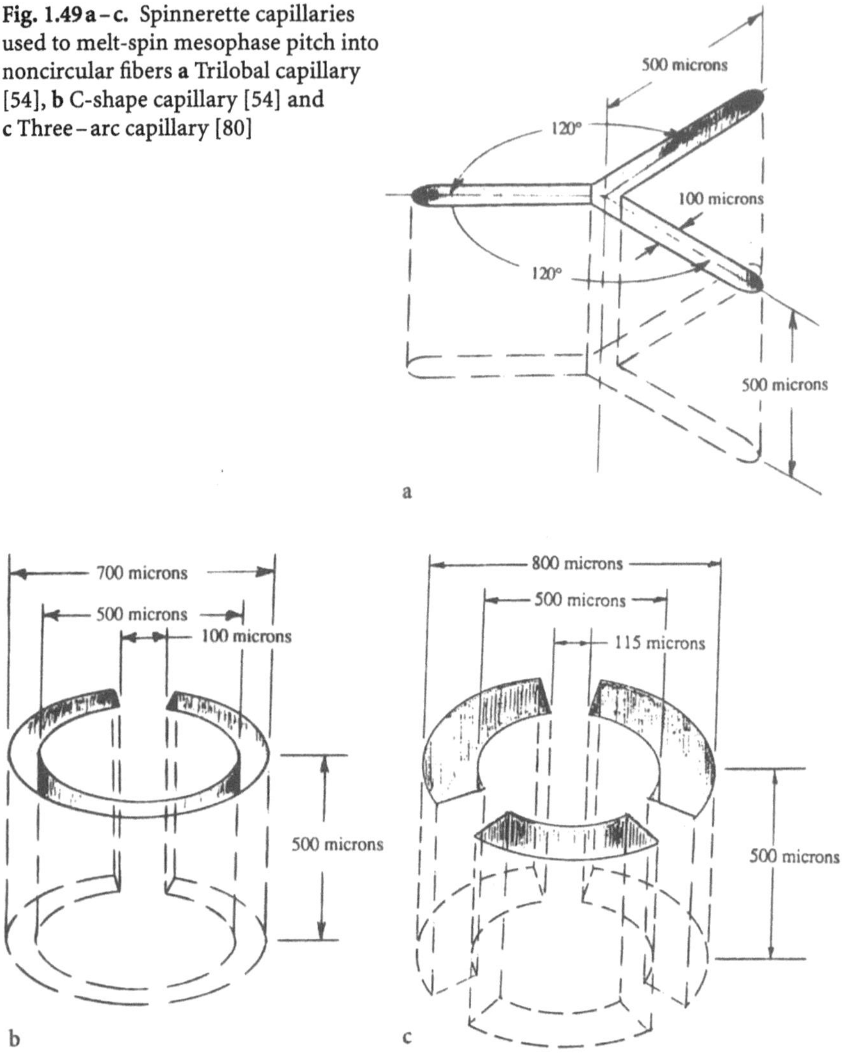

Fig. 1.49 a – c. Spinnerette capillaries used to melt-spin mesophase pitch into noncircular fibers **a** Trilobal capillary [54], **b** C-shape capillary [54] and **c** Three-arc capillary [80]

equivalent during stabilization and heat treatment. The results showed that although the noncircular fibers were significantly larger in cross-sectional area their tensile strengths were nearly equal to those of the smaller round fibers. In addition, the larger noncircular fibers could be spun at higher conversions and heat treated more easily than the smaller circular pitch-based carbon fibers. The increased spinning conversion is the direct result of reduced threadline tension during fiber drawdown. As Edie and Dunham [52] explained, the drawdown step generates high tensions in the pitch fibers. Because of this, even at optimum spinning conditions, the pitch fiber is very close to brittle fracture. One effective method for reducing the threadline tension is to increase the size of the fiber. Since this high threadline stress contributes to the low conversion of current pitch-based fiber processes, noncircular fibers may offer an opportunity to reduce processing costs a critical issue if carbon fibers are to be used in high volume applications.

More recently, Edie et al. [80] have produced hollow-shaped fibers using a spinnerette containing extrusion capillaries such as that shown in Fig. 1.49c. Like other noncircular fibers, this fiber could be easily heat treated. A scanning electron micrograph of a typical hollow fiber is shown in Fig. 1.50b. Hollow carbon fibers may offer advantages in structural applications where either weight or compressive properties are critical. Niederstadt [81] showed that the bending stiffness of composites reinforced with hollow fibers is superior to those reinforced with solid round fibers. Also, their higher moment of inertia make hollow fibers less prone to buckling failure than solid fibers of equal cross-sectional area. However, when subjected to a compressive load, current pitch-based fibers fail in shear. As previously mentioned, disrupted (radial-folded or random) transverse textures appear to inhibit the compressive shear failure of pitch-based carbon fibers. Also, disrupting the transverse structure during melt-spinning can result in improved tensile strength. As Hamada et al. [82] first demonstrated, stir-

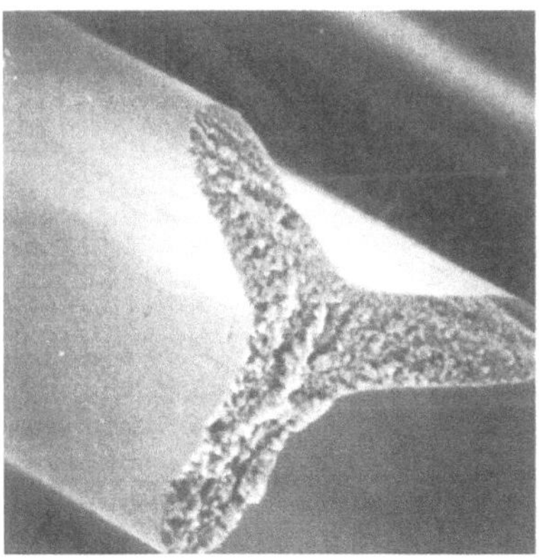

a

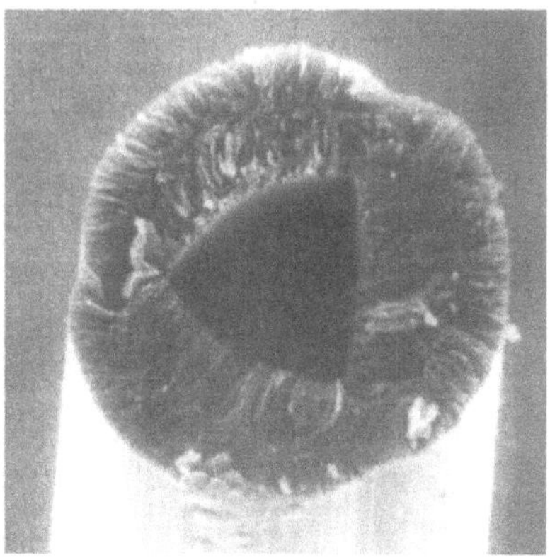

b

Fig. 1.50 a, b. Typical scanning electron micrographs of noncircular fibers. **a** Trilobal fibers (Limb width = 10 microns) [54], **b** Hollow fiber [80]

ring the mesophase prior to extrusion is an effective way to produce pitch-based carbon fibers with a disrupted a transverse texture. Fitzer and Metzler [83] utilized this technique to produce hollow fibers with excellent tensile properties. Like Hamada, they stirred a hydrogenated mesophase pitch, which was from 90 % to 100 % anisotropic, prior to extrusion. However, Fitzer and Metzler used modified c-shape extrusion capillaries, similar to that shown in Fig. 1.51 a. As the figure shows, the c-shape channel has an expansion at one end which served as a material reservoir during extrusion, ensuring closure of the two ends of the "c" during fiber drawdown. Once melt-spun, the mesophase precursor fibers were stabilized over a range of

a

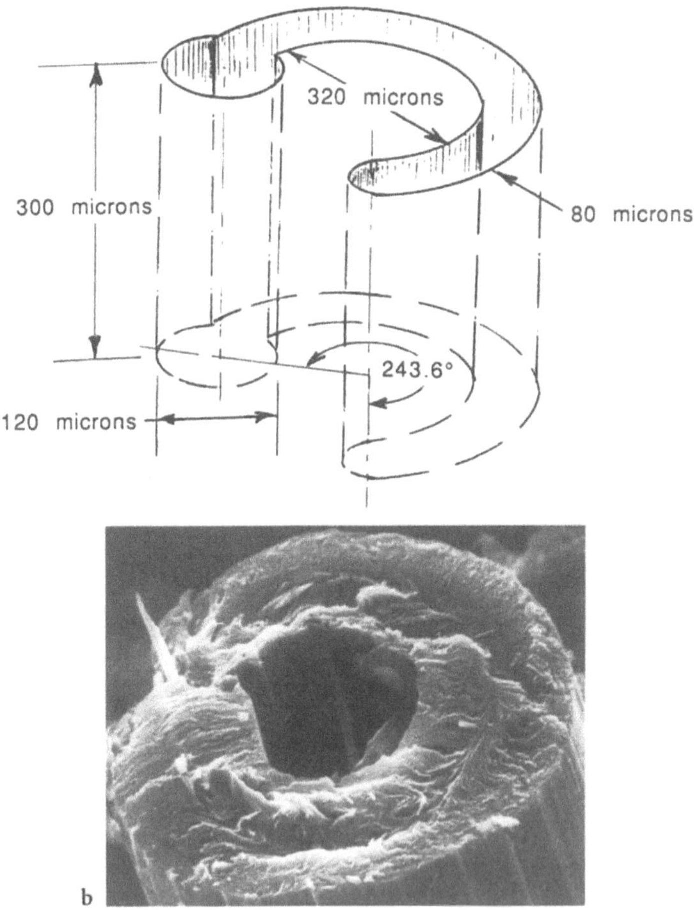

b

Fig. 1.51. **a** C-Shaped capillary used to melt-spin hollow fibers. **b** SEM photograph of typical fiber after stabilization

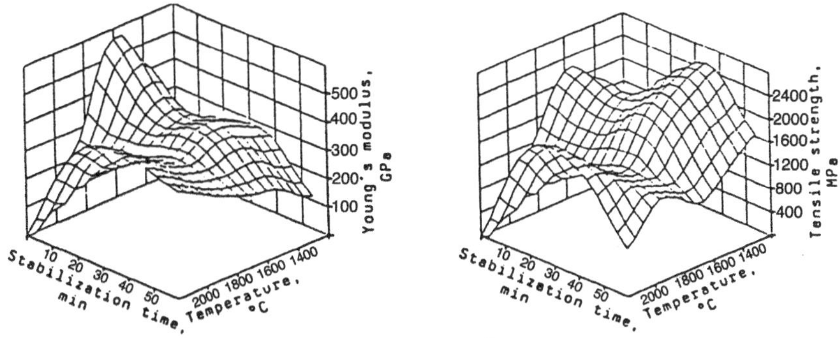

a b

Fig. 1.52a, b. Mechanical properties versus stabilization time and carbonization temperature for hollow pitch-based carbon fibers [83]. **a** Variation in Young's modulus, **b** Variation in tensile strength

conditions. Figure 1.51b shows one of these hollow mesophase pitch-based fibers after stabilization. As in the production of PAN-based carbon fibers, the stabilization step is critical, both in terms of product quality and process economics. Typical results for hollow fibers stabilized at 310 °C for various times and then subjected to carbonization are shown in Fig. 1.52a and 1.52b. Fitzer and Metzler found that when the mesophase was distorted during spinning, the mechanical properties of the hollow fibers were about 50% more superior than those of solid fibers formed from the same mesophase using conventional spinning techniques.

Perhaps the most unique characteristic of pitch-based carbon fibers is their high thermal conductivity. Commercial pitch-based fibers are available with thermal conductivities three times that of copper. Although these fibers should be ideal for high volume thermal management applications such as reinforcing electrical components and integrated circuits, their high price ($ 2000/kg) makes their use impractical. Here again noncircular fibers may well prove to be superior to conventional round fibers. In recent studies, Edie et al. [84] extruded a mesophase precursor through rectangular capillaries, similar to that shown in Fig. 1.53a, to form ribbon-shaped fibers. After stabilization and carbonization, tests showed the electrical resistivity of these relative large ribbon-shaped fibers to be lower than that of equal modulus round fibers. Consequently, these ribbon-shaped fibers promise to exhibit higher thermal conductivity than round fibers. Rogers et al. [85] demonstrated that these ribbon-shaped fibers also graphitize more readily than round fibers which were melt-spun from the same meso-

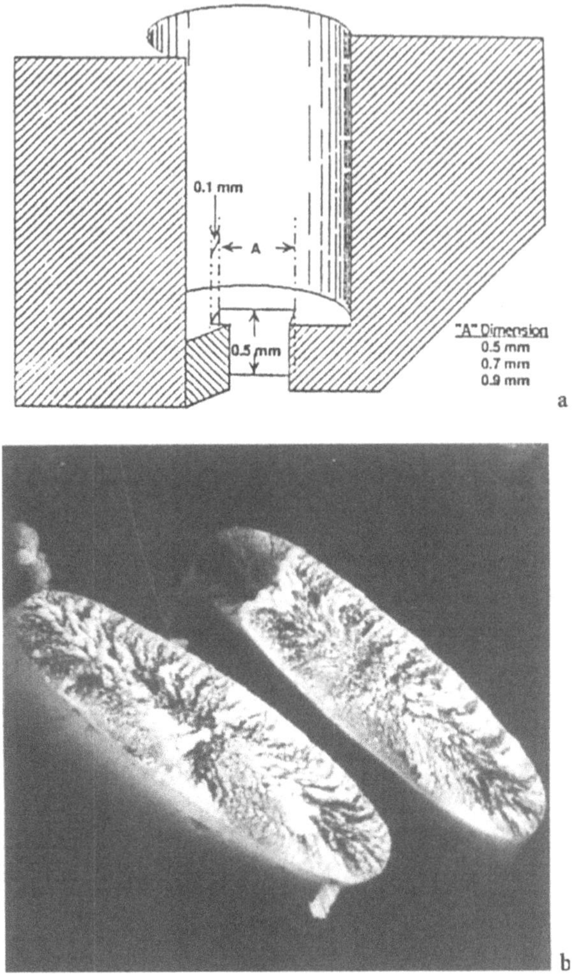

Fig. 1.53. **a** Rectangular capillary used to extrude ribbon fibers. **b** SEM photograph of typical fibers after carbonization (Limb width = 8 Microns)

phase. As Fig. 1.53b shows, these ribbon-shape fibers have what has been termed a line-origin transverse texture. Apparently this texture, which develops during extrusion through the rectangular capillary, is less stressed and, thus, collapses more readily during the high temperature heat treatment. This improved graphitizability, combined with the improved spinnability of the larger ribbon-shaped fibers, give them potential cost advantages over round fibers. Their superior thermal properties and potentially lower cost make ribbon-shaped pitch-based carbon fibers promising future reinforcements for applications ranging from low temperature thermal management to high temperature carbon/carbon composites.

1.12 Adhesion of Carbon Fibers

When a composite structure is loaded, the applied stress must be distributed from the compliant matrix to the load-bearing fibers. Therefore, adhesion between the fibers and the matrix is critical and, in fact, a precondition for proper stress transfer (Fig. 1.54). Both PAN-based and pitch-based carbon fibers, especially the high-modulus fibers, are difficult to wet with resins and are not wet at all with molten metals. Mesophase pitch-based fibers are a particular problem. The degree of adhesion depends on the composition and structural perfection of the carbon layers at the surface of the fibers. Since polymer matrix composites represent the largest market for carbon fibers, all commercial carbon fibers are subjected to surface treatment steps to ensure their compatibility with the resins. While the methods used are proprietary and, thus, not disclosed in the literature, numerous patents provide some insight. The commercial techniques can be broadly classified as: wet oxidation, dry oxidation, and anionic oxidation. The objective of this process step is to increase the number of active surface groups, primarily oxides, and in some cases roughen the surface of the fibers. Numerous surface oxides are created by this process: carboxylic acid, hydroxylic acid, ketones, and even quinonic oxygen and similar structures (Fig. 1.55) [86]. While the exact mechanism is unknown, it is well documented that these carbon fiber surface treatment techniques significantly increase the interlaminar shear strength of polymer matrix composites.

The situation is quite different when we consider reinforcing high-temperature matrix materials, specifically carbon and ceramic matrices with carbon fibers. While good adhesion between the fibers and polymer matrix is desirable for effective load distribution, strong bonding between the fibers and carbon and ceramic matrices is not favored. Therefore, in these applications, non-surface treated carbon fibers are often preferred. How-

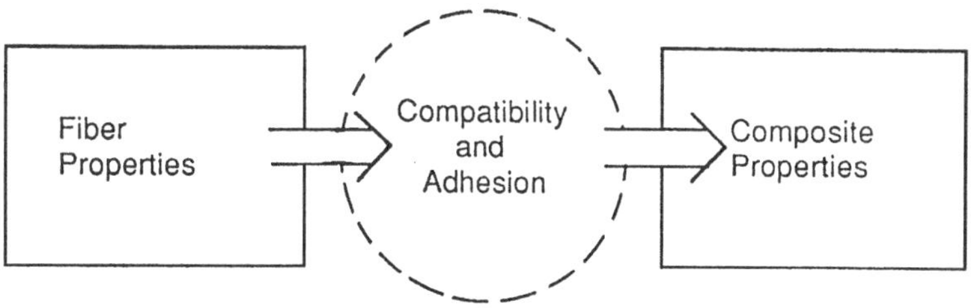

Fig. 1.54. Precondition for the transfer of fiber properties to composite properties

Fig. 1.55. Principal types of oxides on carbon fiber surfaces [86]

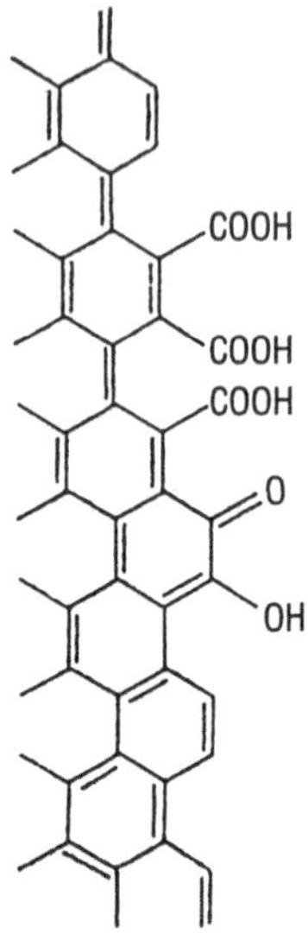

ever, for specific applications with ceramic matrices, carbon fibers are sometimes coated with high-temperature ceramics, such as SiC and TiC [87].

1.13 References

1. Talley CP (1959) J Appl Phys, 30:1114
2. Wawnes FE (1967) Boron Filaments. In: Brautman L, Kersch R (eds) Modern composite materials, Addison-Wesley
3. Lowrie RE (1967) Glass Fibers for High Strength Composites. In: Brautman L, Kersch R (eds) Modern composite materials, Addison-Wesley
4. Gupta PK (1988) Glass Fibers for Composite Materials. In: Bunsell AR (ed) Fiber Reinforcements for Composite Materials, Elsevier Scientific Publishing Co, New York, NY
5. Lubin G (ed) (1982) Handbook of Composites, Van Nostrand Reinhold Co, New York, NY
6. Fitzer E (1989) Carbon 27:621
7. Hillig WB (1988) Verbundwerk '88, Proceedings of International Conference on Reinforced Materials and Composite Technologies, Wiesbaden, Germany, September
8. Kelly B (1981) High Temperatures-High Pressures, 13:245
9. Cooper C (1981) Chemistry and Industry, 18:678
10. Bokros JC (1973) In: Walker PL, Thrower PA (eds) Chemistry and Physics of Carbon, Marcel-Dekker 9:103
11. Bacon R (1973) Carbon Fibers from Rayon Precursors. In: Walker PL, Thrower PA (eds) Chemistry and Physics of Carbon, Marcel-Dekker 9:1

12. Bacon R, Schalamon W (1968) Carbon 6 (2):211
13. Hawthorne H (1971) In: Proceedings of First International Conference on Carbon Fibers, Plastics Institute, London, 81, February
14. Shindo A (1961) Govt Res Inst, Osaka, Japan, Report no. 317
15. Phillips LN, Watt W, Johnson W (1965) British Patent 1110791
16. Otani S (1965) Carbon 3:31
17. Barr JB, Chwastiak S, Didchenko R, Lewis IC, Lewis RT, Singer LS (1976) Appl Polym Symp, No 29:161
18. Singer LS, (1977) US Patent 4005183
19. Diefendorf RJ, Riggs DM (1980) US Patent 4208267
20. Fitzer E, Heine M (1988) In: Bunsell AR (ed) Fiber Reinforcements for Composite Materials, Elsevier, 73–148
21. Shindo A (1960) Japanese Patent 36145
22. Fitzer E, Frohs W, Heine M (1986) Carbon 24:387
23. Johnson W, Watt W, Phillip LN, Moreton R (1966) British Patent 1166251
24. Fitzer E, Müller DJ (1975) Carbon 13:63
25. Fitzer E, Heine M, Metzler W (1986) Carbon '86, Proceedings of the Fourth International Carbon Conference, Baden-Baden, Germany, 853–855, June 30-July 4
26. Bahl OP, Manocha LM (1974) Carbon 12:417
27. Bahl OP, Manocha LM (1975) Carbon 13:297
28. Fiedler AK, Fitzer E, Müller DJ (1971) Proceedings of the 161st ACS Meeting, Los Angeles, p 380
29. Donnet JB, Bansal RC (1984) Carbon Fibers, Marcel Dekker, Inc, New York, NY
30. Manocha LM (1976) Development and Characterization of Carbon Fibers. PhD Thesis, Delhi University, Delhi, India
31. Müller Th (1982) Parameterstudie zur PAN Stabilisierung bei der Kohlenstofffaserherstellung, PhD Thesis, University of Karlsruhe, Germany
32. Clarke AJ, Bailey JE (1978) In: Chemistry in Britian, 6:11 (November)
33. Bahl OP, Mathur RB (1978) Fib Sci Tech 12:31
34. Fitzer E, Gkogkidis A, Jacobson G (1982) Extended Abstracts of the International Symposium on Carbon, Toyohashi, p 284, Carbon Society of Japan, Tokyo
35. Toho Beslon Company, DOS 3132784 (1981).
36. Toray Industries (1975) JP 48228.
37. Bahl OP, Manocha LM (1975) Angewandt Makromol Chem 48:145
38. Morita K et al. (1981) Carbon 18:11
39. Jäger H (1986) PhD Thesis, University of Karlsruhe, Germany
40. Rose PG (1981) Verarbeiten und Anwenden Kohlenstoffverstärkter Kunststoffe, VDI Verlag, p 5
41. Watt W, Moreton R, Johnson DJ (1967) Nature 22:690
42. Brooks JD, G. H. Taylor GH (1965) Carbon 3:85
43. Lewis IC (1984) Journal de Chimie Physique 81:751
44. Fitzer E, Muller K, Shäffer W (1971) In: Walker PL (ed) Chemistry and Physics of Carbon, Marcel Dekker, Vol. 7:237
45. Trendelenburg T (1984) PhD Thesis, University of Karlsruhe, Germany
46. Kompalik D (1986) PhD Thesis, University of Karlsruhe, Germany
47. Diefendorf JR (1983) Extended Abstracts of the 16th Biennial Conference on Carbon, San Diego CA, pp 26–27, July 18–23
48. Mochida I, Shimizu K, Korai Y (1988) Carbon 26 (6):843

49. Mochida I (1991) Mesophase Carbon Fiber and Composites. In: Proceedings of Seventh Annual Conference on Materials Technology, South Illinois University, Carbondale, Illinois, April 10–11
50. Bhatia G, Fitzer E, Kompalik D (1986) Carbon 24 (4):489
51. Edie DD, Diefendorf RJ (1992) Carbon Fiber Manufacturing. In: Buckley JD, Edie DD (eds) Carbon-Carbon Materials and Composites Noyes Publications, Park Ridge, NJ, p 19
52. Edie DD, Dunham MG (1989) Carbon 27:647
53. Stevens WC, Diefendorf RG (1986) Carbon '86, Proceedings of the Fourth International Carbon Conference, Baden-Baden, Germany, pp 37–39, June 30 –July 4
54. Edie, DD, Hayes GF, Rast GH, Fain CC (1990) High Temperatures-High Pressures, 22:289
55. Baker RTK, Harris PS (1978) The Formation of Filamentous Carbon. In: Walker PL (ed) Chemistry and Physics of Carbon 14:83
56. Dresselhaus MS, Dresselhaus G, Sugihara K, Spain IL, Goldberg HA (1988) Graphite Fibers and Filaments, Springer Verlag, New York, NY
57. Yoshida A, Hishiyama Y, Endo M (1985) Extended Abstracts of the 16th Biennial Conference on Carbon, Lexington, Kentucky, p 287
58. Tibbetts GG, Beetz CP Jr (1987) J Phys D 20:292
59. Koyama T, Endo M (1983) Japanese Patent 1982–58966
60. Edie DD, Stoner EG (1992) The Effects of Microstructure and Shape on Carbon Fiber Properties. In: Buckley JD, Edie DD (eds) Carbon-Carbon Materials and Composites, Noyes Publications, Park Ridge, NJ, p 41
61. Johnson DJ (1971) Structural Studies of PAN Based Carbon Fibers. In: Thrower PA (ed) Chemistry and Physics of Carbon, 20, Marcel Dekker, New York, NY, p 1
62. Fourdeux A, Perrett R, Ruland W (1971) Proceedings of the First International Conference on Carbon Fibers, Plastics Institute, London, p 57
63. Guigon M, Oberlin A (1984) Fiber Sc Tech 20:177
64. Reynolds WN (1973) Structure and Physical Properties of Carbon Fibers. In: Chemistry and Physics of Carbon 1:1
65. Fitzer E et al. DFG Report Fi 112/37–2 (1.1 - 30.9.84)
66. Perret R, Ruland WJ (1970) J Appl Cryst 3:525
67. Diefendorf RJ, Tokarsky E (1975) Polymer Engr Sci, 15 :150
68. Johnson DJ (1987) J Phys D: Appl Phys 20 (3):286
69. Guigon M, Oberlin A, Desarmot G (1984) Fibre Sci Tech 20:177
70. Endo M (1988) J Mat Sci 23:598
71. Jones JB, Barr JB, Smith RE (1980) J Mater Sci 15:2455
72. Chawatiak S, Barr JB, Didchenko R (1979) Carbon 17:49
73. Beetz CP Jr (1982) Fibre Sci Tech 16:82
74. Reynolds WN, Sharp JV (1974) Carbon 12:103
75. Kumar S (1989) SAMPE Quarterly 20 (2):3
76. Daumit GP, Ko YS (1986) A Unique Approach to Carbon Fiber Precursor Development. In: Brunsch K, Golden HD, Herhert CM (eds) High Tech – the Way into the Nineties, pp 201–213, Elsevier Science Publishers, BV. Amsterdam
77. Mochida I, Shimizu K, Korai Y, Sakai Y, Fujiyama S, Toshima H, Hono T (1992) Carbon 30:55
78. Liu GZ, McHugh JJ, Edie DD, Thies MC (1992) Carbon '92, Proceedings of the Fifth International Carbon Conference, Essen, Germany, June 22–26, p 795

79. Edie DD, Fox NK, Barnett BC, Fain CC (1986) Carbon 24:477
80. Hayes GJ, Metzler W, Fitzer E, Edie DD (1992) Carbon '92, Proceedings of the Fifth International Carbon Conference, Essen, Germany, June 22-26, p 813
81. Niederstadt G (1981) Z Flugwiss Weltraumforsch 5:30
82. Hamada T, Nishida T, Sajiki Y, Matsumoto M (1987) J Mater Res 2:850
83. Fitzer E, Metzler W (1992) Carbon '92, Proceedings of the Fifth International Carbon Conference, Essen, Germany, June 22-26, p 647
84. Edie DD, Fain CC, Robinson KE, Harper AM, Rogers, DK (1993) Carbon 31(6):941
85. Rogers DK, Robinson KE, Fain CC, Edie DD (1991) Proceedings of the Twentieth Biennial Conference on Carbon, University of California at Santa Barbara, Santa Barbara, CA, June 23-28, p 248
86. Boehm HP (1966) Catalysis 16:179
87. Schuber P (1977) PhD Thesis, University of Karlsruhe, Germany

Carbon/Carbon Composites

2 Introduction

Carbon/carbon composites are a generic class of synthetic, pure carbon materials consisting of carbon fibers reinforced in a carbon matrix. These composites, which were first synthesized in the late 1950s, were only used as an ablative material until the late 1960s [1–4] when space shuttle programmes required thermal protection materials with the following properties: light weight, high thermal shock resistance, a low coefficient of thermal expansion, maintenance of strength at high temperatures, high impact resistance, high stiffness etc. [2–3]. Graphite was known to have most of these properties with the exception of high strength and stiffness [4]. By that time PAN-based carbon fibers had been produced with much higher strength. Concurrently, fiber-reinforced plastic technology had also greatly advanced and new composite fabrication techniques were available [5]. Both of these factors gave impetus to the carbon/carbon technology and these composites emerged as the prospective high temperature structural material for selected critical components of advanced air force exit and re-entry systems. During the 1970s carbon/carbon composites were under extensive development in several laboratories in the US and Europe [6–11]. Research and development efforts were concentrated in all areas of carbon/carbon technology. They included improvements in the properties of constituent materials, i.e., carbon fibers and carbon matrix precursors, and design allowables, i.e., development of multidirection weaving technology and fabrication technology etc. [12]. Technologies from other sectors were utilized effectively to develope carbon/carbon fabrication technology. This included carbon fiber reinforced polymer composite technology to make green composites with thermoset resins which could be pyrolyzed to yield a carbon/carbon composite skeleton; the well known synthetic polycrystalline graphite technology involving the liquid infiltration/carbonization technique; the hot isostatic press (HIP) technology from metallurgy for pressure carbonization of green composites made with pitches, and the chemical vapor deposition (CVD) technology originally developed for producing pyrolytic graphite. The 1970s to beginning of the

1980s was the golden period for carbon/carbon research, both with respect to understanding the science of these composites as well as evolving new technologies [6–14]. However, until then these composites were mainly limited to the aerospace domain where high density and good ablative properties were the main considerations.

For a material to be economically viable, it has to find application in sectors other than the high technology aerospace domain. Therefore, in the mid 1980s certain companies in Europe and in the US started manufacturing carbon/carbon composites for noble metals, glass industry tools, heating elements for semiconductor industry, crucibles etc. [15–16]. The properties targeted for these applications were mechanical properties, specifically high fracture toughness and compatible thermal properties. Fabrication technologies other than those used for aerospace grade carbon/carbon composites were used for these applications. Hence, in the mid 1980s the fabrication carbon/carbon composites received a different direction. By that time high quality pitch-based carbon fibers with very high stiffness and thermal conductivity were commercially produced by the then called Union Carbide (presently Amoco fibers) in the US and various companies in Japan [17]. These fibers initiated another era in carbon/carbon technology. In the late 1980s and early 1990s researchers in the US and Japan focused their efforts on using pitch-based carbon fibers and vapor-grown carbon fibers to develop highly thermal conducting carbon/carbon materials for the first wall of a nuclear reactor, again going back to the high technology area [18–20]. Therefore, over the past three decades, ever since the emergence of carbon/carbon composites, technological developments in this area have undergone many changes.

2.1 Carbon-Carbon Composites and Conventional Graphite

As conventional carbon materials, carbon/carbon composites are high temperature materials par excellence. Carbon is a unique solid substance that can be manipulated to exhibit the broadest variety of different, even controversial, structures and properties. Some carbons can be extremely strong, hard, and stiff, while other forms can be soft and ductile. Many carbons are highly porous and exhibit a large surface area, while some are impervious to liquids and gases. These variations in properties are not achieved through alloying or additions as they might be with metals. Instead, they result from inherent structural characteristics such as amorphous carbon or crystalline graphite content, porosity, type and extent of defects etc.

Stiffness as well as thermal expansion of a material can also be derived from the potential energy curve of that material. Macroscopic elastic strain results from a change in interatomic spacing and is in a given direction equal to its average fractural change in that direction. Young's modulus, therefore is proportional to the curvature of the potential curve at d_0, whereas the symmetry of the curve determines thermal expansion. Symmetrical potential curves give rise to low thermal expansion and asymmetrical curves are associated with high thermal expansion. The bonding or sublimation energy, which also controls melting behavior, is directly related to the depth of the trough. The potential curve of solids with high bonding energy and high melting points have a very deep trough. A high Young's modulus will mean small radius at the bottom. In contrast, the potential curve of solids with only weak bonds between the atoms, such as in polymers, will be highly asymmetrical and flat, indicating a high thermal expansion, low stiffness, and very low melting point. The potential energy of graphite for two principle directions as a function of the distance between the atoms may be predicted as shown in Fig. 2.1 [22]. The different depths of the potential curves for a and c directions and the different curvatures at the minimum of the trough, explain the high degree of anisotropy of Young's modulus and thermal expansion in graphite single crystals [22]. Figure 2.2 shows the thermal expansion behavior of different carbon materials. As is evident from this figure the thermal expansion of pyrolytic graphite approaches that of the single crystal, whereas polycrystalline graphite deviates considerably and exhibits a more or less isotropic thermal expansion behavior [4, 23]. Moreover, in polycrystalline graphite, one cannot use crystallographic a and c directions to describe preferred orientation: rather, one speaks of preferred orientation parallel or perpendicular to the grain.

All synthetic carbon and graphite materials are fabricated by thermal degradation or pyrolysis of carbonaceous precursor materials. During pyrolysis, although formation of chemical bonds with Sp^2 hybridization between carbon atoms is most probable, the pure graphite structure is incompletely formed. The structural defects in pyrolyzed solid carbon residues can be grouped into three catagories as shown in Fig. 2.3 [4]. These are defects within the layers, defects between the layers resulting in stacking faults, and the disclination of parallel layers. These defects can inhibit development of single crystal type structure in carbon even during temperature activated structural reorganization with heat treatment to 3000 °K. Therefore, such materials exhibit isotropic behavior with respect to thermal properties. Those carbon materials formed from pyrolysis of

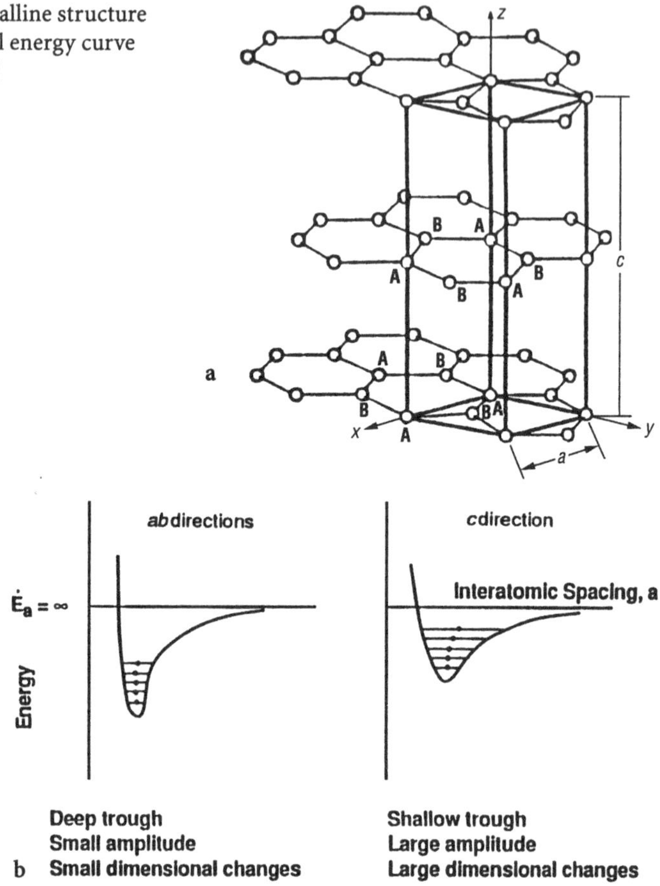

Fig. 2.1 a, b. Crystalline structure a and the potential energy curve b for graphite [22]

liquid precursors, with intermediate formation of liquid crystals, show anisotropic structure in limited areas. The quasi isotropic behavior of polycrystalline carbon/graphite material thus formed is due to statistically distributed slit-like pores generated during the pyrolysis/cooling process.

The type of electron configuration in such disordered graphite, especially for the atoms surrounding the defects or for those which cause some additional bonding between the layers, is not yet clear. Theoretical chemistry cannot exclude the possibility of occurance of sp^3 hybridization. For example, in the case of carbon fibers the bonding forces between the layers are stronger than those in graphite single crystals. This obviously shows that some valence electrons are engaged in cross-linking bonds and do not act only as delocalized π electrons. Microscopically, there are three main

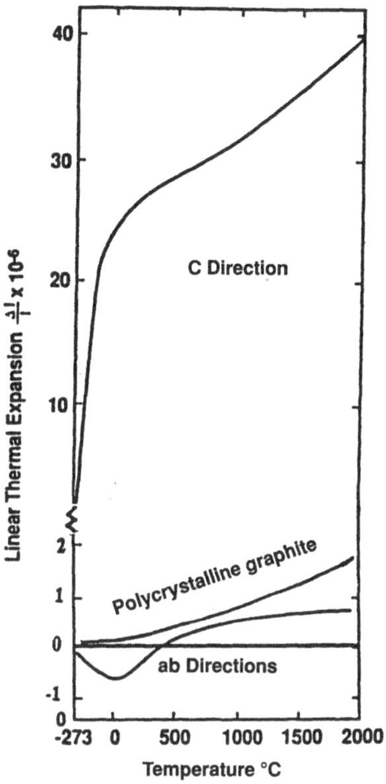

Fig. 2.2. Thermal expansion behavior of single crystal graphite and polygranular graphite

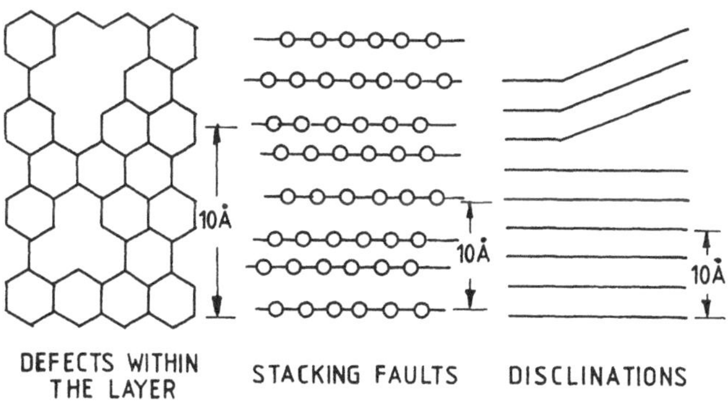

Fig. 2.3. Structural defects in carbon materials

parameters which control the bulk properties of manufactured carbons: (a) preferred orientation; (b) volume and character of porosity; and (c) amount and type of crystalline disorder. The quantum of these parameters in various manufactured carbons is compiled in Table 2.1 [24]. Although the carbon and graphite industry has concentrated its efforts on controlling the influence of these parameters through the use of improvised raw materials and process parameters in manufactured carbons, however, the strength of σ bonds of elemental carbon is only partially utilized. Based on the available data of the best properties of different commercial carbons and graphite products, Fitzer and Huttener have calculated the effectiveness of conversion of bond strength into bulk properties of various heterogeneous carbons (Table 2.2) [24]. As is evident from this Table 2.2, the fraction of the theoretical Young's modulus achieved in polygranular carbons is less than 0.2% while in the newer monolithic forms of carbon it is still less than 0.5%. Only the carbon fibers have achieved more than 50% of the theoretical Young's modulus. Similarly, the fraction of the theoretical strength (10% of the experimental Young's modulus) achieved in common types of polygranular carbons and graphites is negligible compared with that in monolithic carbons and especially in carbon fibers. If one consider the relationship for the bulk strength in terms of the theoretical rather than with the measured Young's modulus, the correlation is even lower. Only carbon fibers show a reasonable correlation between measured and calculated strength. Therefore, in order to utilize the carbon bond strength to the maximum in bulk carbons, high-modulus and high-strength carbon fibers must be incorporated in bulk structures such as carbon/carbon composites.

Table 2.3 shows the parameters which influence the final properties of carbon/carbon composites. In addition to the properties of fiber and matrix, maximum effective utilization of these properties in ultimate carbon/carbon composites are equally important. This can be achieved through optimum choice of geometrical arrangement of the fibers, fiber-matrix interfacial interactions, and processing parameters. All of these are discussed in the following chapters.

Table 2.1. The main parameters: preferred orientation, volume and character of porosity, and degree and type of crystalline disorder controlling the bulk properties of all carbon and graphite materials

	Monolithic carbon			Heterogeneous carbon		
	Glass-like	Pyrolytic	C fiber	C/C composites	Polygranular carbon	
					Ultra-fine grained type	Coarse-grained needle type
Preferred orientation orientation	Minimum; not variable	Maximum; planar; can be partly modified	Maximum; cyclindrical; can be partly modified	Variable up to maximum	Minimum; nearly isotropic	Limited; preferred orientation by grain structure
Porosity (all types of voids exceeding latice defects (> 1 nm))	Minimum; only micropores		Unavoidable; variable in volume and pore characteristics			
Crystalline disorder (defect concentration)	Maximum; only 2-dimendional order; invariable	Minimum obtainable; varies with deposition parameters and HT	Medium; mainly 2-dimensional order; limited variability with HT		Low disorder possible; varies with materials and HT	

HT, heat treatment.

Table 2.2. Mechanical properties of various forms of carbon and graphite materials and yield of theoretical data

	Monolithic carbon			Hetrogeneous carbon		
	Glass-like	Pyrolytic	C fiber	C/C composites	Polygranular carbon	
					Ultra-fine grained type	Coarse needle type
Young's modulus (GN m^{-2})	30	30	300–500	270	10–12	5
Tensile/flexural strength (MN m^{-2})	120	150	2000–4000	1300	10–80	7
Compressive strength (MN m^{-2})	350	450			50–200	20
Effectiveness of theoretical Young's modulus (%)	< 3	< 5	30–50	30	< 0.2	< 0.1
Effectiveness of theoretical strength (%)						
Related to experimental Young's modulus	< 2.5	< 5	10	< 3	< 0.1	0.01
Related to theoretical Young's modulus	0.075	0.1	3–5	< 1	0.002	< 0.001

Table 2.3. Parameters which influence the final properties of C/C composites

Mechanical properties of the components of C/C composites		Interactions between fibre and matrix		
Fiber properties	Matrix properties	Geometrical: volume fraction and fiber arrangement	Interfacial: chemical bonding and physical adhesion	Interaction during application (physical incompatibility)

2.2 References

1. Heym M, Fitzer E (1981) Carbon fiber reinforced Carbon. In: Processing and uses of carbon fiber reinforced plastics, VDI p 85
2. Gill RM (1972) Carbon fibers in composite materials ILIFFE books, London
3. Buckley JD (1988) Ceramic Bulletin 67(2):364
4. Fitzer E (1987) Carbon 25:163
5. Lubin G (1974) Handbook of Composites Materials Reinhold Pub
6. Stoller HM, Frye ER (1971) Processing of C/C Composites an overview paper presented at 73rd Annual meeting, American Ceramic Society, Chicago
7. Kotlensky WV, Pappas J (1969) 9th Biennial Carbon Conf mp 26
8. McAllister LE, Taverna AR (1971) 10th Bienn Conf on Carbon fc-40
9. Fitzer E, Burger A (1971) Ist Int Conf on carbon fibers their composites & Application, London, paper 36
10. Thomas CR, Walker EJ (1978) 5th London Int Carbon Conf p 520
11. Fitzer E, Huttener W, Manocha LM (1980) Carbon 18:291
12. Schmidt DL (1972) SAMPE Journal Vol 8 p 9
13. Savage G (1992) Carbon/Carbon Composites. Chapman & Hall Pub
14. McAllister LE, Lachman WL (1983) In: Kelly A, Milleko ST (eds) Multidirectionally Carbon-Carbon Composites in Handbook of Composites Vol 4. Elsevier Pub, p 109
15. Sigri Elektrographit GmbH, Company Brouchure
16. Schunk Kohlenstofftechnik GmbH Company Brouchure
17. Otani S, Oya A (1976) Petroleum derived carbon p 323
18. Buckley JD, Edie DD (1992) (Ed.) Carbon-Carbon materials and Composites NASA Reference publication No 1254
19. Thomas CR (1993) (ed) Essentials of Carbon-carbon composites, Royal Society of Chemistry (pub)
20. Kimura S, Yasuda E (1995) Advances in research of CF/C composites in Japan, Japan-US CCM-VII, Kyoto, p 1
21. Pierson HO (1993) Handbook of Carbon, Graphite, Diamond and Fullerenes. Noyes Pub
22. Fitzer E, Heym M (1978) High Temperature High Pressure 10:29
23. Bushong RM, Neel EA (1963) Proc 5th Conf on Carbon Vol 1
24. Fitzer E, Huttener W (1981) J Phy D Appl Physics 14:347

3 Carbon Fiber Architecture

The most fundamental aspect of carbon/carbon technology is the substructure or preform which from design considerations may be referred to as fiber architecture. Fiber architecture is the backbone of composites. It not only imparts rigidity to the composites, but also in combination with fiber properties it determines the properties of the composites. Therefore, the choice of preform or fiber architecture is made on the basis of the intended application.

All known types of carbon fibers are used in many different types of architecture: in continuous or discrete form: aligned unidirectionally or as

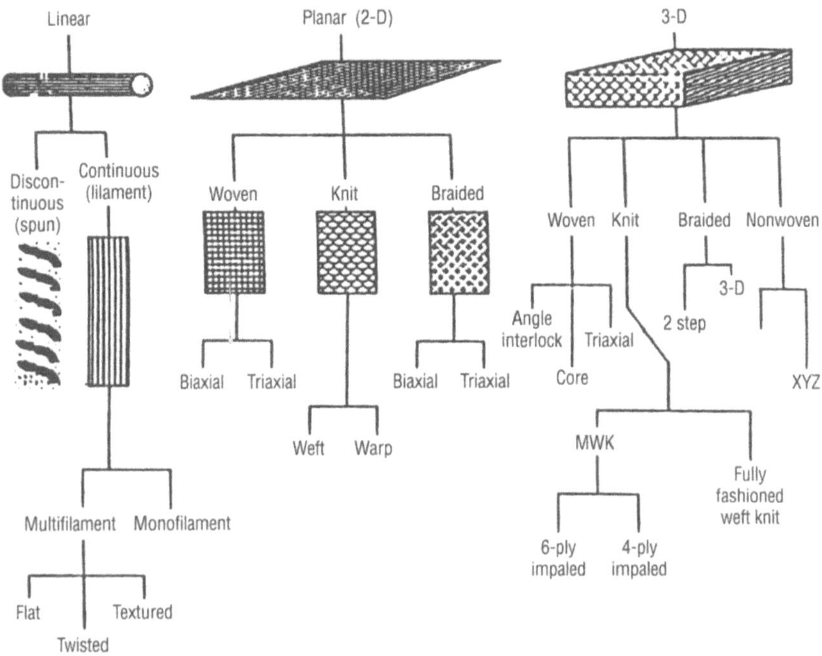

Fig. 3.1. Classification of fiber architecture [3]

various fiber weaves; two-directional (2D) fabrics in stacked, stiched, or pierced configuration; random fibers; braided yarns; multidirectional 3D orthogonal or cylindrical weave or multidirectional weave (4D to 11D) [1–3]. These types of fiber architecture possess a wide spectrum of pore geometries and pore distributions, a broad range of fiber volume fractions and fiber orientation, as well as a wide selection of formed shape and net shape capabilities with design options for tough and reliable carbon/carbon composites [2]. Based on structural parameters and fiber continuity, Frank Ko [3] has classified these types of fiber architecture into four categories: discrete, continuous, planar interlaced (two dimensional), and fully integrated (three dimensional). Figure 3.1 [3] shows some of these types of fiber architectures.

3.1 Discrete Fibers

The first category of fiber architecture, i.e., discrete fibers, is either processed by spray layup or pulp molding. It is difficult to control precisely the orientation of the fibers or pore structure in these forms. Fiber volume fraction and strength translation efficiency in composites made of discrete fiber system are quite low.

3.2 Unidirectional Fibers

The second category of fiber architecture, consisting of linear continuous fibers or the so-called unidirectional system, has the highest level of property translation efficiency. As seen from Fig. 3.2 [4], theoretically, these systems may have the highest fiber packing efficiency, in the order of 90% (although only 50%–60% fiber volume fraction is the optimum), and possess low, controlled porosity. Unidirectional composites are processed either by filament winding or hand layup or by angle ply tape layup. Although these systems are widely used for basic research and for studying processing etc., in actual applications, unidirectional composites are particularly poor in out–of–plane strength.

3.3 Planar (2D) Preforms

Laminar fiber architecture include fiber felts, the first to be used as preform for making carbon/carbon composites and two dimensional woven fabrics

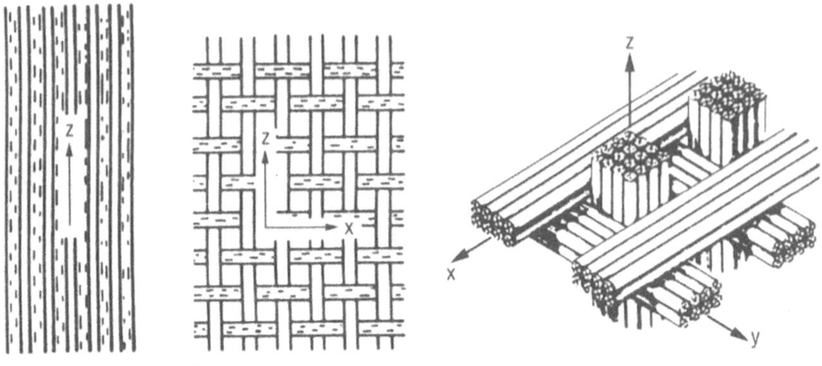

a b c

Fig. 3.2 a–c. Principles of reinforcements with continuous fibers [4]. The maximum fiber volume fractions are calculated assuming direct contacts between neighbouring filaments. **a** Unidirectional; **b** two-directional; **c** three-directional reinforcements. Theoretical maximum fiber volume fractions are 90.6 %, 78.5 % and 67.9 %, respectively

with the fibers oriented in two directions. Figure 3.3 shows the schematic diagrams for the three basic weave geometries: plane, satin, and twill weave. Plain weave, which is the most highly interlaced, is the tightest basic fabric design and the most resistant to in-plane shear movement [3, 5]. Satin weave represents a family of constructions woven as standard four, five or eight harness satin with the least amount of yarn interlacing. In satin weave, a warp yarn skips over a few fill yarns and goes under one fill yarn. Like in eight harness satin, the warp yarn skips over seven fill yarns and goes under one. The satin weave construction offers the advantage of the floating yarns, which contributes to a higher fiber packing density, a high level of fiber volume fraction, and higher strength and modulus translation efficiencies over plain fabrics. Satin fabrics are preferred over plain fabrics also due to their ability to conform to complex contours. Twill weave has interlacing in between plain weave and satin weave, with warps and fills interlacing regularly. Other types of laminar fiber architecture include a biaxial braid with varying braiding angles as shown in Fig. 3.4.

Although the interlaminar failure problems associated with the unidirectional fiber architecture are taken care of in laminar structures, the interlaminar strength is still governed by the matrix properties. The 2D composites also exhibit anisotropic thermal properties.

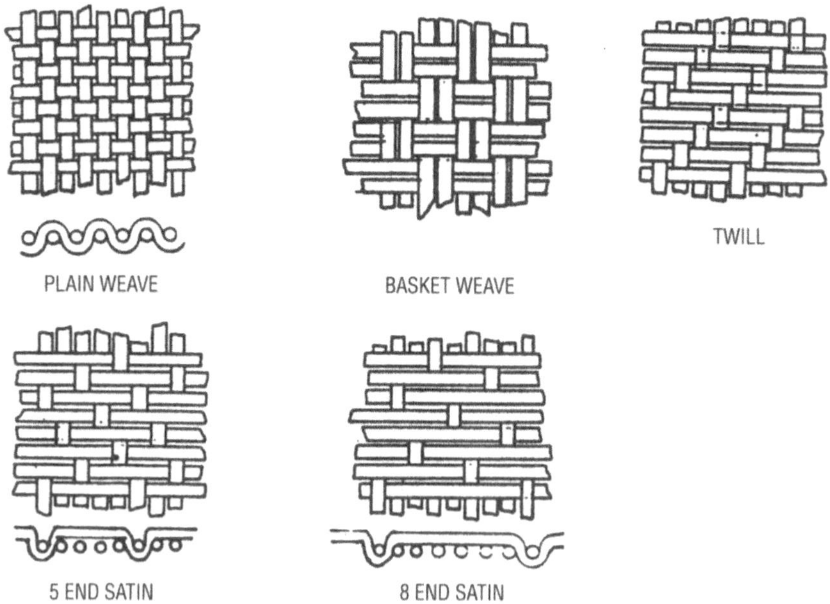

Fig. 3.3. Structural geometry of biaxially woven fabrics

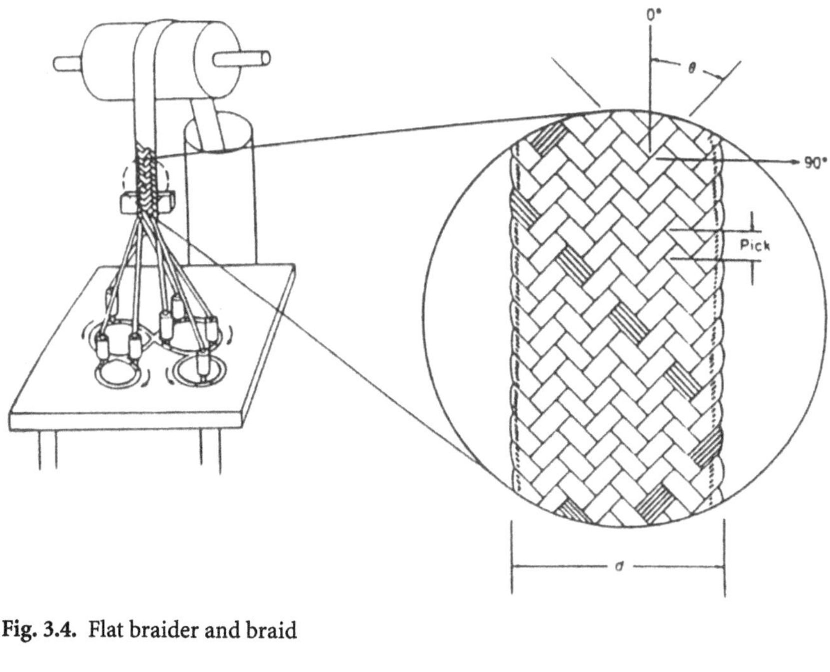

Fig. 3.4. Flat braider and braid

3.4 Three-Dimensional Preforms

The most important and critical component of advanced carbon/carbon composites, especially for aerospace applications, are three-dimensional preforms, since the fiber architecture in these preforms controls the processing as well as the properties of the final carbon/carbon composites. In three-dimensional reinforcements, also termed as through-the-thickness reinforcement, the carbon fibers are laid in three directions using braiding, knitting, weaving, and nonwoven techniques [6]. These reinforcements overcome some of the limitations of 2D carbon/carbon composites, especially the out-of-plane strength and interlaminar shear strength.

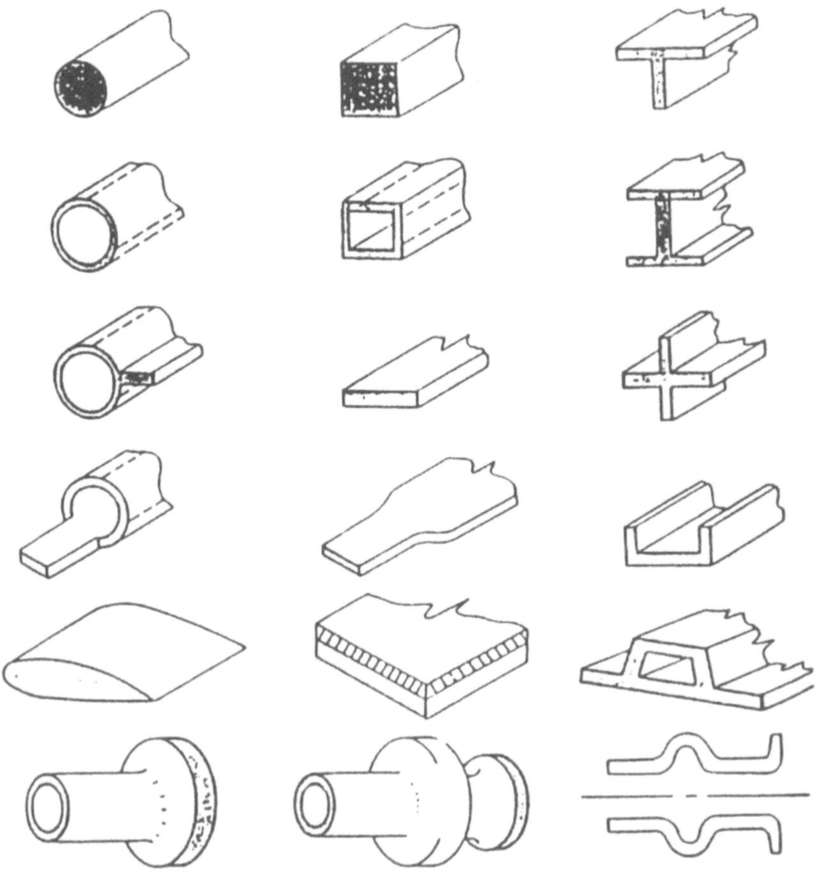

Fig. 3.5. Net shape structures produced by three-dimensional braiding [5]

In three-dimensional braiding, a variety of solid complex structural shapes are produced by intertwining or orthogonal interlacing of three or more yarn systems to form an integral structure (Fig. 3.5) [5]. The knitted three dimensional fabrics are produced by either the weft-knitting or warp-knitting process. An example of one such knitted three-dimensional fabric employed for making carbon/carbon aircraft brakes is shown in Fig. 3.6 [5]. Needle-punched nonwoven fabrics and mats were the early preforms used for making carbon/carbon composites by chemical vapor deposition (CVD) and by a liquid infiltration/carbonization process [5–7].

During the past 15–20 years, 3D woven preforms have gained much attention in the carbon/carbon composite industry. Three-dimensional multidirectional weaving technology provides the ability to orient the desired amount of fibers in preselected directions to accommodate the design loads of the final structural component and, hence, to produce tailored composites. Perhaps one of the most widely used and simplest types of multidirectional preforms for carbon/carbon composite applications is the three-directional (3D) orthogonal weave construction shown in Fig. 3.7 [8]. In such configuration, yarn bundles are laid straight on Cartesian coordinates to obtain maximum structural capabilities of the fiber. The type of yarn, yarn spacing, and number of yarns per site can be varied in all three directions to have a 3D woven structure with the desired preform density and distribution of fiber volume fractions in different directions. In such a preform the fiber volume fraction theoretically achiev-

Fig. 3.6. Carbon fiber braided structure for brake linings [5]

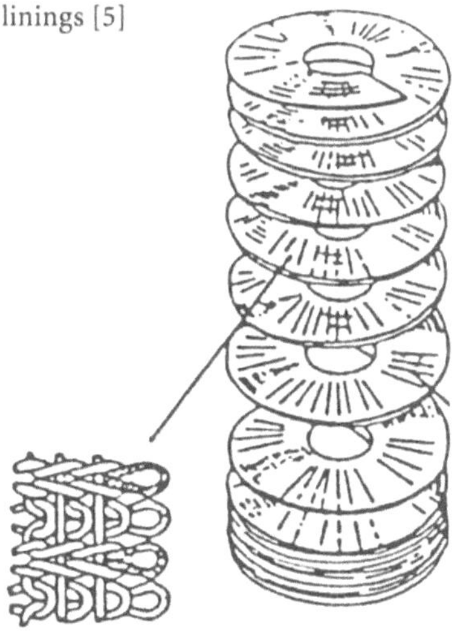

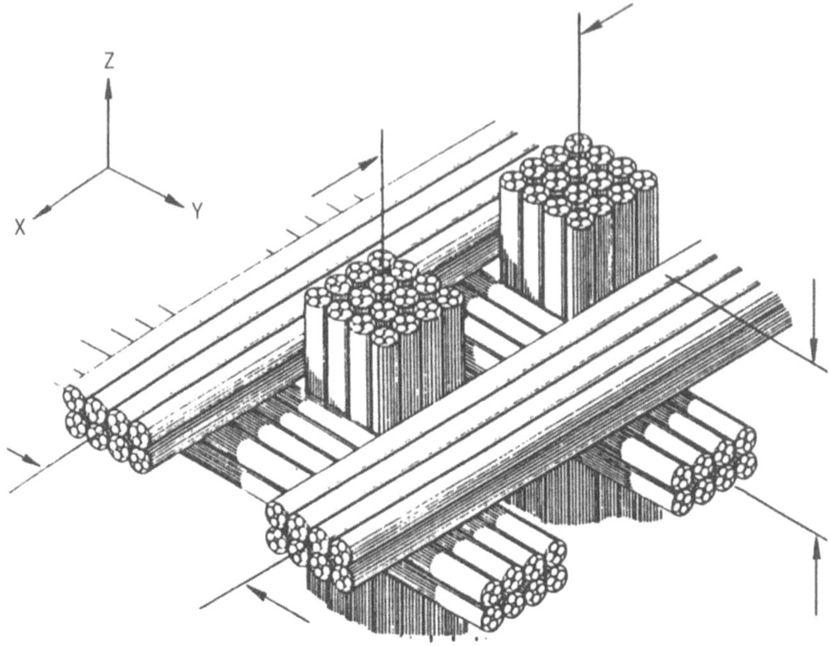

Fig. 3.7. Typical 3D block construction [10]

able has been calculated to be around 70%; however, in practice it is found to be much lower due to inherent void formation as a result of crossover fiber yarns. In order to weave 3D woven preform with small and well-dispersed void cells, yarns with a small diameter are preferred. Multidirectional fiber preforms are fabricated by a variety of techniques including dry yarn weaving, fabric piercing, assembly of rigid pultruded yarn, filament winding, or by a combination of these. Although details of the equipment and procedures used to manufacture multidirectional preforms are proprietary, some information is available in the literature, generally regarding fabrication processes [3, 5–9]. Figure 3.8 shows a schematic diagram of the 3D orthogonal weaving by the dry yarn process and Fig. 3.9 shows the pierced fabric technique [10]. The properties of the 3D preform made by latter technique can also be varied as desired by choosing a proper type of fabric and tows for z-direction piercing. The pierced fabric technique offers a higher fiber volume fraction and, hence, higher preform density, though distribution of fibers in the z-direction is quite lower than in x-y directions.

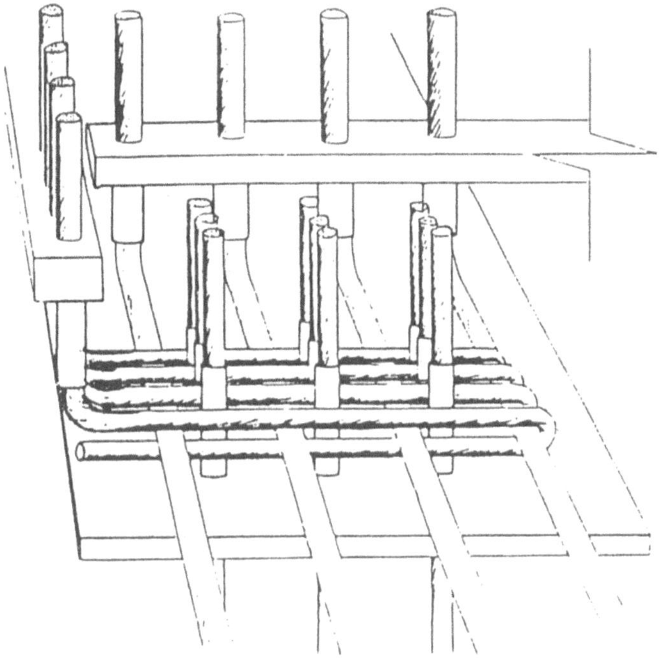

Fig. 3.8. Three-directional weaving [10]

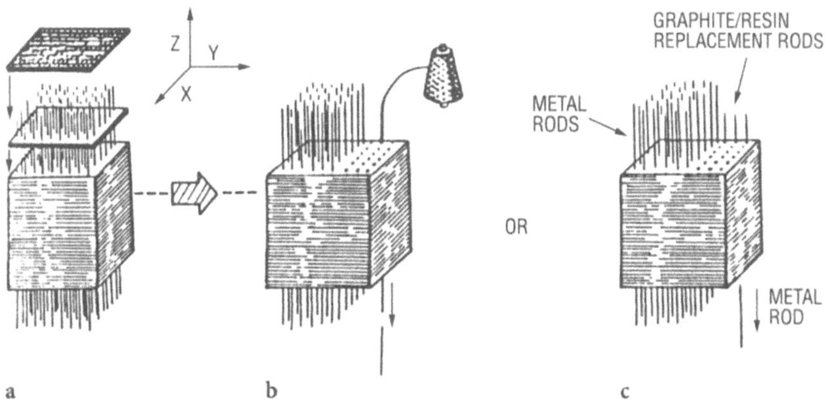

Fig. 3.9 a–c. Pierced fabric block fabrication [10]. **a** Pierced graphite fabric over metal rods. **b** Metal rods replaced with graphite yarns. **c** Metal rods replaced with graphite-resin rods

3.5 Multidirectional Structures

In order to fabricate composites with higher fiber volume fractions and isotropic properties, isotropic woven preforms are desirable. These can be obtained by placing fibers in angular positions or along the diagonals in addition to or in place of three Cartesian coordinates. Accordingly, woven preforms with 4, 5, 7, or 11 directions of reinforcements have been fabricated. Schematic representations of these multidirectional woven preforms are shown in Fig. 3.10 [11]. Four-directional (4D) structures can be fabricated either by placing the fibers across four corners or across four faces of a cube or by modifying the basic 3D orthogonal structure with x- and y-direction fibers at a 45° angle. A five-directional (5D) design consists of a 3D orthogonal structure with the addition of two ± 45° in-plane reinforcements with respect to the x-y fibers along with the axial (z)direction of reinforcement. A combination of 4D construction with 3D orthogonal construction gives rise to seven-directional (7D) woven structures (Fig. 3.11). A combination of the 3D base structure, diagonals across the faces and diagonals across the corner produces a more isotropic woven structure with eleven directions of reinforcement, i.e., the 11D structure [12, 13]. An increase in the number of reinforcement directions ultimately results in lower fiber volume fractions in each in-plane direction of reinforcement with a resultant loss of tensile and compressive properties in those directions. Thus, in general, multidirectional reinforcements lead to enhanced off-axis properties at the expense of on-axis properties. Multidirectional woven preforms, in addition to offering composites with more or less isotropic properties, also help in processing high density carbon/carbon composites. Multidirectional woven preforms possess well-connected pores and voids which are easier to be infiltrated than those made in 3D preforms. The cubic voids of 3D structure share common corners while the voids of the 4D geometry share common sides. Hence, 4D geometry is more permeable than 3D geometry. The 7D structure, a combination of 4D and 3D, will have more permeable channels than either of the two separately and hence will help in fabricating high density composites. In addition to these straight line structures (rectilinear structure), end application of carbon/carbon composites also requires complex shape structures such as that of turbine blades and conical and cylindrical structures. Figure 3.12 [14] shows typical 3D polar coordinate weave constructions with fibers running in axial, circumferential, and radial directions. In 3D polar constructions like in 3D orthogonal structures, fiber amount, types, and spacing can be optimized

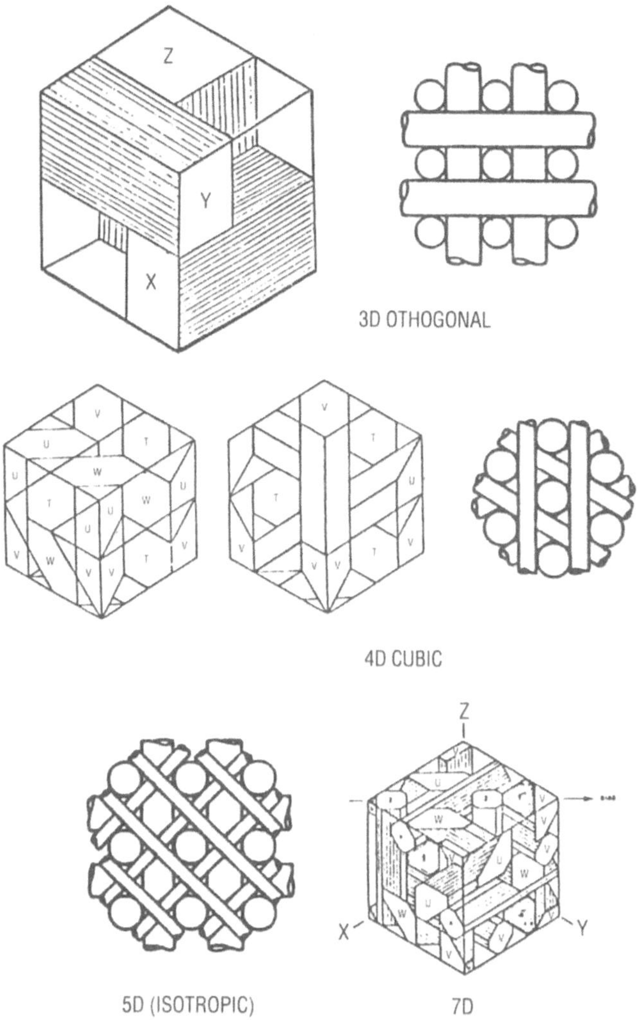

Fig. 3.10. Three-directional (3D orthogonal), 4D (cubic), 5D (isotropic) and 7D fiber structures

in axial, circumferential, and radial directions to obtain a preform with desired characteristics. Cylindrical looms have been developed to produce dry multidirectional polar constructions with desired characteristics.

The multidirectional structures described above can also be produced using nonwoven techniques instead of using prerigidized yarns, i.e., phenolic-coated carbon fibers, or using a combination of filament winding

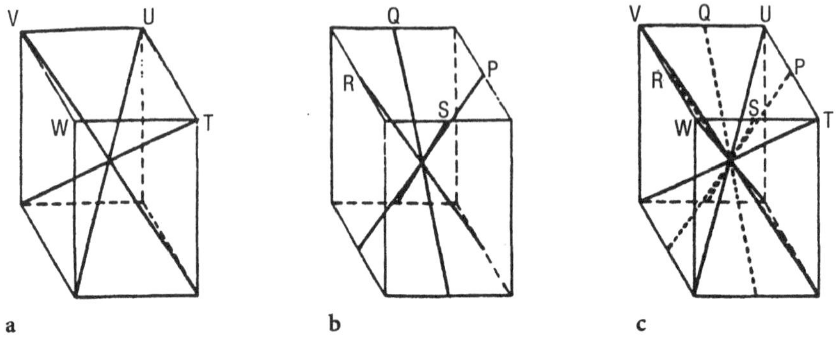

Fig. 3.11 a–c. Directions of reinforcements in multidirectional woven structures. **a** Seven-directional construction – type I; typical "across the corners" diagonals. **b** Seven-directional construction – type II; typical "across the face" diagonals. **c** Eleven-directional construction; typical "across the corners" diagonals plus "across the face" diagonals

Fig. 3.12. Three-directional cylinder construction [14]

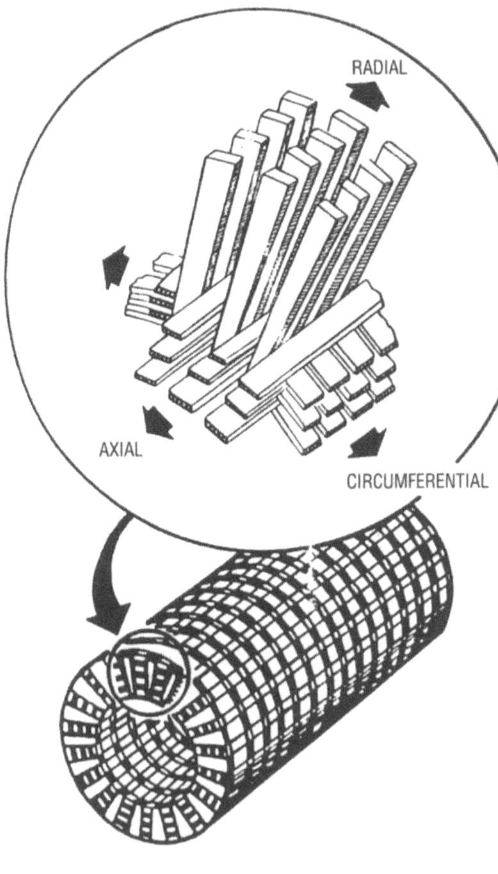

technique and prerigidized yarns, especially for cylindrical structures. Today, carbon fiber preforms of even complex shapes can be woven on automatic machines. Weaving procedures and machines for multidirectional weaving with different fiber configurations have been developed at FMI (USA), Aerospatiale and SEP (France), the Research Institute of Polymers in Tsukuba, and the Research Department of the Industrial Textile Division of Shikibo Ltd. in Osaka, Japan [16]. Figure 3.13 shows a prototype of an automatic 3D weaving machine and Fig. 3.14 shows the woven preform [15, 17].

Textile preforms offer a wide choice of design for tough and reliable structural carbon/carbon composites. However, to take advantage of these attractive features for carbon/carbon technology, fabric geometry models have to be developed which may provide a better understanding of the fiber architecture, matrix processing conditions, fiber/matrix and fiber/fiber interactions, as well as their impact on the ultimate properties of the composites. Ko et al. [5, 18] have developed such a fabric geometric model (FGM) based on a unit cell geometry and have examined the elastic re-

Fig. 3.13. Automatic multidirectional weaving machine

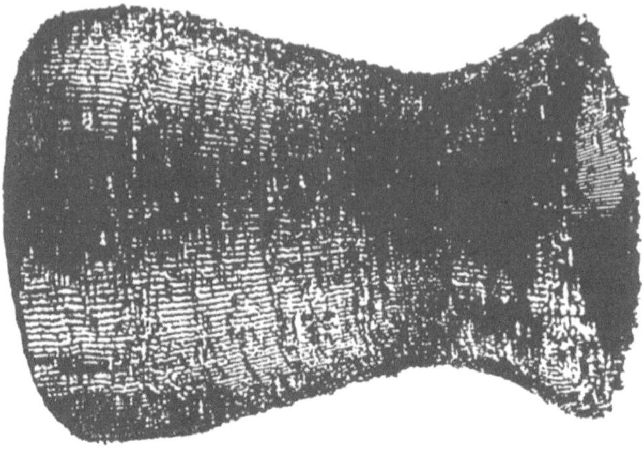

Fig. 3.14. Three-directional woven carbon fiber preform [17]

sponse in different directions of various three-dimensionally reinforced composites. In an xyz orthogonal fabric with an equal distribution of fibers in the x and y directions and a 3D braid reinforced composite with 1 by 1 braided structures, Ez (through the thickness modulus) is found to increase as Ex (uniaxial tensile modulus) decreases. The relationship between Ex and Ez is found to be similar for both of the composites (Fig. 3.15), whereas the two composites are found to have opposite relationships be-

Fig. 3.15. Relationship between longitudinal modulus and through-thickness modulus of 3D composites [5]

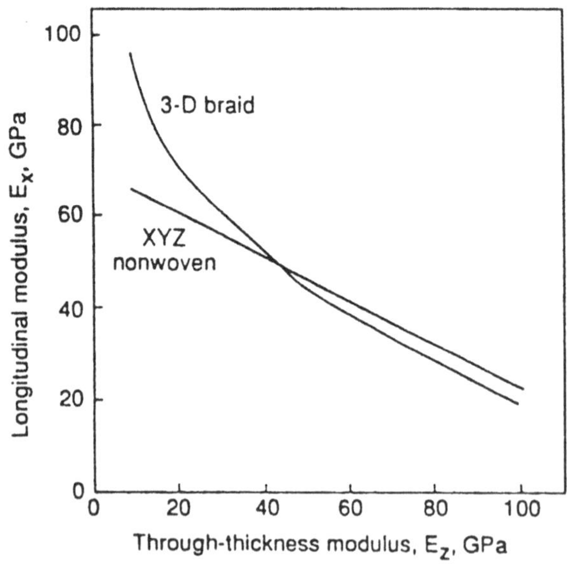

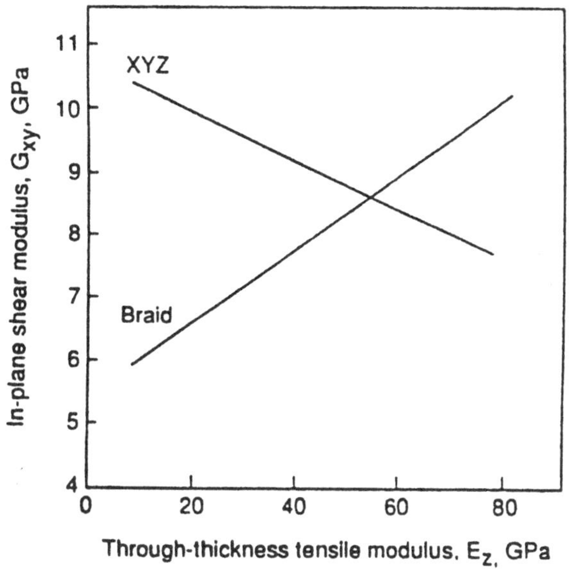

Fig. 3.16. Relationship between in-plane shear modulus and through-thickness modulus of 3D composites [5]

tween Gxy (in-plane shear modulus) and Ez (Fig. 3.16). This also shows the sensitivity of fiber orientation and distribution in various directions in different types of composites.

3.6 References

1. Heym M, Fitzer E (1981) Carbon fiber reinforced Carbon. In: Processing and uses of carbon fiber reinforced plastics. VDI Düsseldorf, p 85
2. Huettener W (1990) Potential of carbon-carbon composites as structural material. In: Figueiredo JL et al. (eds) Carbon Fibers Filaments and Composites, Kluwer Academic Publishers, p 275
3. Ko F (1989) Ceramic Bulletin 68:401
4. Fitzer E, Huttener W (1981) J Phy D Appl Physics 14:347
5. Ko F (1993) Textile preforms for carbon-carbon composite. In: Buckley JD, Edie DD (eds) Carbon-carbon materials and composites. Noyes publication, p 71
6. Sanders LR (1977) SAMPE Quarterly p 38
7. Williams DJ (1978) New knitting methods offer continuous structures. Adv Comp Eng, p 12
8. Lachman WL, Crawford JA, McAllister LE (1978) Multidirectionally reinforced C/C Composites. In: Noton B et al. (eds) Proc Int Conf Comp Materials (Metallurgy Society of AIME, NY)
9. Barton RS (1968) SPE J, 24:31

10. McAllister LE, Taverna AR (1976) A study of composition - construction variation in 3D C/C Composites. In: Scola E et al. (ed) Proc Int Conf Comp Mat (AIME) 307
11. Gebhardt JJ, Stover ER, Mueller W (1976) In: Petroleum Derived Carbon ACS Symp Series 21:212
12. Stover ER, Latva JD (1973) Eleventh Bienn Carbon Conf, p 277
13. Stover ER, Roetling JA (1973) Eleventh Bienn Carbon Conf, p 335
14. Mullen CK, Roy PJ (1972) Fabrication and Property description of AVCO 3D C/C Cylinder material in 17th National SAMPE Symp p III A-2
15. Fitzer E (1987) Carbon, 25:163
16. Meyer RA (1994) Overview of International Carbon-carbon composites research. In: Materially speaking, vol 9 No 3 MTC, Carbondale, Ill USA
17. Aerospatiale les Mureux, France - Product Brochure
18. Ko F, Pastore CM, Lie C, Whyte DN (1987) A Fabric Geometry Model for 3D Braid reinforced FP/Al-Li composites, Competitive Advancements in Metal Proccess. Proceedings, Int SAMPE Metal Conf Aug. 18-21

4 Processing of Carbon/Carbon Composites

4.1 Basic Routes for Fabrication of Carbon Products

The high thermostability of the solid carbon is the basis for the high temperature applicability of all carbon materials. As can be seen from its phase diagram (Fig. 4.1) [1] carbon, or the thermodynamic equilibrium phase graphite, melts only under conditions of high pressure and temperature, i.e., in the order of 100 bars and 4000 K, respectively. As a consequence of this behavior, chemical methods for material fabrication, such as melting

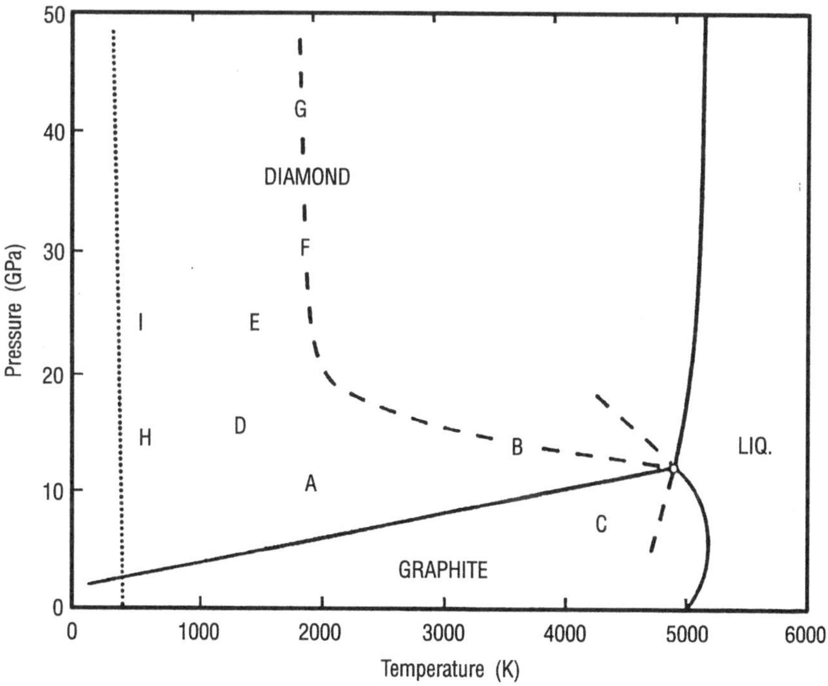

Fig. 4.1. Pressure and temperature phase diagram for carbon [1]

or sintering cannot be applied for fabrication of carbon products. Only in situ methods, i. e., preparation of elemental carbon by thermal decomposition of carbonaceous compounds, are suitable for preparation of solid structural carbon materials.

Hydrocarbons are the basic carbonizable compounds. Alternatively, these may contain additional ligands such as oxygen, sulfur, nitrogen, and halogens. Examples of the most suitable precursor compounds are:

(1) *Pure hydrocarbons:* gaseous low molecular weight compounds with molecular weights between 16 and 50 g/mol or molten pitches with molecular weights from 500 to 1000 g/mol
(2) *Oxygen-containing hydrocarbons:* nonmelting polymers such as phenolics and furans, and oxidized pitches and thermoplastics
(3) *Nitrogen-containing hydrocarbons:* polyacrylonitrile, polyimide, etc.
(4) *Sulfur- and halogen-containing compounds:* modified pitches.

The principle routes for processing carbon/carbon composites by making use of the above carbonizable compounds can be summarised as follows:

4.1.1 One-Step Pyrolysis

The first fabrication process for carbon/carbon composites was the so-called HITCO process. Textiles from rayon were pyrolyzed in hydrocarbon atmosphere. This process is a combination of a solid state pyrolysis of the rayon textile resulting in a carbon fiber framework and gas-phase pyrolysis (chemical vapor deposition; CVD) of hydrocarbon vapor for formation of the pore filling matrix. This process which was developed for nose tip cones of missiles is still used, but only to a very limited extent.

4.1.2 Two-Step Pyrolysis

Solid-state pyrolysis of carbon fiber precursor to carbon fiber is the first step in the process. This is followed by gas-phase pyrolysis of hydrocarbons in an autoclave to form the carbon matrix. This process is widely used today, not only for missile technology but also for carbon/carbon disk brakes. The process requires high investment costs and long residence times for the substrate within the autoclaves (in the order of 1 month). It may be economical only if the necessary equipment is easily available at reasonable costs.

4.1.3 Three-Step Pyrolysis

This process is a modification of the two-step process described above. It involves carbonization of fibrous precursor to carbon fibers by solid-state pyrolysis, formation of a rigid carbon fiber framework by binding with a carbonaceous precursor such as phenolics, followed by pyrolysis of this binder precursor into a porous bulk carbon/carbon material and subsequent impregnation by chemical vapor deposition to fill the pores by pyrolytic carbon or graphite.

The chemical vapor impregnation process (CVI) of the two- and three-step routes was the pioneering process for fabrication of carbon/carbon composites and is still in application in spite of the high fabrication costs. Theoretically considered as a single-step fabrication procedure, it needs various interruptions to remove the preferentially densified surface layers making it de facto a multistep process. This process is advantageous for giving a pyrocarbon finish to carbon/carbon composites and for applying ceramic coatings against oxidation.

4.1.4 Multistep Pyrolysis

In this process, the first steps are the same as those described above in the three-step process, namely, carbonization of the carbon fiber precursor and formation of a rigid fiber framework by means of a binder, usually a phenolic or pitch, and its pyrolysis to yield a carbon/carbon skeleton. The densification of the rigid porous carbon/carbon skeleton is carried out in a liquid impregnation step using phenolic or pitch precursor followed by heat treatment. The densification comprises impregnation and recarbonization cycles repeated up to four times or more to get a carbon matrix around each filament and in between the yarns.

If pitch is used as a liquid carbon precursor for the pore filling, high pressure pyrolysis is advantageous to get a high carbon yield of up to 80%. The use of dehydrogenating chemical additives in pitches in some cases may replace high pressure pyrolysis to get equivalent high yields.

All commercial processes for fabrication of carbon/carbon composites are based on various pyrolysis methods, e.g., gaseous phase or liquid phase alone or in combination [2–5]. Figure 4.2 shows a schematic representation of the two-step, three-step and multistep processes. The details of these methods with respect to control of processing parameters, and their criti-

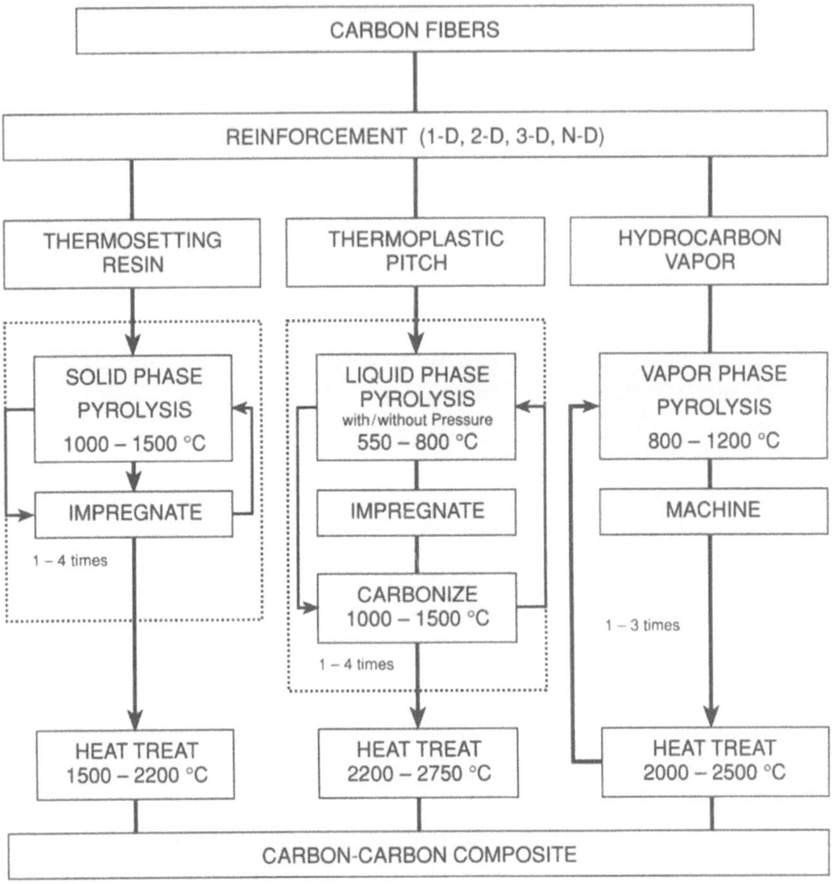

Fig. 4.2. Carbon/carbon composite processing

cality and influence on product properties are dealt with in the following section.

4.2 Gas Phase Impregnation Process

The chemical vapor deposition (CVD) process for formation of carbon uses volatile hydrocarbon compounds such as methane, ethane, propane and benzene as precursors. Thermal degradation is achieved on hot surfaces of the substrate, resulting in a pyrolytic carbon deposit and volatile byproducts, which consist mainly of hydrogen. The process of fabricating carbon/carbon composites through this route is, in fact, analogous to the technique applied to densify highly porous carbon skeleton, i.e. so-called CVI. In this case porous skeleton is replaced by carbon fiber woven preform. Therefore, this process is used for densification of preforms made of either carbon fibers alone or rigidized carbon structures (the two- and three-step processes described above). In the process of CVI, the feed gas undergoes a series of vapor phase interactions as it diffuses into the preform, forming intermediate species which further react to deposit carbon on all heated surfaces in the vicinity of the gas. An important feature of the deposit is its anisotropy and its adherence to the contour of the substrate, so that in the case of fibrous constructions, all of the filaments to a greater or lesser degree are coated with carbon. However, the overall deposition rate is highly sensitive to the surface area (substrate surface) and to the concentration of the cracking gas. Sometimes carrier gases such as hydrogen are also used to control the concentration of cracking gas. In addition, a number of other parameters such as substrate design, temperature, and gas pressure will effect the rate and uniformity of deposition and the matrix characteristics. The CVD processes operating on isothermal, thermal gradient, pressure gradient, and pressure pulsation principles have been applied to fabricate carbon/carbon composites. The first two have been used most extensively.

4.2.1 Isothermal Chemical Vapor Deposition

The isothermal technique is shown schematically in Fig. 4.3 [6]. In this process, a woven preform called a substrate is placed within the susceptor of an induction furnace and hydrocarbon gases are introduced into the susceptor to allow infiltration into the substrate.

The substrate is heated by radiation so that the substrate and gas are maintained at a uniform temperature. The isothermal process is usually conducted at low temperatures (around 1050°–1100 °C) and reduced pressures, with the flow rate of hydrocarbon gas predetermined depending on the substrate surface area [7]. As the cracking gas enters deep into the substrate, its concentration becomes less due to decomposition at the outer layers resulting in the establishment of concentration gradients between the outer surface and the inner depleted zones. This in turn is the driving force for the hydrocarbon gas to continue to enter the substrate from the exterior into interior. As a consequence of this, a deposition gradient is also established within the substrate as a function of distance from the outer surface, depending on the geometry of the substrate. In carbon/carbon

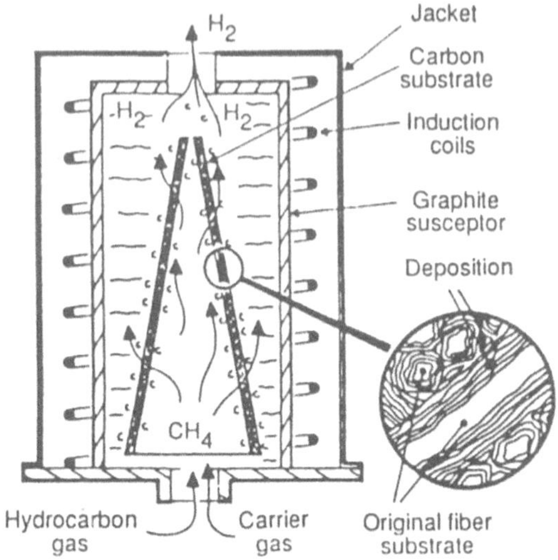

Fig. 4.3. Schematic representation of the isothermal CVD technique

composites fabricated by this technique, the densities of the composites with varying geometry are found to be different although they are processed under similar infiltration conditions [8]. Therefore, in order to achieve high infiltration densities, it is necessary to operate at minimal deposition rates (low temperature) for a long period of time. Moreover, this process produces a crust on the surface of the substrate that must be machined away in between the cycles to achieve more infiltration and a high ultimate density of the composites. Although this process is quite costly and time consuming, it is still used in production because a large number of parts can be densified simultaneously.

4.2.2 Thermal Gradient Process

A second procedure for CVD known as the thermal gradient process comprises maintaining a temperature gradient through the substrate thickness. The outer surface of the substrate, i.e., the entry point for the gas supply is at a lower temperature than the innermost surface. This is accomplished by placing the woven structure over a graphite mandrel which is inductively heated. A schematic diagram of the thermal gradient process is shown in Fig. 4.4 [9]. As can be seen from the figure, the mandrel and induction coil must be tailored to the geometry of the substrate. The surface near the mandrel and furthest from the gas supply is at the highest temperature, i.e.,

Fig. 4.4. Schematic representation of the thermal gradient CVD technique

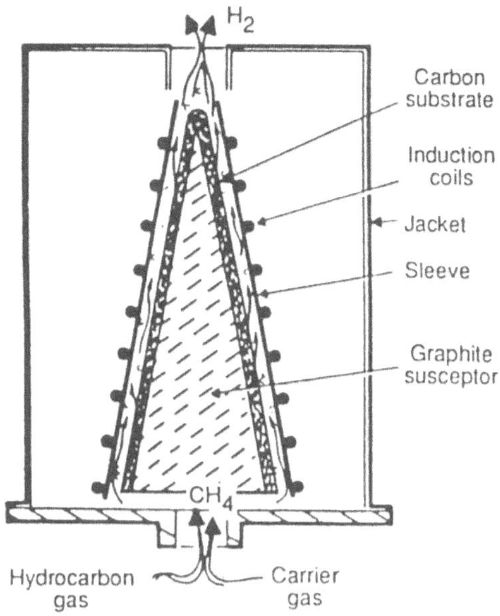

in the order of 1300° – 1400 °C. Therefore, deposition starts at the surface near the mandrel. This densified region exhibits better thermal conductivity than the fibrous preform and starts heating inductively, resulting in migration of the hot section through the thickness of the substrate. Consequently, the deposition progresses radially through the substrate towards the colder surface. In this process, gas velocity is higher than in an isothermal process. This technique prevents the formation of a crust on the outer surface of the preform and thus eliminates the machining step. However, this process is most useful for relatively thin structures and is limited to a single item per furnace. Large cross sections of the specimen cause considerable gas turbulence and uncertain gas-phase pyrolysis.

A modified temperature gradient technique has been developed to densify large-size woven structures. This involves utilizing the conducting property of graphite fibers and, hence, coupling the fiber structure directly with the field placed within an r.f. coil. By using a short coil and slowly drawing the preform through the coil, a complete length of the preform can be heated. During this procedure, the process gas is passed around the preform at high velocity. A large preform can thus be infiltrated and densified uniformly [10]. However, infiltration is greatly influenced by the draw ratio, porosity of the preform, and velocity and concentration of the cracking gas.

4.2.3 Differential Pressure Process

The differential pressure technique is a modification of the isothermal technique. In this technique the substrate is heated to a constant temperature while gas is fed into the middle of the substrate under positive pressure with respect to the furnace chamber. A pressure difference is thus created between the center and outer walls of the substrate and gas is forced to flow through the complete substrate. This process has limited applications.

As discussed above, the preferred deposition of pyrolytic carbon on the surface of the substrate makes filling up the internal pores difficult. The intertow spacings and open porosity is changed into closed porosity due to the bottleneck effect shown in Fig. 4.5 [3, 7]. As in all heterogeneous

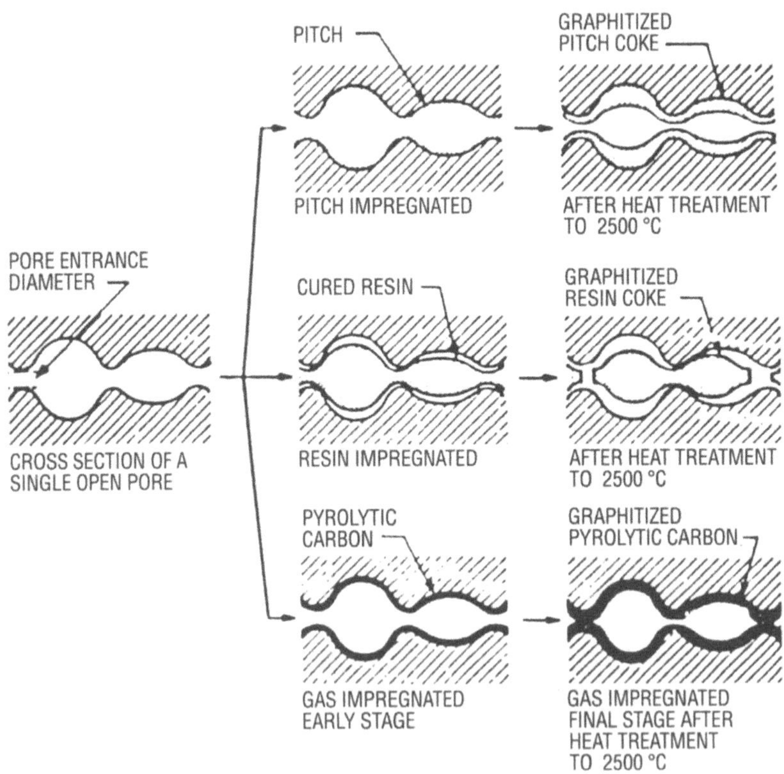

Fig. 4.5. Mechanism of pore filling and pore blocking by liquid impregnation and by CVD [7]

gas–solid reactions, control of the overall reaction rate by diffusion of the reacting gases to the deposition surface will favor the formation of surface crusts and hence must be avoided. In order to reduce cluster formation, chemical reactions on the surface and on the inner surface should control the overall reaction rate [7, 9]. Kinetic studies of this problem indicate that the chemical reaction is much more controlled by the temperature than by the diffusional transport of the reactant. Therefore, the reaction rate can be better controlled by choosing low temperatures. However, with low temperatures very long processing times are required, with the result that autoclaves in which infiltration processes are performed are occupied for weeks for a densification cycle, thus making the process economically unviable.

4.3 Density of CVD Carbon/Carbon Composites

Carbon/carbon composite development work using the CVD technique was started at the Super Temp Company, USA, in 1965 based on their experience with pyrolytic carbons and infrastructure [4, 11, 12]. Initially, composites were fabricated utilizing carbon felt as substrate and methane gas as a carbon source under isothermal conditions. It was observed that the rate of carbon deposition is a function of temperature, pressure, methane concentration, location, and gas residence time.

According to Bauer and Kotlensky [12], the geometrical configuration of the fiber preform and initial porosity have a great influence on the densification process and final density of carbon/carbon composites. The lower the fiber volume content, i.e., the higher the initial porosity, the higher is the ultimate density of the carbon/carbon composites. Figure 4.6 [9] shows a typical example with rayon-based carbon felt with a fiber density of 1.5 g/cc as substrate. Similar results are reported with laminated, 2D fabric and 3D structures.

In the case of cloth layup structures, fiber volumes in the range of 30 % – 50 % appear to be the most suitable for achieving high density in carbon/carbon composites. Gebhardt et al. [9] of General Electric Company have observed that the geometry of the preform, fiber packing density, and organized void structure in multidirectional woven structure all have great influence on matrix formation within the geometric interstices and fiber bundles and consequently on the ultimate average density and density variation in the composites. However, as discussed above, the formation of surface cluster, the final density, and the density distribution in a woven

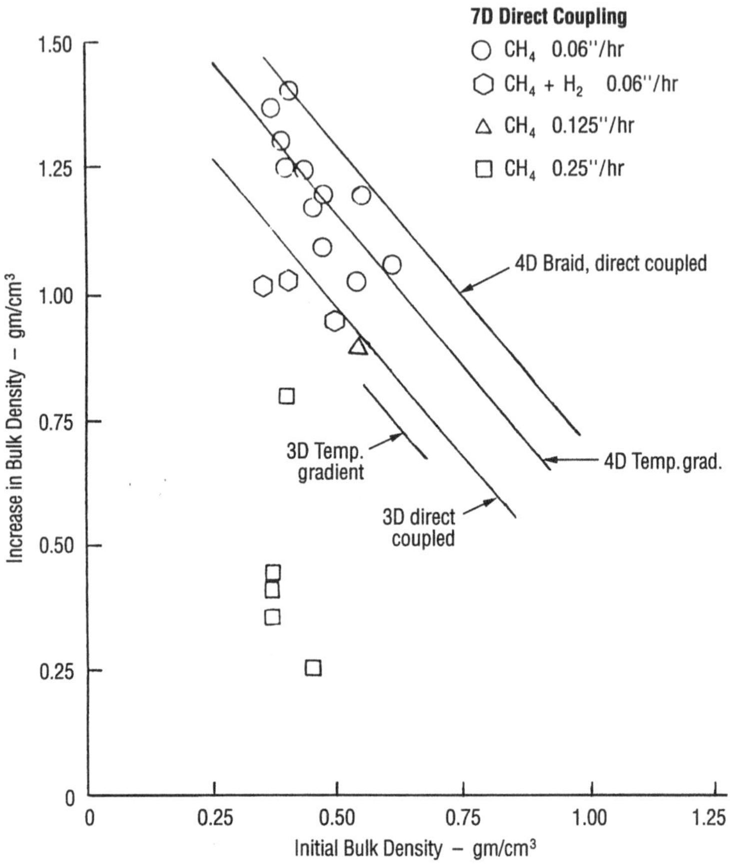

Fig. 4.6. Correlation between initial and final density for various weave geometries and infiltration methods [9]

structure can be improved and controlled by using the temperature gradient technique. If the gradient is quite steep, e.g., 500 °C per cm, with careful control of pressure and gas flow, relatively high-density (~ 1.75–1.8 g/cc) carbon/carbon composites can be fabricated. Using the direct coupling infiltration technique, Gebhardt et al. have demonstrated that available surface area is also equally important in obtaining uniformly densified carbon/carbon composites. Using a fine-fiber 7D structure they found that the fiber bundles get nicely infiltrated under optimum processing conditions (Fig. 4.7) [9].

Fiber overlaps and twisted regions in braided and multidirectional weave structures tend to be the points of blocked voids. Therefore, higher

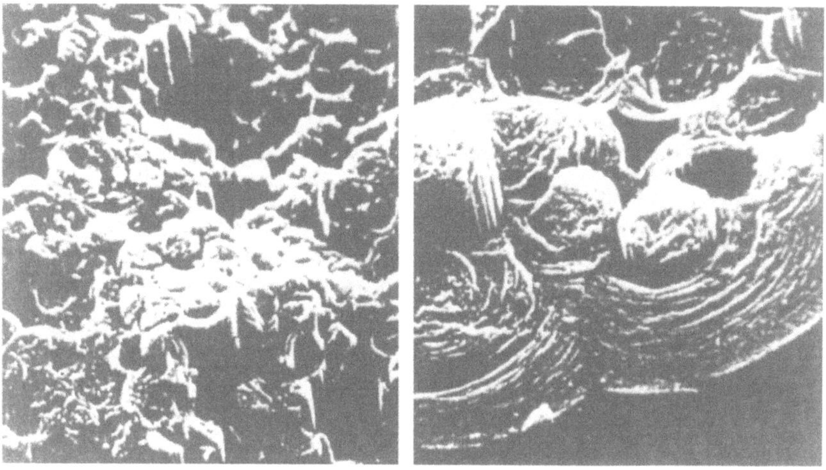

Fig. 4.7. Scanning electron micrographs of pyrolytic carbon deposit within bundles in 7D billet [9]

densities are achievable in 4D and 7D structures than in 3D structures. Even using the direct coupling technique, ultimate higher density carbon/carbon composites can be obtained with low fiber volume preform having relatively open, accessible, and well-connected voids. However, for a particular initial density, the increase in density is greatest for composites made with the direct coupling technique followed by the temperature gradient technique and least for those made by the isothermal process. In all cases the density varies through the thickness. Seven-directional carbon/carbon composites have been produced with densities as high as 1.85 g/cc as compared with 3D structures with a density 1.65 g/cc.

Concurrently with Super Temp, Sandia Laboratories also undertook to develop carbon composites using mainly the temperature gradient technique and carbon felt as substrate. Leeds and Pierson [13] reported the successful development of carbon/carbon composites as full-sized heat shields. The final density of the C/C composites achieved by them is reported to be 1.83 g/cc.

Later researchers worldwide [14–16] used the above mentioned conventional CVD techniques for studying infiltration of pyrocarbon in a carbon/carbon skeleton and the properties of such composites.

Over the past few years, efforts are being made to improve upon the conventional processes. Vaidyaraman et al. [17–19] used forced flow-thermal gradient chemical vapor infiltration process (FCVI) to fabricate carbon/-carbon composites using propylene, propane, and methane. This process incorporates the advantages of both forced flow and thermal gradient processes. In the FCVI process, a temperature gradient in the order of 200° – 500 °C is applied across the preform and the reagent gases are forced to flow through the cold to the hot surface of the preform. In this way, a high reagent concentration at the cold side offsets the lower temperature, and the lower concentration at the hot side offsets the high temperature, resulting in uniform deposition rates. Thus, reagent concentration, temperature, and the flow rate are the critical parameters. Using this technique, high density carbon/carbon composites can be fabricated in shorter times.

Other modifications of the CVI technique which are under investigation for the manufacture of fibrous composites are: pulse CVI technique [20, 21], plasma enhanced CVD [22], and thermal gradient isobaric CVI [23]. In the pulse CVI technique, cycles of evacuation of the preform and introduction of cracking gas are performed repeatedly. The duration of each cycle may vary from a few seconds to a minute or two. Thousands of cycles need to be performed to achieve good densification. Deposition temperature, gas flow rate, and gas introduction time are the most important parameters for obtaining dense anisotropic carbon around carbon fibers.

4.4 The Classical Fabrication Route, as Used for Carbon Ceramics

The classical fabrication methods employing filler and binder phases used for fabrication of carbon products are similar to those used for ceramic processes. However, they differ from the usual ceramic sintering in that the strong covalent bonding that exists between carbon atoms does not allow any significant atom movement at high temperatures and hence thermal sintering is not achieved in carbon particles. Instead, solid particles of pure carbon (primary carbon) are combined with a temporary binder, which then acts as precursor for the secondary carbon formed during the baking, i.e., carbonization treatment. The result is an "all carbon" material with two different phases, namely, the primary carbon as "filler carbon" and the secondary carbon as "binder carbon". The analogy between synthetic granular carbons and carbon/carbon composites is shown in Fig. 4.8 [3]. In carbon/carbon composites, carbon fibers instead of filler grains are used as primary carbon.

Figure 4.9 shows the analogy between the processing of polygranular carbon as a ceramic and the processing of carbon/carbon composites by the multistep pyrolysis route. The impregnation and recarbonization steps in the farbrication of C/C composites are repeated four to six times, where-

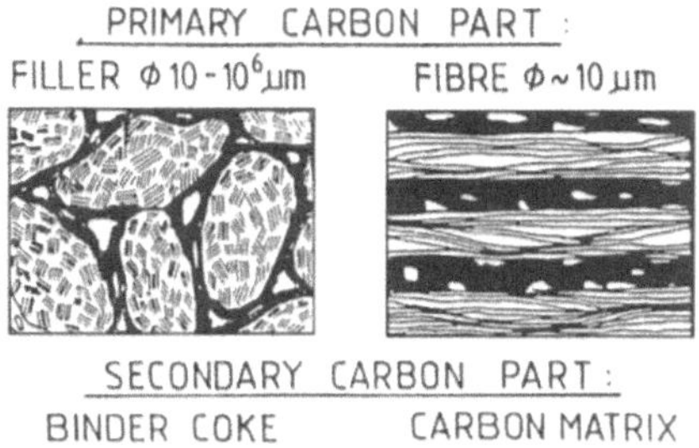

Fig. 4.8. The two phase structures of synthetic polygranular graphites and carbon/carbon composites

as in the industrial production of polygranular carbons, a maximum of two impregnation steps are usually performed. The main characteristics of the ceramic-like processing method for carbon materials fabrication are the mass loss and shrinkage of the matrix during heat treatment. If not controlled effectively, the latter results in a cracked network in the matrix of carbon/carbon composites. One can start with a three-dimensional fiber arrangement and avoid shrinkage by mechanical means. Some bulk shrinkage during carbonization of the "green" composite material, however, is generally tolerated. Carbon/carbon composites are generally highly porous after first production step. They consist mainly of primary carbons, and

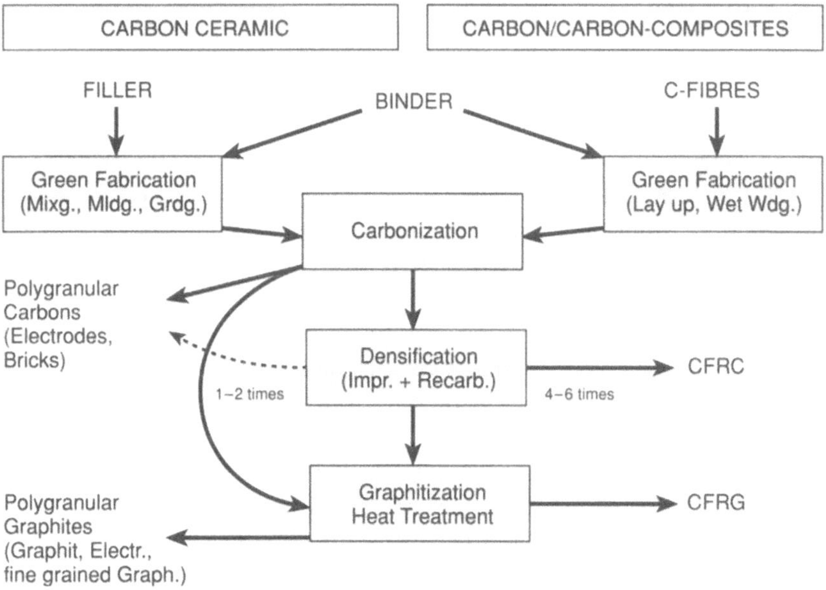

Fig. 4.9. Comparison of production processes for synthetic polygranular graphites and carbon/carbon composites

only small parts of binder bridges constituting secondary carbon. The objective of subsequent process steps is the densification of such porous skeletons. Densification is achieved by impregnation with carbonaceous precursor compounds – liquids or gases – and subsequent recarbonization. Thus, these multiphase "all carbon" composite materials also contain a third type of solid carbon, the impregnated carbon [24].

Densification may be achieved by two different impregnation routes, i.e., gas-phase impregnation and liquid impregnation. However, as is the case for conventional granular carbon technology, the more economical liquid impregnation method has gained importance also for carbon/carbon composites. The process involves vacuum/pressure impregnation of the porous structure with a carbon yielding liquid; thus, the impregnant has access to the innermost fine pores. Therefore, composites with complex shapes in various fiber configurations, such as unidirectional, bidirectional, fabric or multidirectional can be produced with high densities using the liquid infiltration technique. Composites with much higher densities can be processed using a combination of liquid infiltration and CVI since the impregnants in CVD or CVI do get into very small pores resulting from multiple impregnation/carbonization of liquid precursors. These small pores cannot be further impregnated with liquids because of surface tension.

4.4.1 Impregnating Precursors

Organic compounds are the main and almost exclusive raw materials for the production of artificial graphites. By pyrolysis, these carbon compounds are converted to solid carbons and various volatile compounds as by-products. High bulk densities and mechanical strengths are only attainable if the amount of solid carbon formed is much greater than the amount of volatile by-products. The pyrolysis must be directed towards the formation of carbon free volatile by-products such as water, hydrogen chloride, and hydrogen sulfide, avoiding the formation of stable volatile carbon compounds such as carbon monoxide, fluorocarbons, and carbon sulfur compounds. Thus, the carbon yield is influenced by the chemical composition and molecular rearrangement of the starting material as well as by the reaction parameters such as temperature, heating rate, residence time at the pyrolysis temperature, and gas pressure [24, 25]. The morphology of the carbon products is, in addition, influenced by the mass and heat transport phenomenon during pyrolysis and the ease of escape of the volatile by-products.

The ultimate properties of carbon/carbon composites are much dependent on the structure and properties of carbon matrix which in turn are controlled by their precursors. Therefore, the selection of the precursor for the matrix is extremely important and should be considered:

(a) The carbon yield should be very high which means a low weight loss during carbonization.
(b) The precursor should have optimum low viscosity in order to enter the fine pores of the substrate and get contained there.
(c) Controll on wetting of the carbon fibers with the carbon precursor.
(d) A favorable coke microstructure should result after carbonization.

These stringent requirements limit the number of precursors to only a few which can be classified into two general categories: thermosetting resins and thermoplastic pitches.

4.4.4.1 Thermosetting Resins

Thermosetting resins have been in use commercially for a long time as impregnants for various applications because of their ready availability in liquid state with moderate viscosity at room temperature and the ability to polymerize to form a highly cross-linked nonmelting solid when heated approx. at 200 °C. As precursor for carbon products, thermosetting resins additionally should contain an aromatic ring structure in the backbone or be able to form a ring structure during early pyrolysis so as to get a high char yield [24–26]. Carbon yield is determined by the initial carbon content of the precursor and the extent to which volatile-carbon-containing species are released during the pyrolytic reactions. Therefore, along with the molecular structure of the polymer, their pyrolysis behavior is also equally important for obtaining high-density carbon/carbon composites. Polyfurfuryl alcohol and phenolics have been in use for a long time in carbon industries as binders and impregnants [24–27]. These resins polymerize at 150°–250 °C to form highly cross-linked, nonmelting solids which undergo solid-state pyrolysis [28, 29] and are therefore well contained within the filler particles. These resins have also been used in fiber-reinforced polymer matrix composites for ablative applications. In fact, the first carbon/carbon composite was formed as a result of charring a carbon-fiber-reinforced thermosetting resin in the absence of air. Because of the importance of thermosetting resins for carbon products and carbon/carbon composites, there has been great interest in the development of new resin systems which

Table 4.1. Carbon yield of various thermoset precursors

Precursor	Carbon yield (%)	Pyrolysis efficiency (%)
Polyacetals	85	> 90
Polyphenylene	85	> 90
Polyimides	60	85
Phenolic resins	55–65	75–80
Furan resins	55–65	75–85
Polybenzimidazole	74	95
Resorcinol furfuryaldehyde	60	90
Oxidized polystyrene	55	85
Polyacrylonitrile	45	65
Polysulfone	48	66
Epoxidized phenol-formaldehyde	50	63
Cellulose	20	NA
Epoxy resin	5	13

NA, not available.

give higher and higher char yields and in reaching an understanding of the pyrolysis behavior of these resins [30–31]. The carbon yield of some of the thermosetting resins are compiled in Table 4.1. The char yield of most of the resins lie in the range of 45%–60% by weight whereas polyphenylene and polyacetal resins exhibit a char yield as high as 85%. Figure 4.10 shows the chemical structures of some potential precursors for carbon matrix in carbon/carbon composites. A thermogravimetric analysis of these resins is shown in Fig. 4.11.

During pyrolysis of these resins, condensation, oxidation, dehydration, and decomposition reactions may occur as both simultaneous and sequential reactions. Figure 4.12 shows the sequence of probable chemical reactions occuring during pyrolysis of phenol formaldehyde and furfuryl alcohol polymers; their thermogravimetric and gas analyses are shown in Fig. 4.13 [29]. As stated above, during pyrolysis, various dissociation/condensation reactions take place concurrently. Therefore, weight loss of the polymers during pyrolysis is accompanied by volumetric shrinkage. In many cases the extent of the two are found to be in the same proportion. Matrix shrinkage during pyrolysis has a dominating effect on the microstructure and performance of carbon/carbon composites. Excessive shrinkage during carbonization generates stresses at the fiber/matrix interface and often results in matrix cracks and fiber damage. A systematic

PHENOLICS:

RESIN A

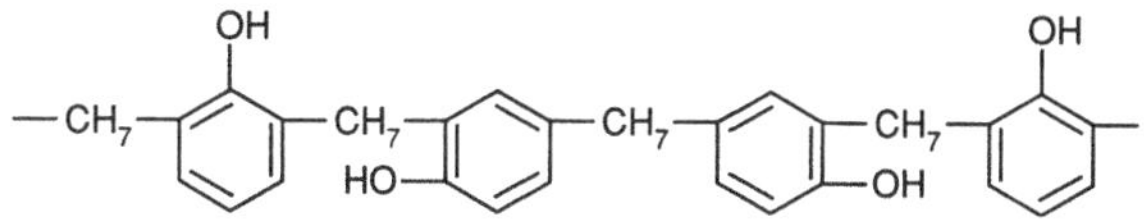

+ HEXAMETHYLENTETRAMIN (HARDENER)

RESIN B

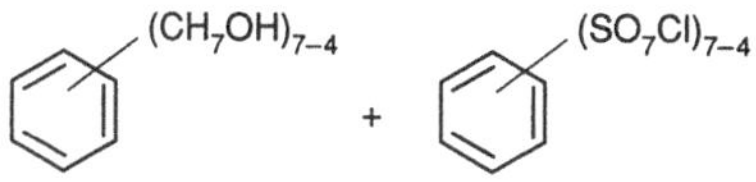

POLYIMIDES:

KAPTON

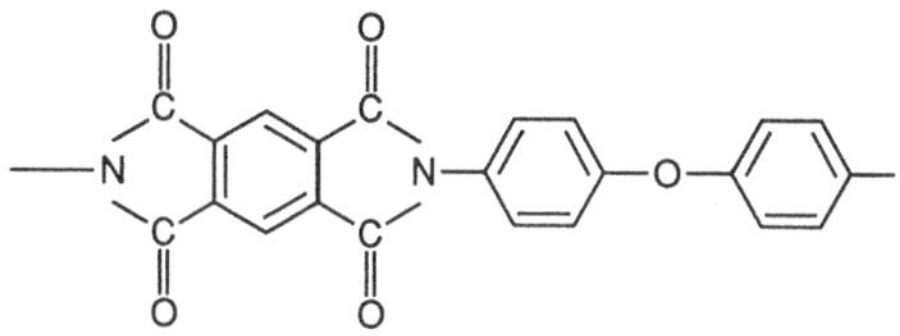

QX 13

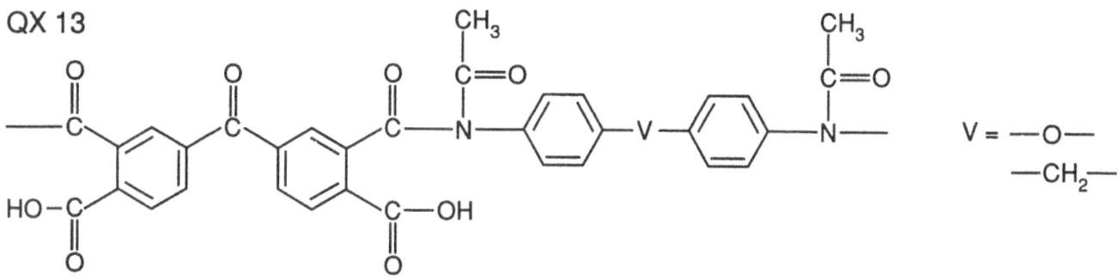

POLYPHENYLENE:

HA 43

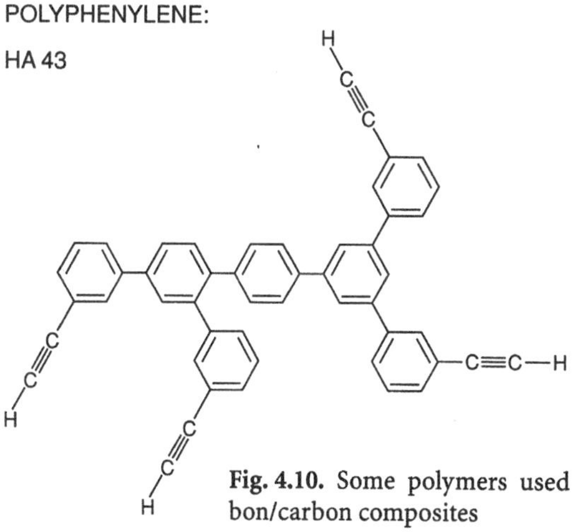

Fig. 4.10. Some polymers used as matrix precursors for carbon/carbon composites

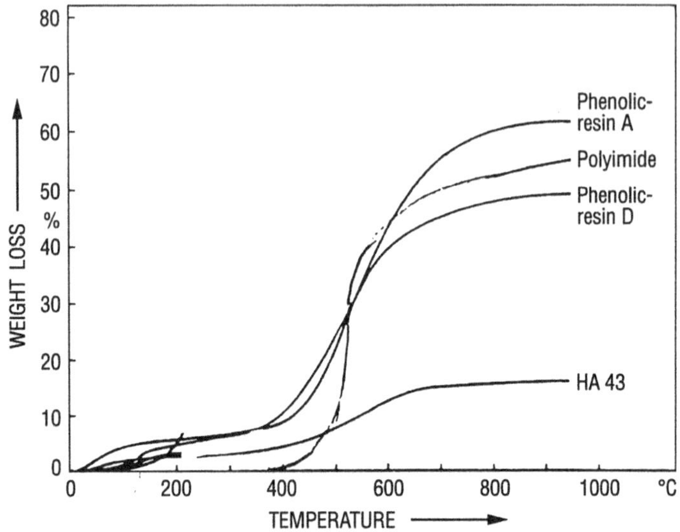

Fig. 4.11. Weight loss during pyrolysis of precursor cured resins

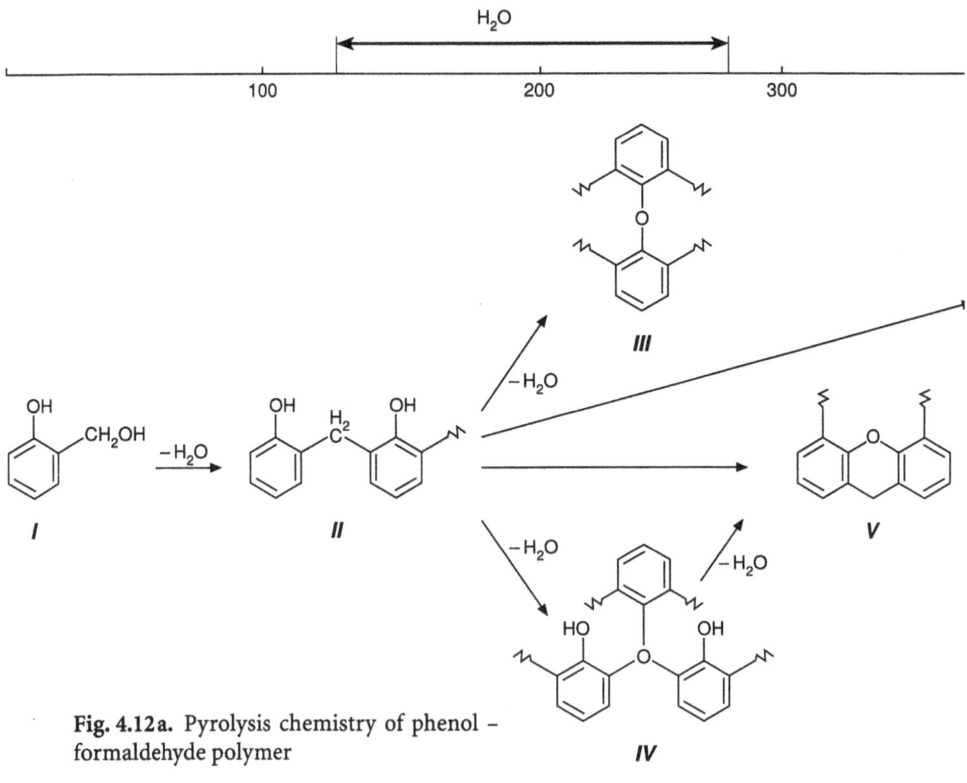

Fig. 4.12a. Pyrolysis chemistry of phenol – formaldehyde polymer

analysis of the pyrolysis mechanism and thermogravimetric analysis of resin matrixes help in choosing optimum carbonization conditions, especially the rate of heating in particular temperature region.

The solid-state pyrolysis of highly cross-linked polymers results in turbostratic carbon with aromatic systems angularly displaced, similar to their previous arrangement in the cross-linked polymers, since there is not sufficient mobility of the molecules for ordered rearrangement. These resins yield low density nongraphitizable carbons with closed porosity, which limits the density of carbon/carbon composites to lower values (1.5–1.7 g/cc) [29].

Shimizu et al. [32] synthesized various highly cross-linked, high molecular weight phenolic resins in order to increase char yield and decrease carbonization shrinkage. By using linear high ortho type phenolic resin, they found an increase in char yield of about 25% and a reduction in carbonization shrinkage of 10%. These resins are found to contain anisotropic carbon to some extent.

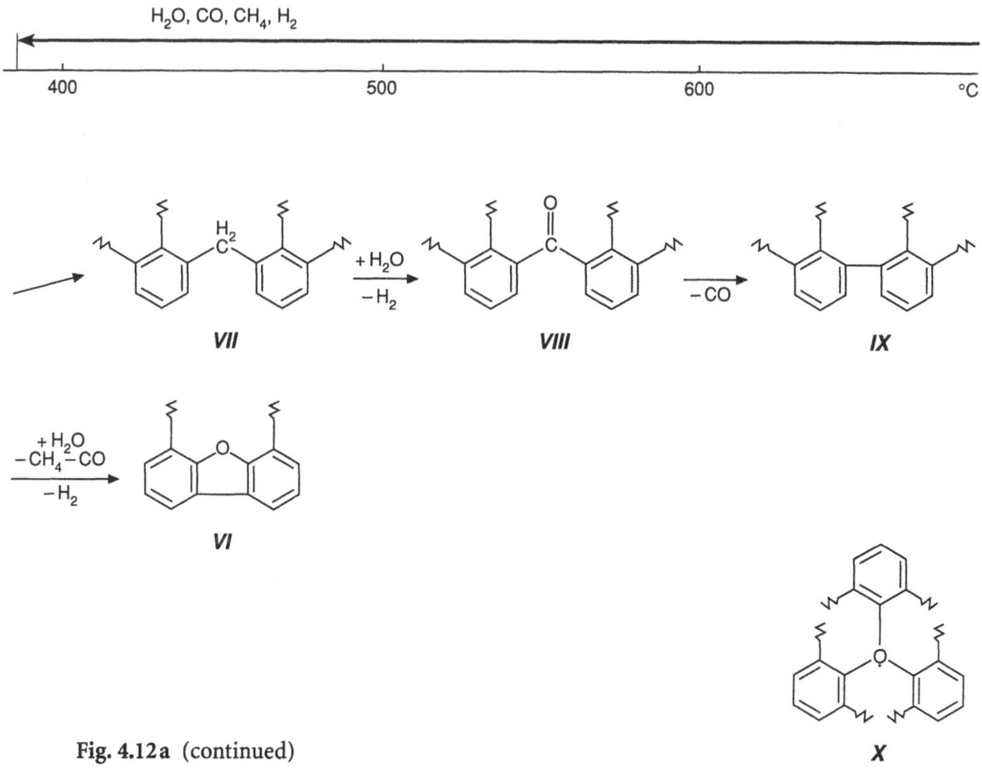

Fig. 4.12a (continued)

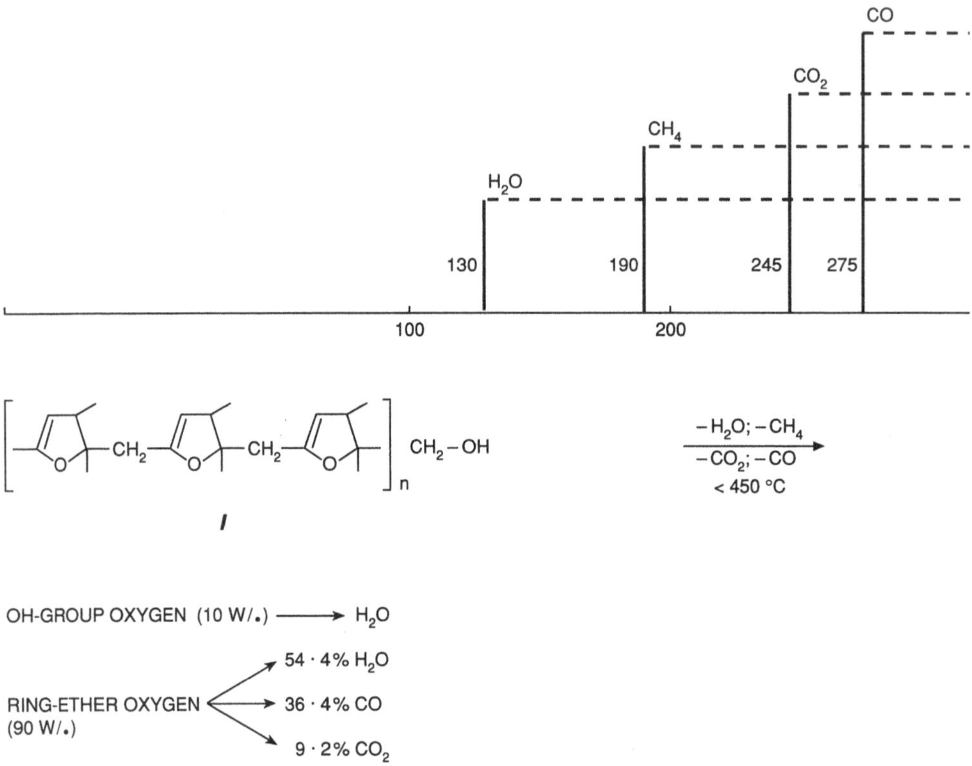

Fig. 4.12b. Pyrolysis chemistry of polyfurfuryl alcohol [24]

In addition to the commonly used phenol formaldehyde and polyfurfuryl alcohol resins, a wide range of other new types of resins have also been tried as carbon/carbon matrix precursors. As can be seen from Table 4.1, polyimide gives a carbon yield of approximately 60% [33]. Like other resins, these resins are also commercially available. However, the studies carried out by various authors on the use of polyimides as a matrix precursor for carbon/carbon composites do not speak in favor of this resin over phenolic or furan resins because of their high cost combined with the problems associated with their processing, especially during prepreging and polymerization. Moreover, polyimides exhibit excessive shrinkage during carbonization. Similarly, poly *p*-phenylene resin, whilst exhibiting high char yields is costly and associated with hazardous problems in the

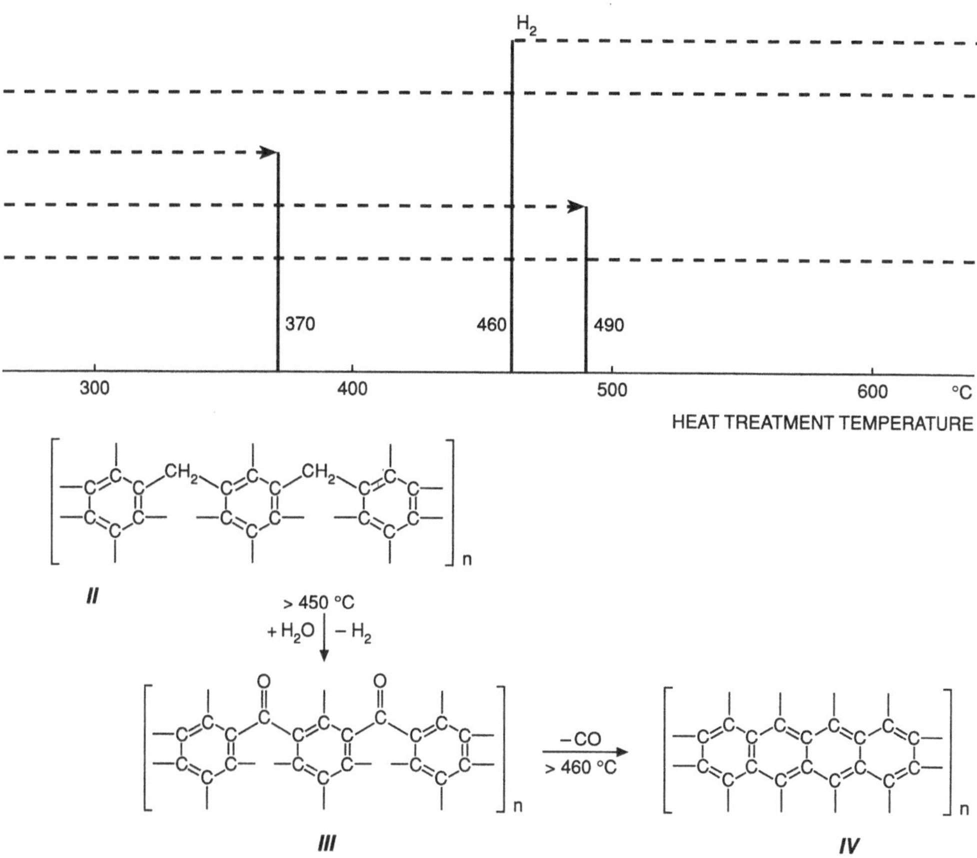

Fig. 4.12b (continued)

processing of composites, making it prohibitive for application as a matrix precursor. In general, it is observed that the resins giving high char yields have exorbitant prices and pose problems in the manufacture of polymer composites. Therefore, phenolic and furan resins are still the most popular matrix precursors for making carbon/carbon skeletons and/or as liquid impregnants.

4.4.1.2 Pitches

Pitches are an essential constituent in the manufacture of more than 80% of carbon products. They are used both as binder and for subsequent im-

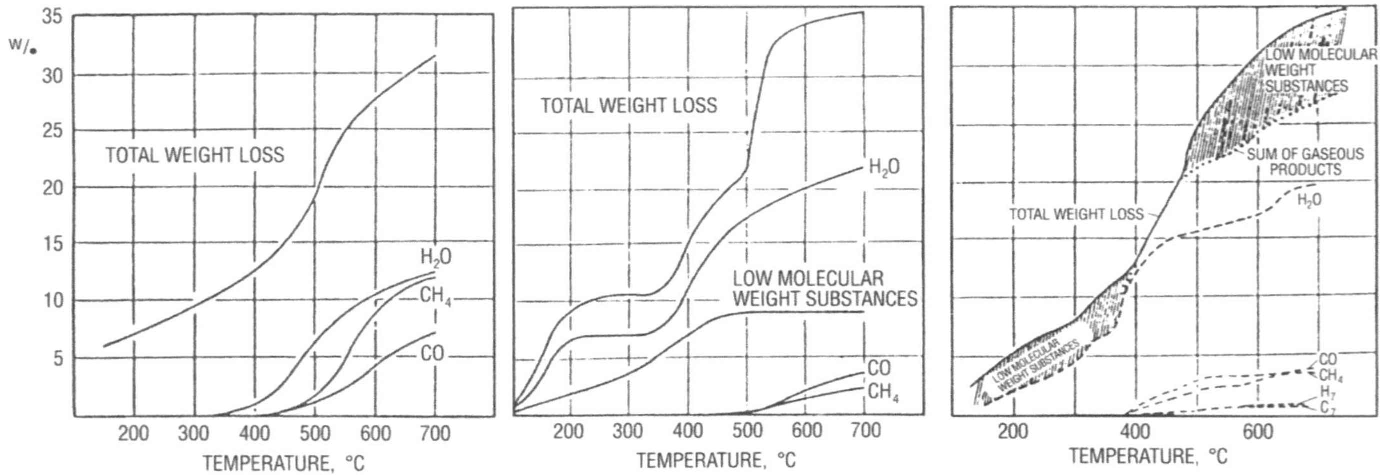

a b c

Fig. 4.13 a–d. Thermal analysis and gas analysis of **a** highly cross-linked phenol-formaldehyde resin, **b** weakly cross-linked phenol-formaldehyde resin, **c** commercial phenolic resin

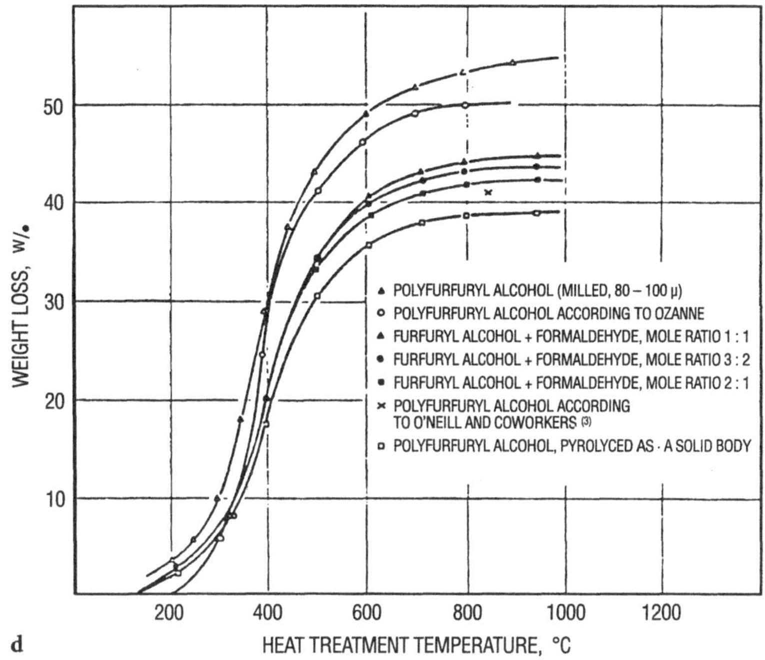

d

Fig. 4.13 (continued) Thermal analysis and gas analysis of **d** polyfurfuryl alcohol resins

pregnation and densification purposes [34–36]. Therefore, a large database exists from graphite industries on their processing and conversion to carbon. High carbon yield, low cost, and graphitizability in addition to controllable viscosity are the attractive characteristics of pitches. Depending on their origin, pitches are classified as coal tar pitches, petroleum pitches, and synthetic pitches [36–39].

Coal tar is a by-product obtained during coking of bituminous coals. Pitches are obtained from the coal tar by distillation and heat treatment processes. These are complex mixtures containing a range of aromatic and heterocyclic organic compounds consisting predominantly of carbon and hydrogen with small amounts of nitrogen, oxygen, and sulfur in specific cases. The precise composition and properties vary according to the source of the tar and the method of removal of low molecular weight fractions [37–41].

Petroleum pitches are obtained from the residues at the bottom of catalytic crackers of naphtha, crude oil distillation plants, or refineries by distillation, heat treatments, or oxidation, or by a combination of these processes. Like coal tar pitch, the chemical and physical characteristics of these pitches are dependent on the source of the petroleum crude and processing conditions. Primarily, the pitches are characterized by coke yield and coke structure produced from the crude petroleum, which in turn depend on the aromaticity or C/H ratio of the pitches. Coal tar pitches are usually more aromatic than petroleum pitches. However, from the point of view of fundamental understanding of the pitch processing, its chemical characteristics and their influence on pyrolysis behavior, solvent fractionation has widely been used to divide the pitch into a broad range of constituents. Though coal tar and petroleum industries have tended to use different solvents and nomenclature, the carbon industries have used a system of nomenclature for many years which still prevails.

Quinoline or Pyridine Insolubles (QI)

The fraction which is insoluble in quinoline or pyridine represents high molecular weight aromatic compounds and solid impurities such as free carbons. These are known as C_1 components or QI or α resins.

Benzene or Toluene Insolubles

Fractions soluble in quinoline or pyridine but insoluble in benzene or toluene are known as C_2 components, TI, or β resins.

Petroleum Ether or n-Hexane

Fractions soluble in benzene or toluene but insoluble in petroleum ether or *n*-heptane are known as resinoid fractions, and are identified with asphaltenes; the soluble material is the crystalloid fraction.

The solubility spectrum of a pitch can be useful in its characterization. Many of the molecules that are present in typical pitches are mesogenic in character and can be extracted and used to make specialized pitches for specific applications with the best possible mechanical and thermal properties of carbons [38–42]. Soluble fractions can be analyzed by vapor phase osmometry for average molecular weight, and an indication of the molecular weight distribution can be obtained from gel permeation and high pressure liquid and gas chromatography. The aromatic and aliphatic character of fractions can be determined by protons and ^{13}C nmr spectroscopy. In addition, a number of other analysis techniques such as IR and UV spectroscopy, mass spectroscopy, X-ray diffraction, and electron spin resonance have been used to study the molecular species in the pitch and as tools to modify the chemical structure of the pitches for specific end application and to study the reaction mechanism during carbonization of pitches [43–45].

Commercially, pitches are available under two categories: binder pitches and impregnating pitches. Since, pitches are one of the major precursors of most carbon artifacts, their behavior on carbonization has been the main subject of interest for many carbon scientists. Pitches are comprised of a large variety of organic compounds differing in chemical composition and number of aromatic rings; hence, it is difficult to study and develop carbonization models. Therefore, various workers have studied the carbonization behavior of model compounds representative of certain fractions [39, 42, 46, 47]. It is believed that carbonization basically consists of aromatic growth and polymerization to two-dimensional networks which may ultimately attain the three-dimensional order of graphite. The major chemical processes involved during pyrolysis may be thermal scission of aliphatic side chains and C–H and C–C bond cleavage to form reactive free radicals; recombination of reactive free radicals to form planar aromatic ring structures of higher molecular weight, molecular rearrangement; thermal polymerization; and aromatic condensation with elimination of side chains and hydrogen leading to three-dimensional ordering during graphitization. All of these processes may occur sequentially or simultaneously, depending on the amount and nature of organic species present in the original pitch. Most of the chemical reactions and polymerization to two-dimensional ordering take place in liquid phase below 450 °C. These reactions make the building blocks for the ultimate graphitic structure in carbon. Therefore the development of liquid crystal phase (mesophase) during liquid phase pyrolysis of the pitch is the most significant feature of pitch carbonization.

The polynuclear aromatic molecules of the mesophase have a disc-like shape and tend to align parallel to an applied magnetic field [48]. The amounts of different pitch fractions in a pitch determines the ultimate structure of the coke produced from it since each fraction result in coke with different microstructure [49, 50, 51]. Carbonization of the benzene solubles produces a strongly anisotropic, coarse-flow microstructure, whereas β resins carbonize to a coarse mosaic texture.

The QI fraction produces an isotropic carbon. The mobility of these polycyclic aromatic molecules in the liquid phase during carbonization (shearing) also influence the orientation of the basal planes in the resultant carbon, suggesting that these will be influenced by the viscosity of the pitches. The viscosity of the pitches as such can be controlled by fractionation and the mixing technique.

Pitch is preferred as matrix precursor for advanced carbon/carbon composites due to the high coke yields obtainable from them. The yield depends very much on the composition of the precursor pitch and pyrolysis conditions. It is very strongly influenced by the proportion of β resin and quinoline insolubles, generally increasing with an increasing amount of these fractions. The carbon yields of QI and β resin fractions are found to be 95% and 85% respectively, almost constant, independent of the origin of the pitch [48, 50]. The carbon yield of the benzene soluble fraction varies between 30% and 55% depending on the average molecular weight of this fraction present in the pitch with respect to the yield one may be tempted to have a pitch with higher QI and β resin contents. However, these fractions raise the viscosity of the pitches and also reduce the wettability of the pitches with the fillers, thereby not only affecting the final macrostructure of the product but also creating difficulties during processing, i.e., infiltration in the innermost fine pores. Therefore, a balance of different pitch fractions has to be arrived at, keeping in mind the viscosity, wetting characteristics, and coke microstructure .

The pitches, from their softening point onwards, exhibit continuous change in viscosity with temperature, decreasing to an extremely low value. The minimum viscosity and temperature of least viscosity depend on the elemental composition and thermal history of the pitches [50]. Pitches attain minimum viscosity over a small temperature range, and knowledge of this behavior is of great importance for processing carbon/carbon composites. Rand [51] has proposed T_g to be taken as a reference temperature since the pitches are found to attain minimum viscosity at a temperature of about 100° – 150 °C above T_g. The viscosity of various pitches and pitch fractions with increasing temperature, is shown in Fig. 4.14 [50].

The change in viscosity from the softening point to least viscosity is associated with melting and fractionation of isotropic pitch, whereas the increase in viscosity is associated with devolatization of low molecular weight compounds, polymerization, and structural ordering of planar aromatic molecules to a liquid crystalline state called mesophase. Mesophase spheres of about 0.1 μm diameter similar to those shown in Fig. 4.15 [3] begin to appear at temperatures greater than about 400 °C and consist of polyaromatic molecules with a broad molecular size distribution and average molecular weight in the order of 1000. Such a spherical mesophase behaves like nematic liquid crystal and tends towards coalescence, forming large areas following further pyrolysis. The structural arrangement of the polyaromatic in the liquid crystal determines the micro and ultrastructure of the resulting carbon, which is formed after splitting off the residual hydrogen. Figure 4.16 shows a schematic representation of the molecular arrangement within the mesophase [48]. The chemical composition of the pitch, reaction temperature, and heating rates control the kinetics of meso-

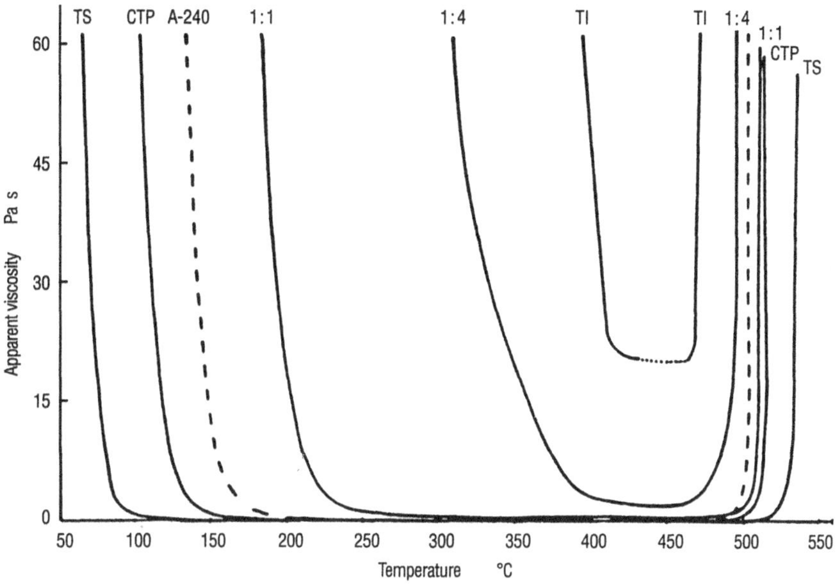

Fig. 4.14. Viscosity of various pitch fractions [50]

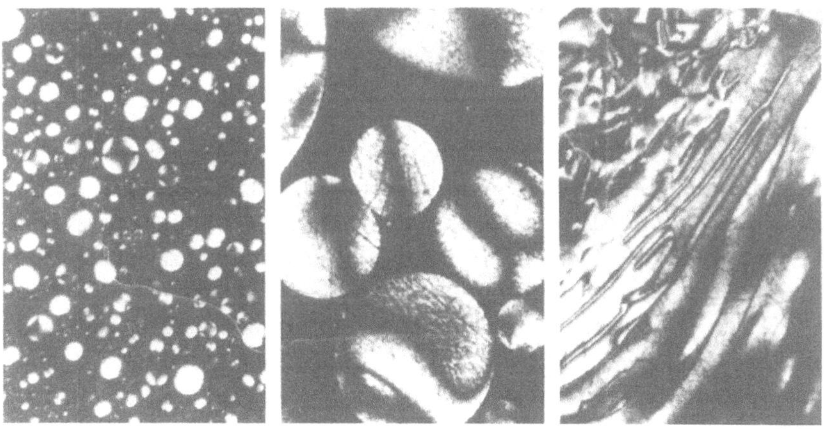

Fig. 4.15. Growth of carbonaceous mesophase [3]

phase formation and determine the microstructure of the resulting carbon solids.

It is well known that while most of the pitches form mesophase structures, those with a reactive side chain structure result in more or less isotropic carbons. Chemically modified pitches mostly lead to isotropic carbons

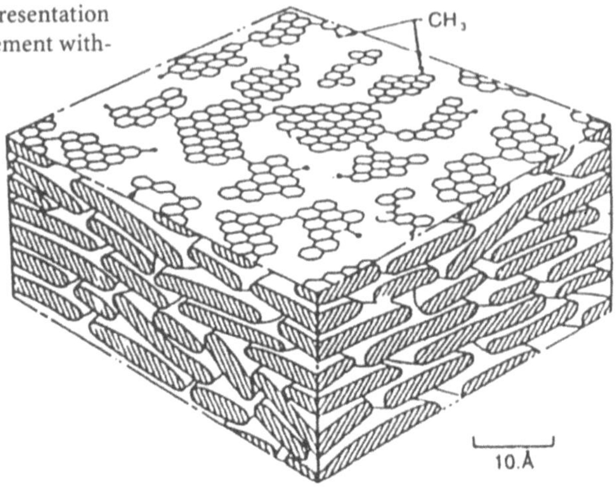

Fig. 4.16. Schematic representation of the molecular arrangement within the mesophase [48]

[51–54]. The coalescence of mesophase is also very sensitive to the presence of particulate matter (usually 1 µm) within the pyrolyzing system which normally adheres to the surface of the spheres and prevents coalescence. These particulate impurities normally known as quinoline insolubles are found to be contained in coal tar pitches while petroleum derived pitches do not contain quinoline insolubles. Therefore, the latter pitches result in anisotropic needle-like cokes. Coking pressure and temperature also influence mesophase formation and microstructure of carbons from pitch. According to Burns and McAllister [55], as the carbonization pressure is increased the coke microstructure changes from a needle-like structure to a coarse mosaic structure. At still higher pressures of about 200 MPa, coalescence of the mesophase does not occur resulting in still coarser and more isotropic carbons.

Coal tar and petroleum-based pitches on normal pressure carbonization yield about 50% coke by weight, the same as that found for phenolics and furfural resins. The coke yield of coal tar pitches can be increased either by using chemical additives such as sulfur or by slow carbonization or by pressure pyrolysis at 550 °C. Figure 4.17 [2] shows the effect of pyrolysis pressure on coke residue. The results of pressure carbonization of coal tar and petroleum pitches indicate that pressures in the range of 6.9 and 68.9 MPa are sufficient to achieve high coke yields. By dehydrogenation of coal tar pitch with 10% sulfur, the three-dimensional network of coke increases, resulting in glass-like carbon with poor mechanical and thermal properties [52, 54].

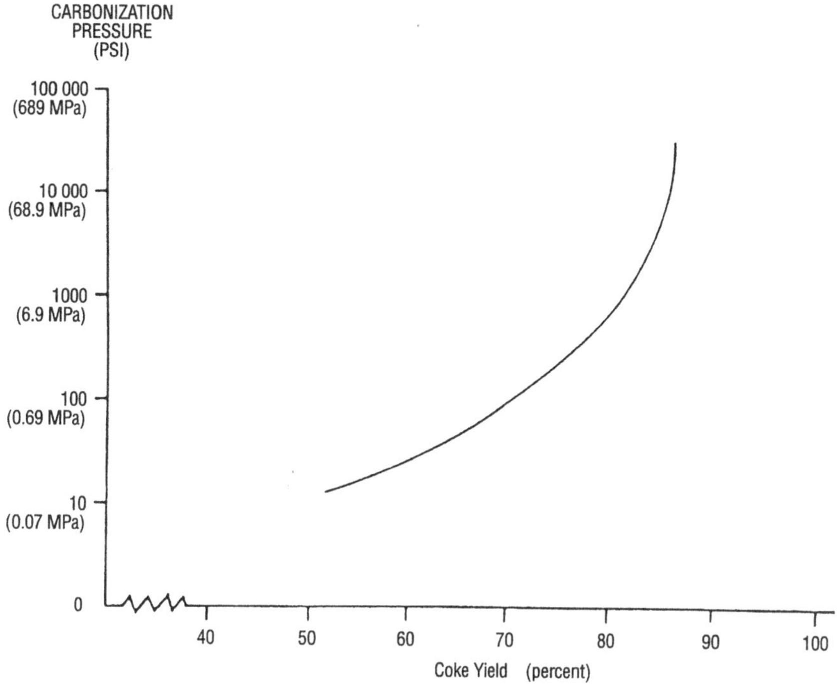

Fig. 4.17. Effect of pressure carbonization on coke yield for petroleum pitch [2]

Synthetic pitches with high coke yields have also been developed by thermally reacting quinones with other quinones or with unsaturated fused ring polycyclic hydrocarbons [38, 47, 51]. These quinone compositions exhibit very high carbon yields (as high as 77% – 88%), low melt viscosity, easy processing, and low carbonization shrinkage.

4.5 Processing of Composites with Thermosetting Resins as Matrix Precursors

In undirectional (UD) and 2D composites, the laminates with thermosetting resin are laid up as per predetermined configuration, uniaxially pressed, vacuum bagged, autoclaved, or press cured and postcured depending on the shape and geometry of the composites. The products in polymer forms may also be made by the filament winding technique. Polymer composites so formed are carbonized in a controlled way. During carbonization, the polymer matrix decomposes, resulting in the release of vapor by-products and weight loss which, if not controlled properly, may erupt the structure of the composites. Therefore, usually very slow heating rates, 5° - 10 °C/h, depending on the thickness of the composites, are used during carbonization of polymer matrix composites. As discussed in Sect. 4.4, simultaneous decomposition, condensation, and rearrangement of molecular chains during carbonization of polymers lead to the shrinkage of the char matrix in all directions [3, 55]. This will obviously cause shrinkage in the composites as well. However, bulk shrinkage in the composites is constrained by the presence of carbon fibers. In a UD composite, although no shrinkage takes place in the fiber direction, appreciable changes are observed in the direction perpendicular to the fibers. The trend of shrinkage in composites in the direction perpendicular to the fibers is almost similar to that observed during carbonization of unreinforced poly-

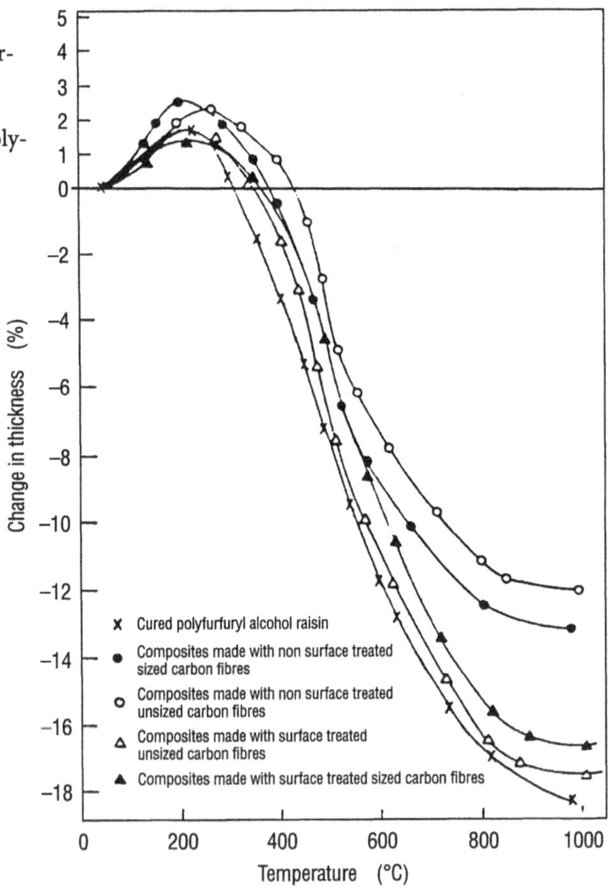

Fig. 4.18. Dimensional changes in direction perpendicular to the fibers during pyrolysis of carbon-fiber-reinforced polymer composites [56]

mers (Fig. 4.18) [56]. However, in composites, the temperature of onset of shrinkage is lower by 50° – 100 °C. Fiber reinforcement influences both the onset of shrinkage and the rate of shrinkage. The maximum shrinkage takes place between 450 °C and 650 °C. Therefore, due care must be taken with the rate of heating in this temperature range. Between 650 °C and 800 °C, the rate of shrinkage is found to be reduced by 50 %. Shrinkage continues up to 1100 °C, although at a very low rate. Suppression of shrinkage in an anisotropic way results in extensive microcracking and void formation in the composites (Fig. 4.19) [57]. The extent of shrinkage and cracks depend on the fiber-matrix bonding in the polymer composites which are in turn controlled by the type and amount of surface chemical groups on the carbon fibers [56–58]. Carbon fibers made by the commercial process of heat treatment at a temperature between 1200° – 1500 °C (high-strength type or intermediate-modulus (IM) type) are found to possess an active carbon surface and more surface groups than high-modulus (HM) type carbon fibers. Consequently, polymer composites made with the latter fibers exhibit lower shrinkage during carbonization than those made with

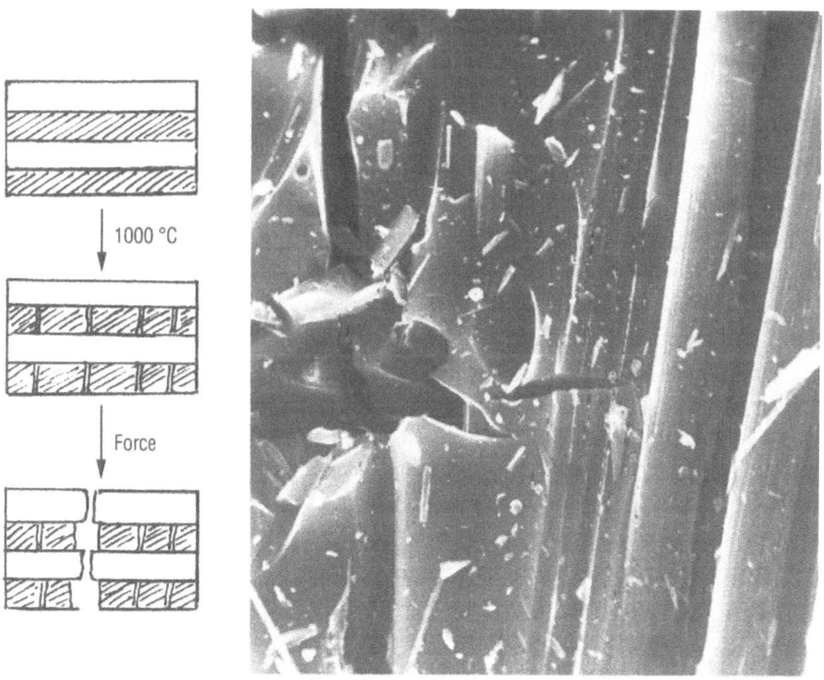

Fig. 4.19. Intramatrix crack formation during first carbonization of polymer composites [57]

Fig. 4.20. Cross-sectional shrinkage during pyrolysis of composites made with different carbon fibers

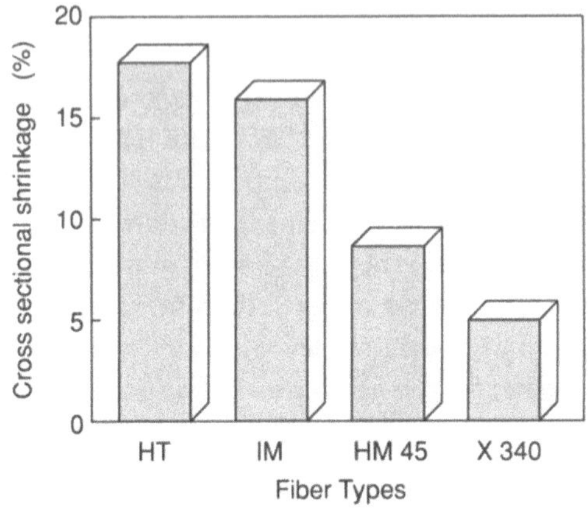

the former fibers. This is evident from Fig. 4.20, which shows shrinkage in a direction which is normal for the fibers of model UD polymer matrix composites made with different carbon fibers [58]. Composites made with high-strength carbon fibers and surface-treated HM carbon fibers exhibit higher shrinkage than those made with non-surface-treated HM carbon fibers [57–60]. In the former composites the matrix seems to have shrunk onto the fibers, whereas in the latter, the matrix shrinks away from the fibers. Consequently, in composites made with non-surface-treated HM carbons fibers, slit-type pores and cracks running parallel to the fiber axis predominantly at the fiber/matrix interface are observed in the charred matrix. These look like annular gaps around carbon fibers (Fig. 4.21). In composites made with surface-treated HM carbon fibers, the cracks are found to run both in the fiber direction and perpendicular to it. If the fiber/matrix bonding in polymer matrix composites is too strong, the shrinkage cracks running perpendicular to the fibers may even damage the fibers (Fig. 4.22). Recent studies have shown that the surface energetics on carbon fibers are a function of the heat treatment temperature of the fibers. If commercial high-strength and IM carbon fibers are post-heat-treated to around 1800°–2000 °C, the resulting carbon fibers form intermediate bonds with the polymer matrix. When such composites are carbonized, a lower pyrolysis shrinkage and better structural integrity results. In contrast to UD composites, very little bulk dimensional changes are observed in woven or multidirectional fiber-reinforced composites because of the presence of fibers in multidirections [57, 61]. However, the matrix shrinks in the usual way, resulting in development of a system of cracks in different directions.

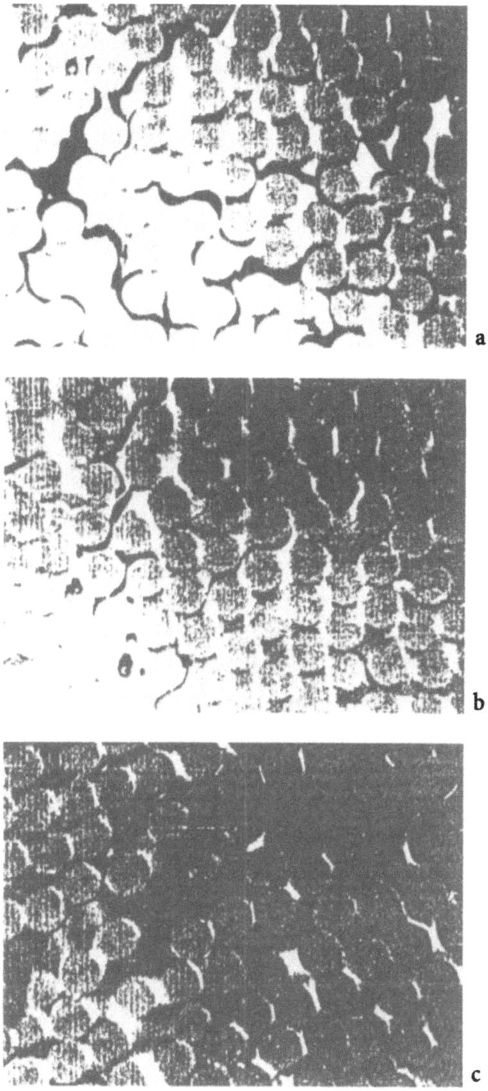

Fig. 4.21. Optical micrographs of the composites pyrolyzed to 600 °C. **a** composites made with no-surface treated unsized carbon fibers. **b** composites made with non surface treated sized carbon fibers. **c** composites made with surface treated unsized carbon fibers [56]

After initial carbonization, the pores and voids in carbon/carbon materials which may comprise up to 20%–30% of the volume are filled either by CVI or by liquid infiltration using thermosetting resins or pitches as impregnants. These impregnants have normally lower viscosities than those used in the first fabrication. Liquid impregnation is usually conducted under vacuum and pressure to achieve maximum pore filling. Impregnation is followed by carbonization at slow rates and sometimes by graphit-

Fig. 4.22. Carbonization shrinkage crack running perpendicular to the fibers

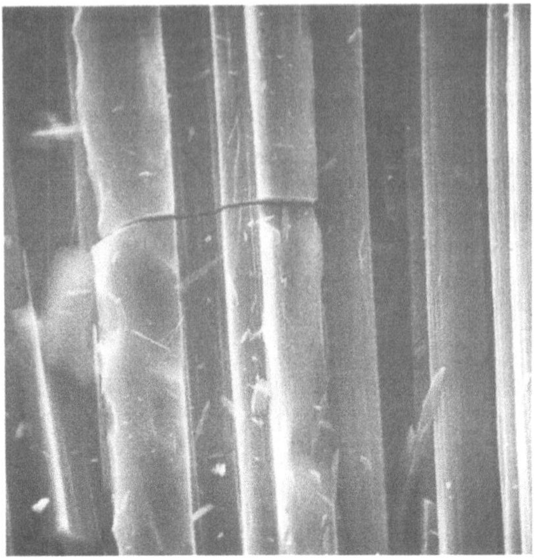

ization. Heating rates for all process cycles may vary depending on the size and shape of the composite being processed. To achieve high density, impregnation/carbonization cycles are repeated until the desired density or maximum density is achieved. The increase in density with densification cycles is dependent on the size and accessibility to infiltration of the pores and therefore varies with impregnant, with the type, dimensions and processing of the composites, and with the macrostructure of the carbonized skeleton which is controlled in addition to the above factors by the type and volume of fiber used. Carbon/carbon composites made with carbon fibers having surface functional groups or strong fiber–matrix bonding in the polymer stage exhibit a smaller increase in density on redensification than those made with non-surface-active carbon fibers (Fig. 4.23) [56, 58, 62]. In composites made with low surface energetic carbon fibers, i. e., with low or intermediate fiber/matrix bonding at the polymer stage, slit-like pores and an intermediate crack network is formed during carbonization, which can be filled during impregnation with thermosetting resin and coal tar pitches. Therefore they exhibit a higher densification efficiency (increase in density). In contrast, composites made with high surface energetic (strongly surface-treated) carbon fibers, although they also form cracks during carbonization, they are not interconnected and, hence, are not available to be filled in. Therefore, these composites exhibit very little increase in density during subsequent densification. Carbon/carbon composites made with

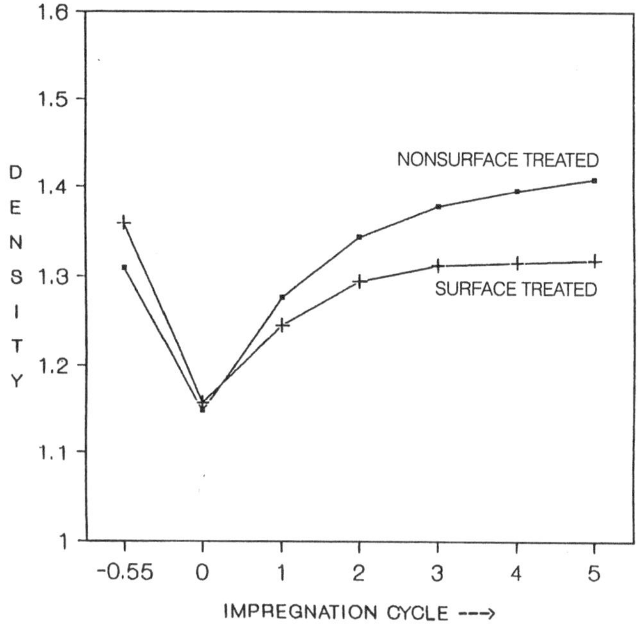

Fig. 4.23. Variation in density with impregnation cycles of the composites made with different types of carbon fibers •–•– Nonsurface treated, –+– surface treated

post-heat-treated carbon fibers are observed to yield very high density carbon/carbon composites after redensification cycles [3].

Fabrication of carbon/carbon composites through the thermosetting resin pyrolysis route may be preferred for general purpose carbon/carbon composites if the excessive shrinkage during carbonization and fiber/matrix bonding in carbonized composites can be controlled. There are some precursor materials available based on polyimides which exhibit preferred anisotropic shrinkage and result in excellent fiber strength yield after the first carbonization. However, the polyimide conversion to carbon results in the formation of closed pore systems which prevent redensification. Another means of controlling carbonization shrinkage is through the addition of fine graphite powder, coke, or carbon black (1–2 µm) to thermosetting resins so that these enter into the interfilament regions and hinder carbonization shrinkage [57, 62–65]. Figure 4.24 shows the effect of addition of graphite powder to phenolic resin on linear shrinkage during carbonization. As can be seen from this figure ultimate shrinkage at 1000 °C is proportional to the amount of graphite powder added. The same is found with composites containing resin plus coke filler as matrix precursor. Graphite

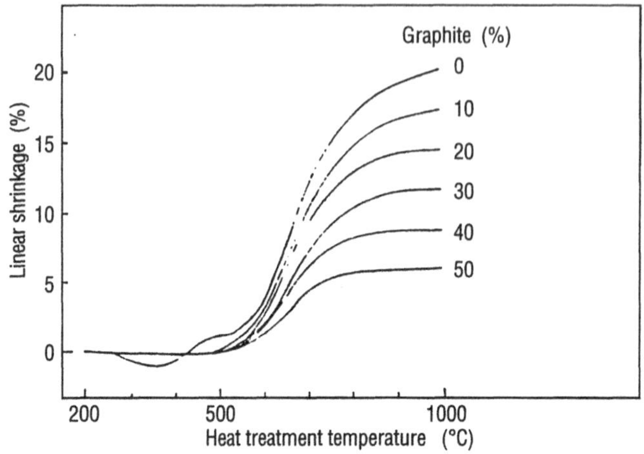

Fig. 4.24. Decrease in carbonization shrinkage of cured phenolic on addition of ultra-fine graphite powder [63]

powder addition to phenolic resin is found to decrease the net carbonization shrinkage of the composite by about 50 %. The decreased shrinkage in filler modified resin based composites is attributed to a decrease in the overall content of resin and a preferential adhesion of the resin to the fillers as compared with the fibers. Under these circumstances, the resin matrix still shrinks during carbonization but it does so on the graphite powder, reducing fiber–matrix interactions. These fine powder fillers have a higher surface area than the reinforcing fibers. Consequently, filler-modified composites made with even surface-treated carbon fibers after carbonization, posses accessible pores and cracks which can be filled in during the densification cycle and exhibit an increase in density on subsequent densification cycles.

In the case of 2D composites made with fabric, the space accessible to liquid matrix infiltration additionally depends on the type of weave pattern and size of the tow. Accordingly, the density of the composites after the first carbonization and on subsequent infiltration cycles will be governed by the weave pattern. Carbon/carbon composites made with 8H satin cloth are reported to posses higher ultimate density than those made with plain carbon cloth [61].

4.6 Composites with Pitch as Matrix Precursor

In contrast to thermosetting resins, composites made with pitches have to be carbonized in a very careful manner to prevent liquid pitch exudation during carbonization. The ultimate density of the carbon/carbon composites processed by pitch impregnation routes is controlled by (i) initial porosity of the preform, (ii) physical changes taking place in the fibers and the preform such as expansion etc. due to high temperatures, and (iii) densification efficiency of the pitch, which, in turn is a function of the initial liquid impregnation efficiency, pitch exudation, and carbon yield. Since pitches undergo liquid-phase pyrolysis, vapor percolation through the hot liquid matrix must also be curtailed. This is done through pressure carbonization. Pressure carbonization of pitches not only prevents percolation of gases through the matrix but also helps the polymerization of low molecular weight hydrocarbons, resulting in a high coke yield and, hence, high densification efficiency [2, 51–65]. Therefore, carbon/carbon composite processing with pitch as matrix precursor can be classified into two broad categories:

(a) standard process (low pressure carbonization under pressure up to 100 bars)
(b) high-pressure process (pressure impregnation carbonization (PIC) under pressure up to 1000 bars).

Both in normal pressure as well as high pressure processing routes, the pressure must be applied in particular temperature ranges in order to have its maximum beneficial effects on the pyrolysis of pitches. Figure 4.25 gives a schematic overview of the temperature/pressure/time conditions employed during fabrication of carbon/carbon composites using different processes. To avoid an exudation of liquid pitch during carbonization, composites are preferably carbonized in sealed metallic vessels of the same geometry as that of the composites [2].

4.6.1 Standard Process

The standard process is the most widely used process for making unidirectional, bidirectional, and multidirectional fiber-reinforced medium density composites. For UD and 2D composites, prepreg technology with a high softening point (120°–150 °C) pitch, called preforming pitch, can be

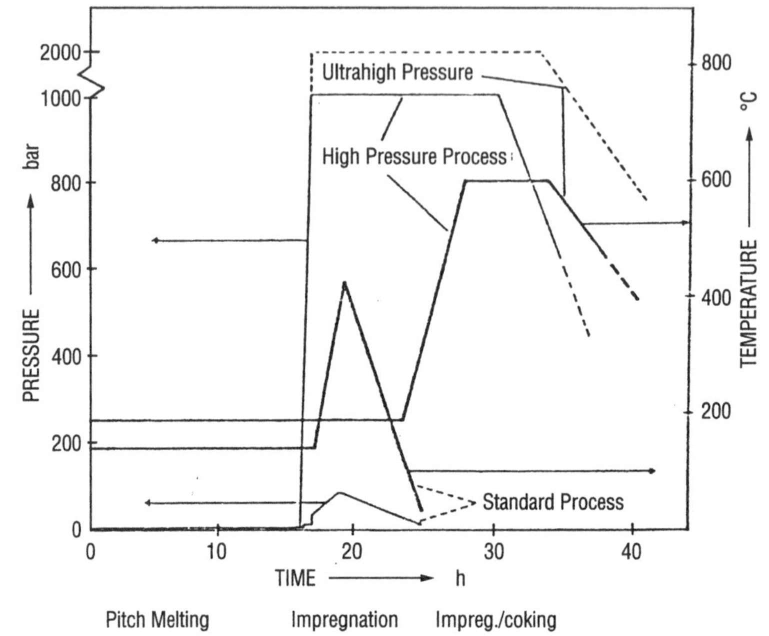

Fig. 4.25. Schematic representation of temperature/pressure relation employed during the processing of pitch-based carbon/carbon composites

used similar to that with thermoset resins [66–70]. Composite molding techniques as used with thermoplastics, are employed for making pitch polymer composites from the prepregs. Multidirectionally woven preforms are impregnated with preforming pitches using a vacuum/pressure impregnation technique. Polymer composites so fabricated are carbonized in a controlled way either at normal pressure or up to a pressure of 60–100 bars in nitrogen. These carbonized composites, so called carbon/carbon skeletons initially have extremely low density (1–1.2 g/cc) and well connected pores and voids with a wide pore size distribution. The carbon/carbon skeletons after evaluation are impregnated with medium softening point pitch (SP = 70°–80°) and pressured to 10–30 bar for 5–10 h. Pitch-impregnated composites are then carbonized to 1000°C under inert atmosphere. The impregnation cycles are repeated normally 4 to 16 times depending on the end density requirements. The impregnation cycles can be cut short by different ways. One way is to carbonize the impregnated composite at 550 °C under inert gas adiabatic pressure up to 60–100 bar followed by carbonization to 1100 °C under normal pressure [71]. Figure 4.26

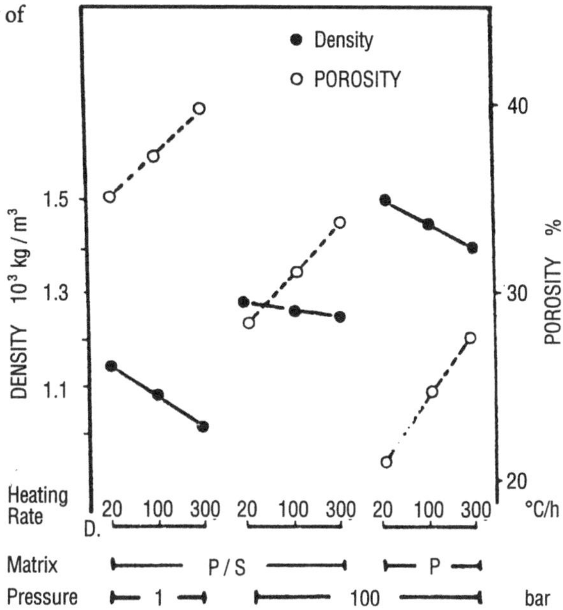

Fig. 4.26. Change in density of carbon/carbon composites with reimpregnation cycles using medium pressure carbonization technique [71]

shows an example of increasing density with impregnation cycles at different carbonization pressures. Composites may be graphitized to 2600° – 2800 °C at intermediate levels in order to densify the matrix, thereby opening up more and more pores for subsequent impregnations. The advantage of intermediate graphitization in density pickup is evident from Fig. 4.27 [72] . Here, each time the pitch impregnation is done at 25 atm and carbonization is carried out under normal atmospheric pressure. The graphitization is carried out at intermediate stages after three to five impregnation cycles. Using this technique UD and 2D composites have been processed with a density of 1.7 – 1.85 g/cc. In multidirectional composites however, and especially in big size composites, achievable densities are limited to 1.6 – 1.8 g/cc .

It is to be expected that pore filling requires multiple reimpregnation and recarbonization steps. During the impregnation process, generally good wetting is considered essential for filling the fine pores. In carbon/carbon composites, adhesion of the impregnant, however, should be minimized to avoid pore blocking as shown in Fig. 4.4. The residual impregnating liquid should shrink away from the pore surfaces in order to open new pore entrances. More important is the surface activity of the filler, in this case the fibers. Therefore, adhesion becomes a relative term depending on the type of filler, the fibers, and the impregnants, i. e., thermosetting resin

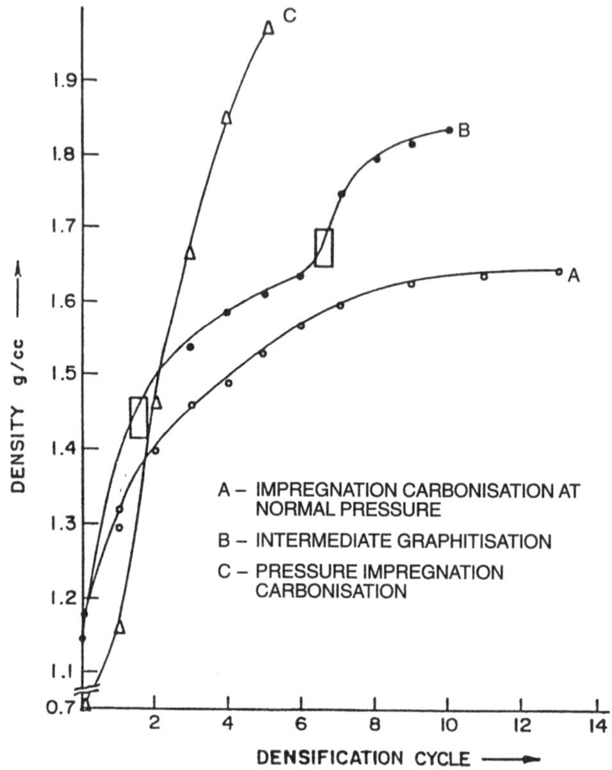

Fig. 4.27. Change in density of carbon/carbon composites with reimpregnation cycle, processed under different pressures [72]

or thermoplastic pitches. Accordingly, good adhesion on pore walls for pitches and poor adhesion for resins is required.

4.6.2 High Pressure Processing

The main aim of an efficient densification process, especially for multidirectional carbon/carbon structures, is to achieve full impregnation with pitches and to derive maximum coke yield from impregnated pitches. As discussed above, the coke yield of pitches can be increased from 50% at ambient pressure to over 85% at 700 bars. This forms the basis of high pressure processing. In this process the carbon/carbon materials are impregnated with pitch under pressure as in the standard process, and carbonized under high pressure. Therefore this process is also called pressure

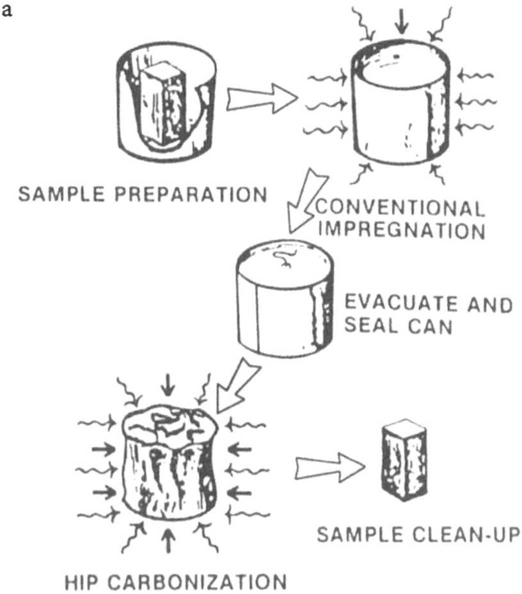

Fig. 4.28. Sequence of events in a PIC cycle [2]

impregnation carbonization (PIC). This process, which is shown in Fig. 4.28 [2, 73], is most suitable for densification of multidirectional woven preforms which are contained in metal cans and impregnated with pitches. These cans containing impregnated performs and some extra pitch, sealed under vacuum, are placed in a pressure vessel and heated to a temperature of 180°-200°C. The pressure is increased to 1000 bars. After about 8-10 h at 200°C, the temperature is increased to 600°-800°C at a programmed rate. The pressure is transmitted to the pitch through the very thin walls of the can behaving like a flexible bag. After the process is complete the can is opened, extra pitch coke is removed, and the composite further heat treated to a temperature of 1500°-2500°C. The complete process cycle is repeated until the required density is achieved. In the PIC process, pressure not only increases coke yield but also helps the impregnation of pitch into small pores with tortuous paths and prevents exudation of this impregnated pitch from the pores during carbonization. Figure 4.29 [2] shows the increasing density of composites with impregnation cycles using the PIC process. For composites requiring multidirectional fiber reinforcements and high carbon matrix packing, the high pressure impregnation carbonization route is preferred. However, this process is quite costly due to autoclaving etc. Therefore, extensive research programs have been undertaken to evolve different ways to build up dense, high carbon containing composites.

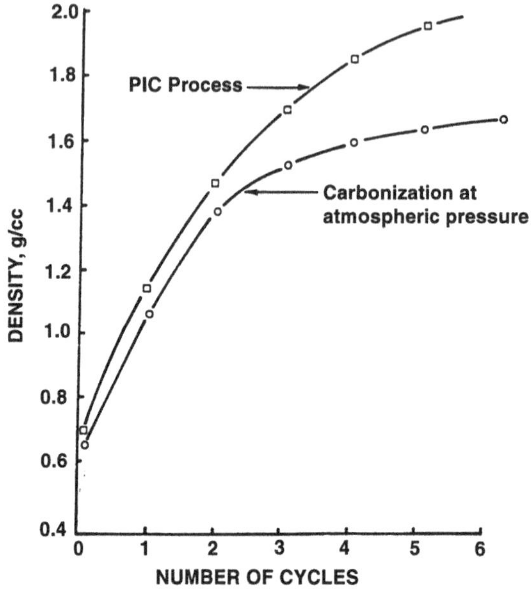

Fig. 4.29. Densification of fine weave multidirectional carbon/carbon composites

Over the last two decades, extensive research has also been carried out on the development of mesophase pitches, and these pitches are now commercially available. These pitches when used as a binder with carbon fibers may result in high-density carbon/carbon composites which is supported by the fact that over past few years numerous reports have appeared on processing of carbon/carbon composites with mesophase pitches [2, 51, 68, 69, 74–77]. In addition to the development of synthetic and high-char-yielding mesophase pitches, concerted efforts are being made to develop improved, less expensive techniques to process pitch-based composites at relatively lower pressures. Bruckmann [75] has evolved a method for fabricating all carbon composites with a mesogenic pitch as carbon precursor which involves heat treatment of the green body to form mesophase from the binder followed by hot pressing to 800 °C. Even after one densification cycle, the composites are found to possess a porosity of less than 10% which can easily be filled up with a relatively low number of subsequent impregnations. In this way, UD carbon/carbon composites have been produced with bulk densities up to 1.78 g/cc and mechanical properties which could otherwise only be attained after numerous impregnation steps. Considering the progress which has been made in developing appropriate mesophase pitches and their commercial availability, this process may cut down the processing time and cost. Rhee et al. [78] have developed UD composites with mesogenic pitch as matrix precursor and mesophase pitch fibers as reinforcement using the prepreging hot molding process shown in Fig. 4.30.

Composites so prepared exhibit low volume shrinkage (~3%–4%), high skeleton density, and well-connected open bulk porosity which can be filled in by the classical impregnation procedure. Mochida et al. [79] and Hajk and Glogar [80] have developed new techniques for the room temperature molding of a blend of carbon fibers and mesophase pitch followed by carbonization without and with (30 bar) pressure. Bulk density of carbon/carbon composite is found to increase with mesophase content. As expected, pressure carbonization was found to be effective in increasing the bulk density of carbon/carbon composites.

White and Shaeffer [81] have suggested bloating of mesophase pitch during carbonization to be the main cause of the low density of pitch-derived carbon/carbon composites. Therefore, in order to have dense carbon/carbon composites with controlled desired matrix microstructure, they emphasized the need to stabilize the matrix in place after the mesophase transformation but prior to carbonization. An oxidation process, similar to that used in the stabilization of mesophase pitch fibers, is used to render the mesophase matrix in the composite infusible and to allow carbonization without disturbing the layered microstructure [77, 82]. Carbon-fiber-reinforced pitch matrix composites are oxidized in oxygen for periods of 50 to 100 h at a temperature of around 220 °C followed by carbonization at

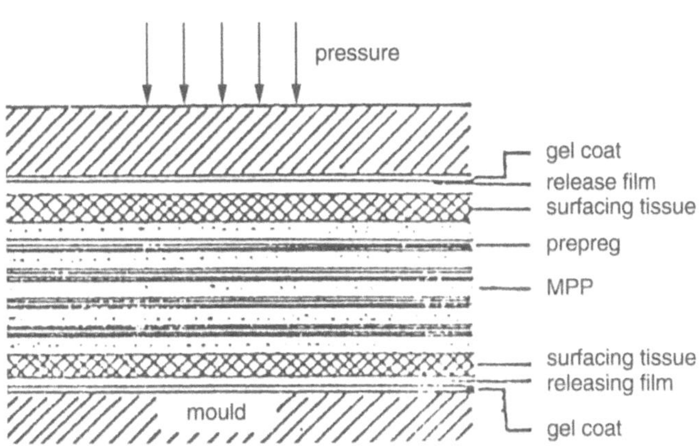

Fig. 4.30. Schematic representation of laminated construction for hot press moulding and carbonization technique using mesogenic pitch

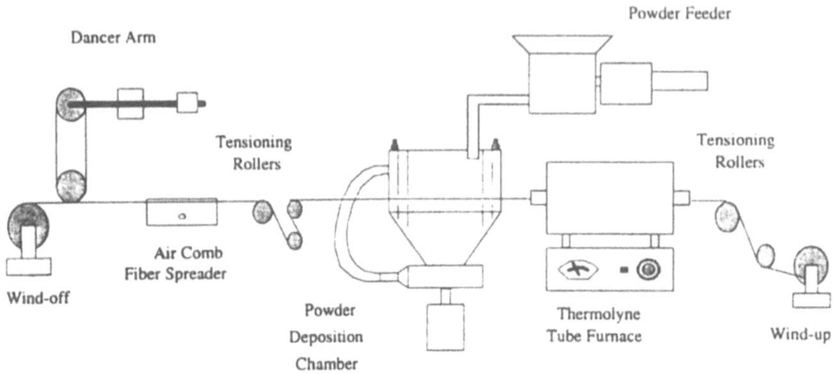

Fig. 4.31. Powder coating process [83]

1100 °C in the presence of an inert gas. The oxidation stabilization technique not only eliminates the need for high-pressure autoclaves in composite processing with mesophase pitch as matrix precursor but also has potential for the use of a wider variety of matrix microstructure ranging from glassy to fully graphitizable mesophase microstructure. Edie et al. [83] have developed a continuous powder coating process to produce flexible, preimpregnated pitch matrix tow prepreg to fabricate UD carbon/carbon composites. This process, which is based on resin prepreging technique, is quick and economical. Figure 4.31 shows an overall scheme of the powder coating process. This essentially comprises tow spreading, powder deposition, and powder fusion. Figure 4.32 shows a pressurized air comb setup for tow spreading and Fig. 4.33 shows the fluidized bed chamber for powder coating. The powder-coated tows are heated for powder fusion resulting in pitch-coated flexible tow prepregs. These prepregs are then used for making UD composites by the compression molding technique. This process can be extended for making multidirectionally reinforced composites.

Another new process developed for manufacturing carbon/carbon composites is the "powder sintering method" [84]. In this process, a mixture of fine pulverized coke and a binder material, used as matrix precursor, is deposited on the carbon fibers or cloth followed by hot pressing and carbonization. Chang and Okura [85] used bulk mesophase pitch as binder material for coke and carbon fiber impregnated with resin as reinforcement. Optimization of the mesophase binder content and hot press pressure and temperature is needed to get maximum desired properties. Sakagami et al. [84] have also used a binder-less technique. Fine self sinterable carbon powder is electrodeposited on the carbon fiber or cloth and hot pressed for

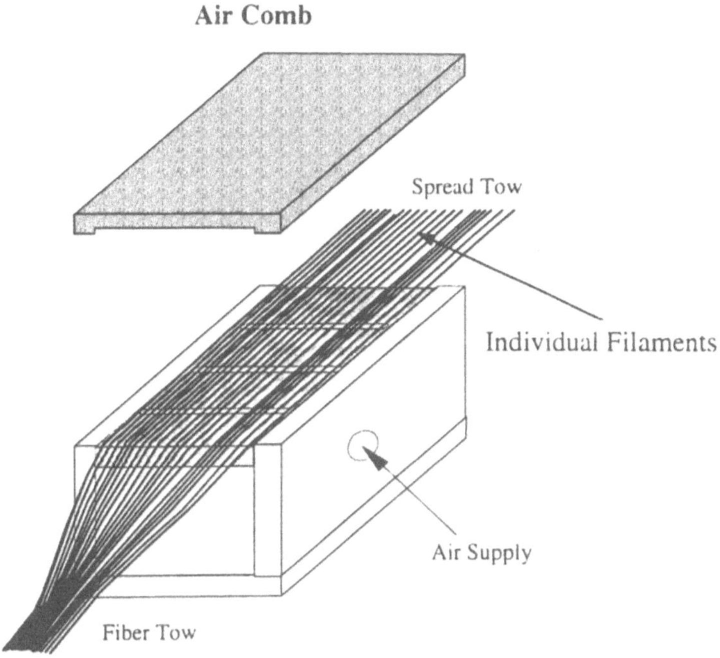

Fig. 4.32. Air-comb fiber spreader [83]

sintering and compacting. Using this technique, they claim to have achieved a density of 1.75 g/cc by processing once at 2000 °C. Yasuda and Tanabe [86] propose treatment of pitch with iodine to increase the coke yield and improve processability of pitch as matrix precursor for carbon/carbon composites. They have found the carbon yield to increase from 56 % to 76 % as a result of iodine treatment for 20 h. The optical texture of the coke is also found to change from a flow type to a mosaic type.

Rand [51] has compared the merits and demerits of the three conventional processes used for manufacturing carbon/carbon composites. These are compiled in Table 4.2. Based on the analytical relationship and experimental data available in the literature, Rellick [87] has calculated the densification efficiency for different carbon/carbon systems. According to him, the main factor for multidirectional (MD) billet densification is the efficiency of the pitch to transform to carbon matrix which inturn depends on various reinforcements and precursor parameters discussed in the beginning of this section. The theoretical limit of pitch densification efficiency is estimated to be about 55 %. However, under normal pressure

Table 4.2. Comparison of characteristic features and properties of carbon matrices from vapor phase, pitch, and resin precursors [51]

Characteristic feature or property	Type of carbon matrix		
	CVD	Pitch	Resin
Density	High, approx. 2000 kg m^{-3},	High, increases with Carbonization pressure	Low 1300 – 1600 kg m^{-5}
Carbon yield	C directly deposited – no further thermal degradation	Varies according to pitch composition 50 % – 80 % w/w	About 50 % for phenolics, increasing to 85 % for polyphenylenes.
Porosity	Low, except for isotropic form; laminar fissures	Macro-sized gas entrapment pores plus shrinkage and thermal stress fissures	High microporosity (pore diameter < 1.0 nm); becomes closed above 1000 °C. Macrovoids may be evident due to vapor evolution during the curing of resin
Microstructure (in the bulk state)	Varies from isotropic to highly orientated laminar forms	Macro-domains (1 – 100 µm) showing preferred orientation developed from the mesophase state	Isotropic-except on the nanoscale, i. e., BSUs are randomly orientated
Orientational effects within fiber preforms	Strong orientation of laminar matrix with fiber surfaces; orientation also on crack surfaces within other matrix types if subsequently subjected to CVI	Preferred orientation of lamellae with fibre surfaces increases as HTT develops the graphitic structure; modified by pressure pyrolysis	Preferred orientation at fiber surface – but to a much lower extent than other precursors; increases with HTT
Graphitizability/ crystallinity	Laminar forms highly graphitizable, can have L_c and L_a > 200 nm after 3000 °C HTT	Highly graphitizable L_a, and L_c usually less than 100 nm	Normally nongraphitizing

Table 4.2 (continued)

Characteristic feature or property	Type of carbon matrix		
	CVD	Pitch	Resin
Purity/composition control	Controlled by gas phase composition – enables other elements to be incorporated into the deposit (e.g., Si, B)	Controlled by source – can be of high purity not easy to incorporate other elements except as powders	As for pitch
Reactivity to oxidizing gases	Very low reactivity for highly oriented pyrographite	Low reactivity decreases with HTT	Usually high reactivity due to micropore network – decreases with HTT but still relatively high
Thermal expansion	Depends on preferred orientation; can be highly anisoptropic approaching values for the crystal in the two major directions	Depends on domain orientation; expansion partly accomodated by lamellar cracks; $1-5 \times 10^{-6}\,K^{-1}$	Isotropic in bulk; $\sim 3 \times 10^{-6}\,K^{-1}$
Thermal and electrical conductivity	Determined by preferred orientation – approaching single crystal graphite values	Depends on domain orientation, HTT and internal porosity, increasing with HTT	Isotropic in bulk
Young's modulus	7–40 GPa depending on the structure	5–10 GPa depending on grain size, porosity, and degree of graphitization but up to 14 GPa for very fine (1 μm) grain size materials	10–30 GPa

Table 4.2 (continued)

Characteristic feature or property	Type of carbon matrix		
	CVD	Pitch	Resin
Strength	Depends on microstructure and degree of graphitization; 10 – 500 mPa	Depends on porosity and pore geometry; 10 – 50 mPa for most polycrystalline carbon/graphites rising to 120 mPa for very fine grain (1 µm) size materials	Approximately 8 – 150 mPa for glassy carbon.; Lower for resin carbons, depending on porosity
Failure strain	0.3 % – 20 % depending on structure; Higher values at highest deposition/HTT	Up to about 0.3 % depending on grain size and degree of graphitization	Up to about 0.4 % (largest values for glassy carbons)

L_c, stack height; L_a, stack width; HTT, heat treatment temperature; BSUs, basic structural units.

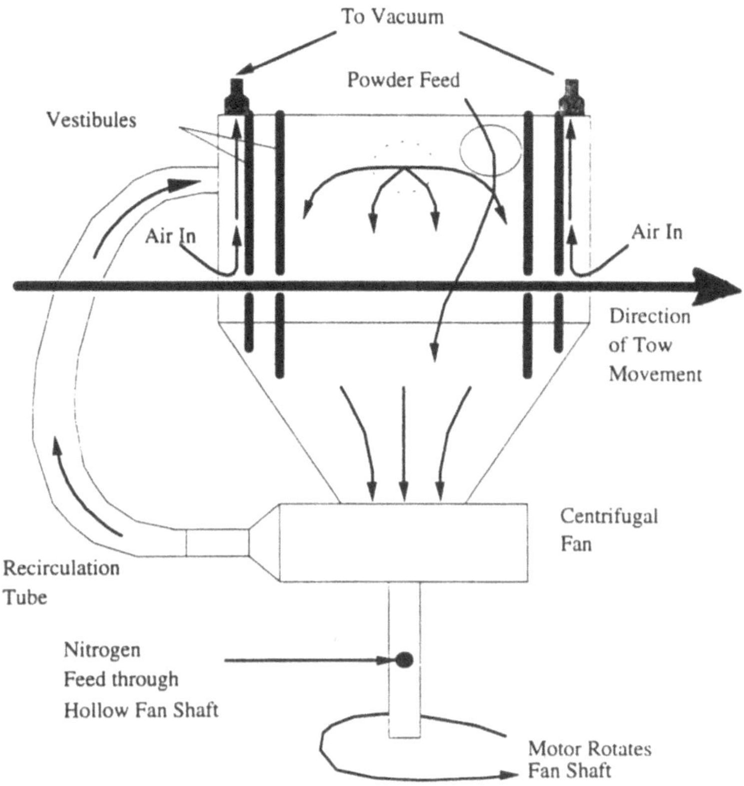

Fig. 4.33. Detail of powder deposition chamber [83]

densification conditions, it is found to be only 15%–20% and could be increased to about 35% in the first cycle using HIPIC (High Pressure Impregnation Carbonization) processing at 1000 atm and to about 50% in later cycles. The same order of densification efficiency can also be achieved using heat-treated or mesophase pitches with the oxidation stabilization technique.

4.7 References

1. Bundy FP et al. (1996) Carbon vol 34 p. 141
2. McAllister LE, Lachman WL (1983) Multidirectionally Carbon-Carbon Composites In: Kelly A, Milieko STM (eds) Handbook of Composites Vol 4 p 109, Elsevier
3. Fitzer E (1987) Carbon, 25 p 163

4. Savage G (1992) Carbon-carbon composites. Chapman & Hall Pub
5. Pierson HO, Northrop DA (1975) J Comp Materials, vol 9 p 118
6. Stoller HM, Frye ER (1972) SAMPE Quarterly Vol 3 p 10
7. Kotlensky WV (1973) In: Walker PL jr (ed) Chemistry & Physics of Carbon. Vol 9 p 151
8. Schmidt DL (1972) SAMPE Journal vol 8 p 9
9. Gebhardt JJ, Stover ER, Mueller W (1976) In: Petroleum derived carbon, ACS Symp series 21, p 212
10. Granofl B, Pierson HO, Schuster DM (1973) Carbon 11 p 177
12. Bauer DW, Kotlensky WV (1972) SAMPE Quarterly Vol 4 p 24
13. Leeds DH et al. (1971) Carbon vol 9 p 237
14. Kimura S, Yasuda E, Takase N, Kayusa S (1980) High Temp High Pressure 13:193
15. Oh SM, Lee JY (1988) Carbon 26 (6) p 763
16. Balik K, Glogar P, Tomanova A (1990) Fachbricate der DKG 9 p 200
17. Vaidyaraman S, Lackey WJ, Agarwal PK, Freeman GB (1996) Carbon vol 33 p 1211
18. Vaidyaraman S, Lackey WJ, Agarwal PK, Miller MA (1996) Carbon vol 34 p 347
19. Vaidyaraman S, Lackey WJ, Agarwal PK, Starr TL (1996) Carbon vol 34 p 1123
20. Dupel P, Bourrat X, Pailler R, Naslain R (1995) Carbon vol 33 p 1193
21. Jeong HJ, Park HD, Lee JD, Park JO (1996) Carbon vol 34 p 417
22. Levesque O (1989) Ph D Thesis, University Bordeaux
23. Golecki I, Morris RC, Narasimhan D, Clements N (1995) 23 Bien Carbon Conf, p 90
24. Fitzer E et al. (1970) In: Walker PL jr (ed) Chemistry & Physics of Carbon Vol 7 p 368
25. Madorsky SL (1969) Thermal Degradation of Polymers. Interscience NY
26. Billimeyer FW jr (1970) Text book of polymer science. Wiley Interscience NY
27. Jenkins GM, Kawamura K (1976) Polymer Carbon, Carbon fibers, Glass & Char, Cambridge University Press
28. Fitzer E, Schaffer W, Yamada S (1969) Carbon 7 p 643
29. Fitzer E, Schaffer W (1970) Carbon 8 p 597
30. Zaldivar RJ, Rellick GS, Yang JM (1991) SAMPE J vol 27 p 29
31. Huettener W (1980) Ph D Thesis, University of Karlsruhe
32. Shimizu K, Murdie N, Hinckley C, Wright M (1991) Proc Seventh Annual Conf Materials. Technology Center SIU p 149
33. Burger A, Fitzer E, Heym M, Terwiesch B (1975) Carbon 13 p 149
34. Mantel CL (1967) Handbook of Carbon and Graphite Interscience Pub
35. Newman JW (1986) Petroleum Pitches – New routes to advanced solid materials, Proc Third Annual Conf, Materials Technology Center SIU p 118
36. Marsh H (ed) (1990) Science of Carbon Butterworth (Pub)
37. White JL (1986) The carbonaceous mesophase and its role in controlling the structure of carbon products. Proc Third Annaul Conf, MTC, SIU p 182
38. Mochida I (1991) Mesophase, carbon fiber and composites. In: Seventh Annual Conf Materials Tech Center, SIU Carbondale p 3
39. Rand B (1987) Fuel p 1491
40. Mochida I, Shimidzu K, Korai Y (1990) Carbon vol 28 p 311
41. Kimura S, Yasuda K, Yasuda E, Tanabe Y (1987) Tanso 128 p 30
42. Riggs DM, Diefendorf RJ (1983) 16th Bien Carbon Conf p 24
43. Esumi K et al. (1993) Carbon vol 31 p 1358
44. Azami K et al. (1990) Int Conf Carbon Tsukuba, Japan, p 218
45. Yokono T, Oka N, Sanada Y (1984) Carbon vol 22 p 614
46. Mochida I (1988) Carbon vol 26 p 843

47. Mochida I (1994) Recent progress of mesophase pitch. Tanso 163 p 150
48. Zimmer JE, White JL (1982) Adv in Liq Crystal 5:157
49. Bhatia G, Agarwal RK, Bahl OP (1987) J Mat Sc 22 p 3847
50. Bhatia G, Fitzer E, Compalik D (1983) 16th Bien Carbon Conf
51. Rand B (1994) Matrix precurssors for carbon-carbon composites. In: Thomas CR (ed) Essentials of carbon-carbon composites. Royal Soc chem p 67
52. Rhee B, Fitzer E, Heym M (1976) High Temperature, High pressure 8 p 307
53. Yasuda E, Kajiura H, Tanabe Y (1995) Tanso 170 p 286
54. Tillmans H (1975) Ph D Thesis, University of Karlsruhe
55. Jain RK, Manocha LM, Bahl OP (1991) Indian Journal Of Technology, vol 29 p 163
56. Manocha LM (1988) Composites Vol 19 p 311
57. Fitzer E, Huttener W, Geigl KH (1980) Carbon 18:265
58. Jain PK, Bahl OP, Manocha LM (1992) SAMPE (Q) p 43
59. Manocha LM, Yasuda E, Tanabe Y, Kimura S (1988) Carbon 26 p 333
60. Manocha LM, Bahl OP, Singh YK (1989) Carbon vol 27 p 381
61. Manocha LM, Bahl OP (1988) Carbon vol 26 p 13
62. Fitzer E, Geigl KH, Manocha LM (1978) Proc 5th London Int carbon conf p 405
63. Yasuda E, Tanabe Y, Manocha LM, Kimura S (1988) Carbon 26 p 225
64. Menendoz R et al. (1995) 22nd Bien Carbon Conf pp 78
65. Inagaki M, Washiyama M, Goto K (1987) J Mat Sc 22 p 3315
66. Meyer RA (1982) Int conf Carbon, Toyohashi, Japan
67. Diefendorf RJ (1970) 10th Annual Symposium of ASME, New Mexico p 125
68. Manocha LM, Bahl OP (1989) Tanso vol 140 p 255
69. Matzinos PD, Patrick JW, Walker A (1996) Carbon vol 34 p 639
70. Manocha LM, Bhatia G, Bahl OP (1982) Proc Ist Indian Carbon Conf, Delhi p 325
71. Fitzer E, Huettener W, Manocha LM (1990) Carbon vol 18 p 291
72. Bahl OP, Manocha LM et al. (1991) J Sc Ind Res vol 50 p 533
73. Chard W, Conaway M, Nesz D (1974) Petrol Der Carb ACS Symp Series Vol 21 p 155
74. Christ K, Huttinger KJ (1993) Carbon vol 31 p 731
75. Bruckmann H (1979) Dr Ing Thesis University of Karlsruhe
76. Diefendorf RJ, Tokarsky FM (1971) AFML-TR-133 Part I & II
77. Dillon K, Thomas KM, Marsh H (1993) Carbon 31 p 1337
78. Rhee BS et al. (1987) High Temp -High Pressure vol 19 p 677
79. Mochida I, Fujiura R, Korai Y (1992) Tanso, No 155 p 398
80. Hajk M, Glogar P (1995) Proc Int Symp Brittle Matrix Composites, Warshaw p 689
81. White JL, Shaeffer PM (1989) Carbon vol 27 p 697
82. Liedtke V, Huttinger KJ (1996) Carbon, vol 34 p 1057
83. Klett JW, Edie DD (1995) Carbon vol 33 p 1485
84. Sakagani S, Iwata K, Kawase M, Wakamatsu S (1989) C/C Composites by Carbon Powder Sintering method. Ist Japan SAMPE Symp Chiba Japan
85. Chang T, Okura A (1986) Fabrication Methods for C/C Composites & Their properties. Japan US CCMIII Tokyo p 425
86. Yasuda E et al. (1996) Application of Iodine treatment on fabrication of pitch based carbon/carbon composites in "Applications of Carbon and Carbonaceous Composite Materials" Palmer KR, Marx DT, Weight MA (Eds.) World Scientific (Pusb.)
87. Rellick GS (1990) Carbon vol 28 p 589

5 Microstructure of Carbon/Carbon Composites

The use of carbon/carbon composites for thermostructural applications requires a basic knowledge of the structural characteristics of the composites that control their engineering properties. These include macroporosity, microstructure of the matrix including interbundle and interfilament matrix, microcracking and interfaces at different levels, fiber orientations, demaged fibers etc. By understanding details of the composites' structure and mechanism of microstructural development, one can control the matrix microstructure and improve its mechanical properties, especially the fiber strength utilization. Microstructural features of carbon/carbon composites may be classified according to various scales of magnification. At the largest useful scale, the millimeter scale, microstructure relates to composite architecture. It comprises the distribution of fibers and matrix, especially interbundle matrix, voids, and pores. At the microscale, it reveals features such as interfilament matrix and at the nanoscale it relates to fiber/matrix interface. Figure 5.1 shows a broad classification of some of the characteristics. For 2D and MD carbon/carbon composites, Jortener [1] classified the former as minimechanical features and the latter as micromechanical features. The origin and characteristics of these depend on (i) the reinforcement configuration, (ii) type of carbon fiber used, (iii) matrix precursor, (iv) fabrication routes, and (v) processing conditions. The macrostructure (constituting bulk characteristics such as porosity, matrix distribution) and microstructure characteristics (microcracks, matrix microstructure at different positions, interfaces, preferred orientations of the graphitic matrix etc.) of the composites may vary significantly depending on the reinforcement configuration, from model unidirectional composites to cloth-reinforced or multidirectional carbon/carbon products due to the presence of fiber crossover regions, waviness of the fibers, stacking sequence etc. Though the exact microstructure of carbon/carbon composites will depend on various factors stated above, some typical features observed in model composites are discussed below. Some of these may be typical of carbon matrixes derived from CVD, thermosetting resins,

Nature of cracking	Scale	Location	Effect on Properties
Micromechanical cracking	• Fiber diameter scale	• Cracking in Matrix • Fiber-matrix interphase	• Load transfer among fibers • Transverse properties of fiber bundles
Minimechanical cracking	• Cloth layer thickness scale	• Fiber bundle • Interface between fiber bundle • Interyarn pocket of unreinforced matrix • Interface between fiber bundle and a matrix pocket • Matrix-matrix interphase	• Porosity • Load transfer among fiber bundles and cloth laminates • Major influence on mechanical properties

Fig. 5.1. Macro- and microstructural features of carbon/carbon composites

or thermoplastic pitches and some may be common in the composites made by these routes.

5.1 Microstructure of Carbon Matrix Produced by Chemical Vapor Deposition

Carbon can assume a variety of microstructures, depending on condensation, stacking, and orientation of hexagonal rings. It is obviously expected to be more variety arising in complex processes such as chemical vapor deposition (CVD) where there are a number of additional parameters in addition to those stated above which control the deposition of carbon within the fibrous structure and the microstructure of the deposited carbons. These parameters include:

1. cracking temperatures
2. total pressure in the crackers as well as partial pressure of cracking gas
3. C/H ratio of cracking gas
4. flow rate
5. fiber type, arrangement, and volume percentage in the preforms.

Table 5.1. Crystallite dimensions of carbon matrix in as-deposited and heat-treated carbon/carbon composites made by CVD routes

Material	Condition	Density (g/cc)	d_{002} (nm)	L_c (nm)
RL-PAN	As-deposited	1.83	0.344–0.345	8.2–9.2
	Heat-treated	1.86	0.337	31–46
SL-PAN	As-deposited	1.70	0.346–0.347	5.8–6.3
	Heat-treated	1.58	0.337–0.342	9.1–15.2
ISO-PAN	As-deposited	1.47	0.344–0.345	5.4–9.1
	Heat-treated	1.45	0.341–0.345	7.0–11

RL, rough laminar; SL, smooth laminar; ISO, isotropic; PAN, polyacrylonitrile; d_{002}, interlayer spacing; L_c, stack height.

The effects of deposition temperature on the crystalline dimensions of the carbon matrix in as-deposited and heat-treated carbon/carbon composites can be seen from Table 5.1.

Depending on the optical activity of carbon structures under polarized light and texture roughness, the microstructure of carbon deposited from a hydrocarbon gas is classified into three broad categories:

1. smooth laminar (SL)
2. rough laminar (RL)
3. isotropic (ISO)

Figure 5.2 [2] shows typical examples of these microstructures which always follow a specific order. For example, SL structure is always followed by RL with a distinct boundary and not by ISO structure. In contrast, the change-over from RL to ISO structure is always associated with diffused boundaries. These microstructures are sensitive to deposition parameters and can vary from ISO to extreme anisotropic. Figure 5.3 [3] shows the corre-

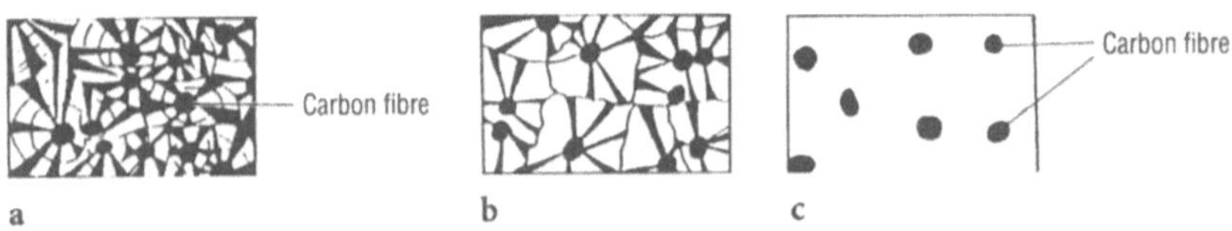

Fig. 5.2. a–c. CVD-derived carbon microstructures. **a** Rough laminar. **b** Smooth laminar. **c** Isotropic

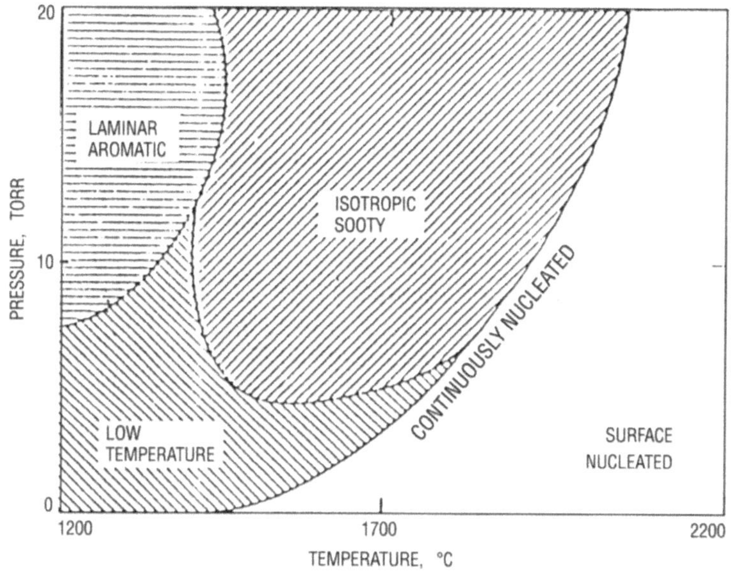

Fig. 5.3. Dependence of pyrocarbon microstructure on deposition temperature and gas pressure [3]

lation between the deposition parameter and the development of microstructure in a methane-derived carbon matrix.

RL deposits formed between 350 °C and 1400 °C are characterized by high optical activity and numerous irregular extinction crosses. SL deposits have a large growth cone microstructure and a lower degree of anisotropy than RL. These are formed at an intermediate temperature (1100°–1300 °C) and high pressure at a high surface/volume ratio and larger contact time. Isotropic deposits have a fine-grained structure with no preferred orientation or gross structural details. Isotropic depositions form at a low contact time and high temperature (~1500 °C). The anisotropic nature which is a consequence of graphite structure is further corroborated by crystallographic parameters of matrix structures. The densities of various carbon structures increase with increasing anisotropy ($d_{RL} > d_{SL} > d_{ISO}$), consistent with crystallographic parameters. Lieberman and Pierson [2] attributed different microstructure types to relative quantities of gaseous acetylenic and aromatic species, i.e., C_2H_2/C_6H_6 molar ratio in cracking gas typically as SL < 5; 5 < RL < 70; I > 70.

It has also been observed that SL and ISO microstructures form under a wide variety of conditions, while RL material apparently forms only under a narrow range of conditions. These deposited structures are free of any large cracks. Only a few circumferential cracks are found to occur within an individual laminar matrix sheath, concentric with the fiber. The bonding between the isotropic carbon matrix layer and the fiber seems to be quite good and continuous, independent of fiber type, i. e., PAN or pitch-based carbon fibers [4]. Upon heat treatment to 3000 °C, the SL matrix develops extensive microcracks due to anisotropic nature of the matrix. These cracks form circumferentially within a matrix sheath and at intersections between matrix sheaths. This also results in decreased density of SL structured matrix on heat treatment. However, due to local variation in temperature and the C_2H_2/C_6H_6 molar ratio caused by the temperature gradient across the thickness of the substrate and cracking of infiltrating gas, a mixture of microstructures are commonly observed [5].

In the case of a thermal gradient process, the temperature of the substrate felt is lower near the outer portion than near the inner one. The C_2H_2/C_6H_6 molar ratio will also be low near the outside surface. Accordingly, one may expect a structural transition from the inside to the outside of the felt to occur in the order of ISO $\rightarrow$ RL $\rightarrow$ SL. This has also been confirmed by Ehrburger et al. [6] through optical and TEM studies on CVI carbon/carbon composites. According to them, the pyrolytic carbon matrix around carbon fibers is composed of several sheaths of varying brightness with alternating dark and bright sheaths. Bright layers are aligned parallel to the fiber axis and surface. Their onion-like texture has irregular surface aspects and numerous fine extinction crosses and, hence, are analogue to RL texture. Dark sheaths correspond to ISO matrix.

In a typical carbon/carbon composite with a density of 1.85 g/cc, the RL component can be as high as 78 %. Upon heat treatment of such composites to a temperature of about 2600 °C, the texture coarsens and some optical anisotropy develops in ISO dark layers particularly near its inner surface. On heating further to 3000 °C the ISO layers become thin and the interface between the innermost dark and bright layers becomes diffuse. Cohesion between various layers is found to be quite strong and existence of micro-fissures have not been observed in such composites. However, in another case, in a composite with a density of 1.78 g/cc, all three types of texture, RL, SL, and ISO, have been observed [7]. Cracks are detected at the interface between the rough and smooth textures. Upon graphitization of these composites, the fissures become more pronounced. It has also been found that SL structure is more brittle than RL and ISO textures [8]. Electron microscopic studies made on these composites have revealed that for the same high heat treatment temperatures, the latter composites with a density of 1.78 g/cc having rough, smooth, and laminar textures exhibit better

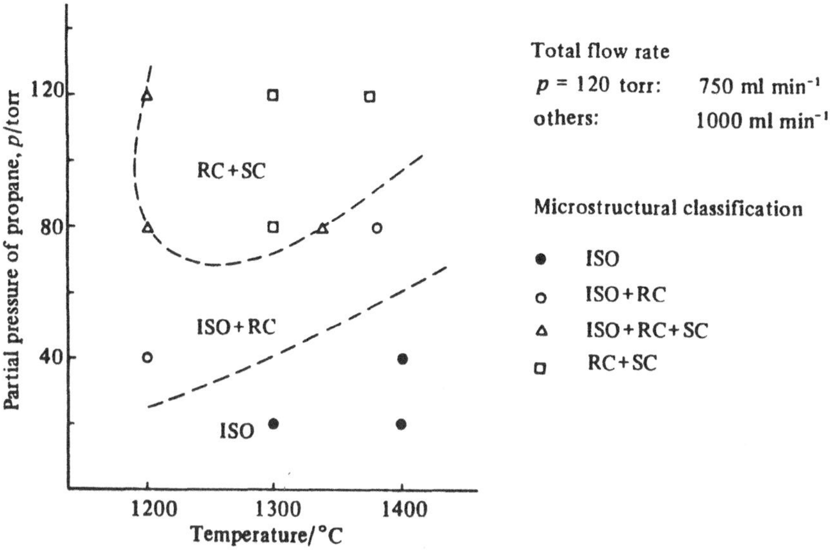

Fig. 5.4. Effect of temperature and partial pressure of propane on microstructure of deposited carbon. *RC* rough column, *SC* smooth column [11]

graphitization ($d_{002} = 0.339$ nm, $L_0 = 17$ nm) than the former composites with a density of 1.85 g/cc ($d_{002} = 0.341$ nm, $L_0 = 16$ nm) [8–10].

The type of hydrocarbon has also been observed to influence the texture of carbon in CVD carbon composites. Kimura et al. [11] have observed cone-like columnar structures in a propane-derived pressure-controlled carbon matrix, the dependence of which on processing parameters is shown in Fig. 5.4. At low partial pressures ($p < 20$ torr), the microstructure of matrix is always observed to be ISO over the temperature range of 1200°–1400 °C. At intermediate values of partial pressure and temperature, ISO and RC carbon are deposited together. Kimura et al. proposed the change of microstructure to be in the order ISO → RC → SC as the temperature is decreased and/or as the partial pressure is increased. This order agrees well with the order stated earlier. However, Oh and Lee [12] are reported to have observed four predominant microstructures in the carbon matrix derived from propane. These have been termed as columnar, SL, transition, and ISO structures. High optical activity and distinct growth cones are the characteristics of columnar structure while SL structure is characterized by high optical activity and distinct extinction crosses and ISO structures is characterized by the lowest optical activity. Transition structure is the one between ISO and SL. They did not observe RL struc-

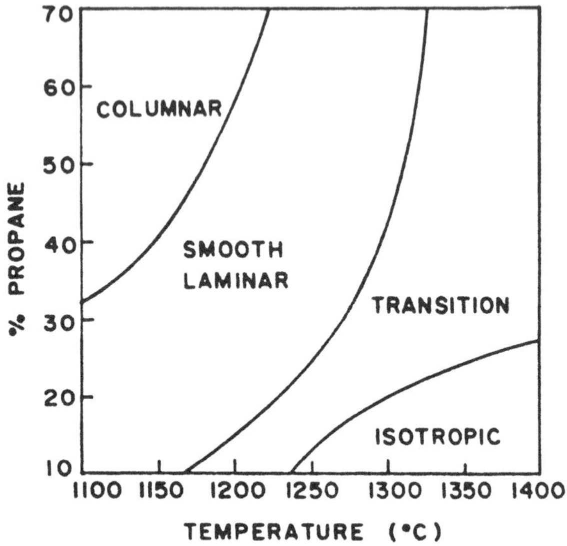

Fig. 5.5. Effect of deposition conditions on microstructure of pyrolytic carbon matrix deposited from propane

ture in their studies, which is characteristic of methane-derived CVD carbon. The deposition conditions for these microstructures are analogous to those for their counterparts in methane-derived carbons, and are shown in Fig. 5.5. The composites exhibit a wide range of matrix density (1.55–2.10 g/cc) and crystallite size 23–33 Å depending on the propane concentration and processing conditions. With a high concentration of propane (around 70%) the density and crystallite size of the matrix seems to be almost constant at all deposition temperatures. In contrast, at low propane concentrations, the density as well as crystallite dimensions of the matrix, characteristics for different textures dveloped in the matrix, decrease with an increase in deposition temperatures as shown in Figs. 5.6 and 5.7 [12].

Figure 5.8 shows some of the typical structures of CVD-derived carbon/carbon composites. The newly developed techniques of forced flow thermal gradient CVI [13] and pulse CVI [14] also result in the uniform deposition of carbon along the thickness of the composites. In these cases also the classic pyrolysis textures of RL, SL, and ISO are observed in the carbon matrix.

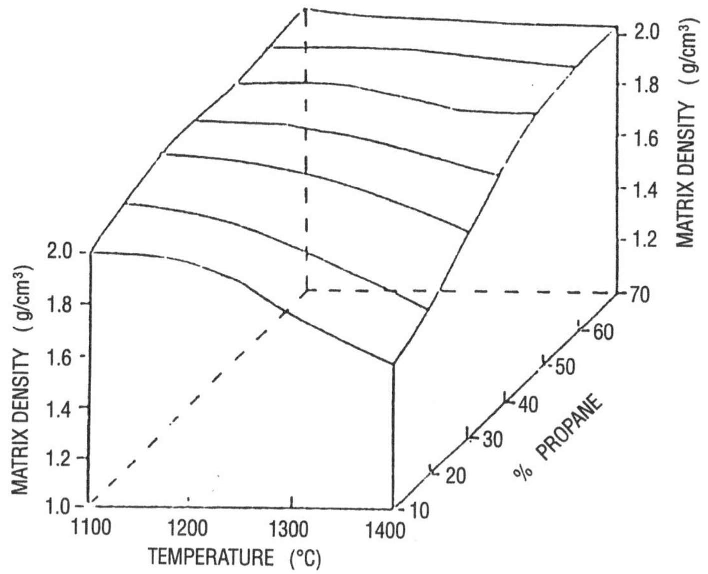

Fig. 5.6. Relation between density of pyrolytic carbon matrix and deposition conditions [12]

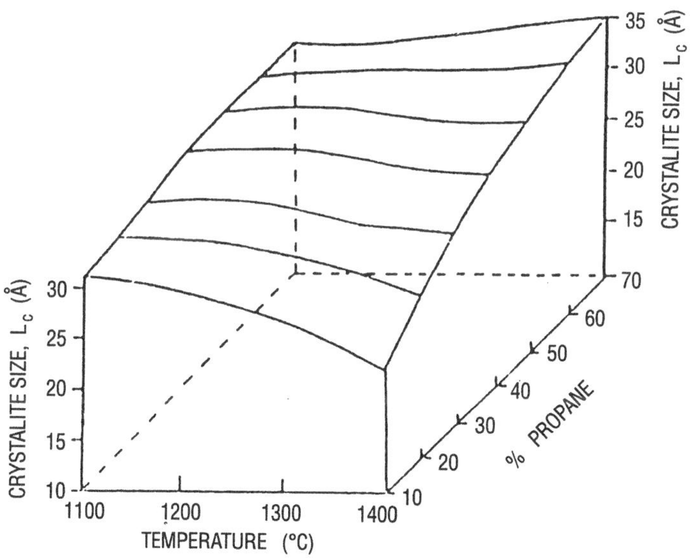

Fig. 5.7. Effect of deposition conditions on crystallite size of pyrolytic carbon matrix [12]

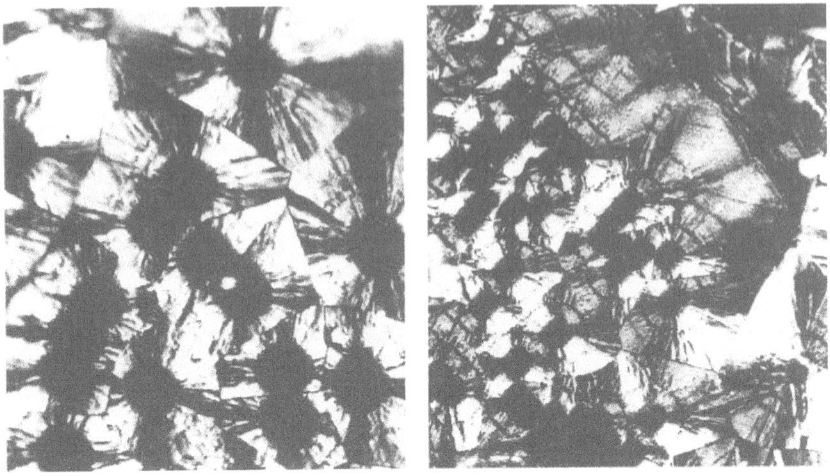

Fig. 5.8 a, b. Typical microstructures of CVD-derived carbon/carbon composite

5.2 Carbon/Carbon Composites Fabrication by the Liquid Infiltration Route

Detailed optical microscopy, XRD, SEM, and TEM studies are reported in the literature on the matrix microstructure derived from thermosetting resins and thermoplastic pitches [15, 16]. However, only quite restricted studies have been carried out on the macrostructure and microstructure of the composites as a whole, especially on the origin, characteristics, and effects of voids, pores, matrix rich regions, etc. These features are equally important and control the properties of the composites.

5.2.1 Macroscopic Features

With respect to minimechanical features, carbon/carbon composites are a build-up of fiber bundles containing filaments, thin interfilament matrix, interbundle-matrix-rich regions, and macropores. The latter includes the voids formed as a result of imperfect fiber stacking or processing, bubble pores due to evolution of gaseous byproducts during carbonization, dry zones and shrinkage cracks etc. [1]. Figure 5.9 is a schematic representa-

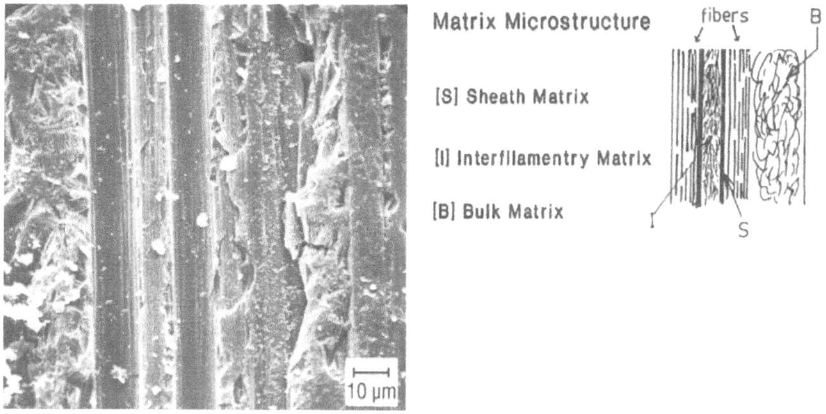

Fig. 5.9. Schematic representation of different matrix domains present in carbon/carbon composites made by multiple liquid impregnation technique

tion of different types of matrixes present in carbon/carbon composites. End-product applications largely determine the fiber architecture. However, the fiber bundle characteristics such as the size of the fiber tow, the fiber surface characteristics and composite processing conditions influence the spreading of fiber tow and infiltration of the liquid matrix in the fiber bundle. This in turn decides the fiber volume contents particularly in those areas which may influence the properties of the end product. Interfilament spacing in a bundle may also effect the microstructure of the matrix in that region.

5.2.1.1 Unidirectional Composites

Carbon/carbon composites as unidirectional model composites are made in research laboratories using prepreg technology either with thermosetting resins or thermoplastic pitches as matrix. If prepregged, stacked, and compressed properly, these composites will not exhibit gross matrix-rich pockets, dry zones, or voids. However, the size of the interbundle matrix will be larger than the intrabundle matrix. This results in variation in microstructure of the carbon matrix over the cross section. In the case of thermosetting resin derived carbon/carbon composites, the surface properties of carbon fibers are found to have controlling influence on the fiber/matrix bonding in the composites even after carbonization [17–21]. In composites with strong fiber/matrix bonding at the polymer stage, the carbon matrix remains well bonded to the fiber even on subsequent heat treatments, resulting in macro- as well as microcracks in the unreinforced matrix (interbundle matrix).

The composite macrostructure consists of lumps of fibers bonded together, separated by large voids. The size of lumps and voids depend on the relative amounts of fibers and matrix at the polymer stage and/or on the surface treatment and type of the carbon fibers. Annular gaps are observed between the fiber and matrix in composites with poor fiber/matrix bonding at the polymer stage, i.e., made with non-surface-treated high-modulus carbon fibers [Fig. 5.10]. However, in both cases extensive matrix cracking occurs in the direction perpendicular to the fibers due to matrix shrinkage during carbonization.

In carbon/carbon composites made with pitches as matrix precursor by low pressure processing, voids, bubble pores, and matrix macrocracks are commonly observed. These are partially filled up during subsequent densification cycles and exhibit regions of distinct primary and secondary carbon matrixes (Fig. 5.11).

a

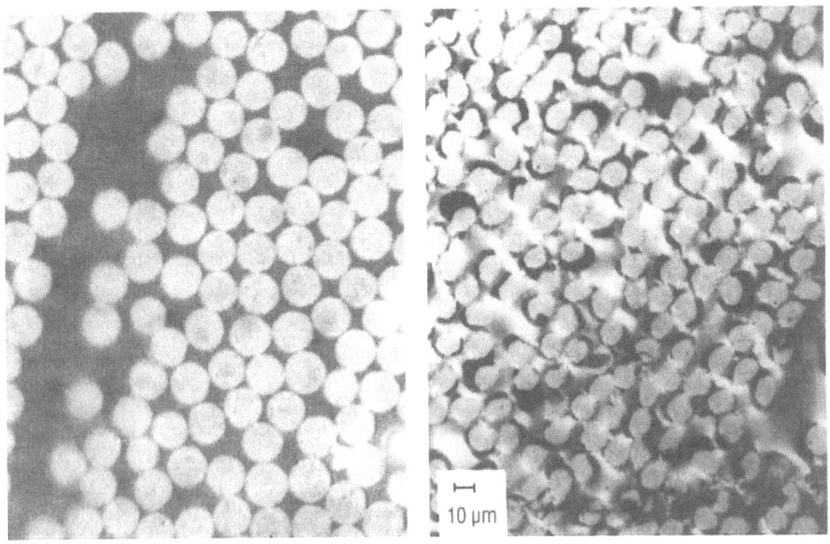

b

Fig. 5.10 a, b. Microstructure of composites made with non-surface-treated high-modulus carbon fibers. **a** Polymer composites. **b** Carbonized composites

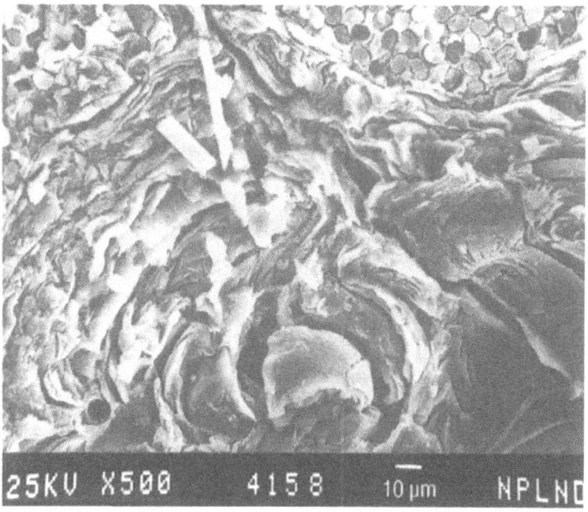

Fig. 5.11. Macrostructure of carbon/carbon composites made with pitch

5.2.1.2 Cloth-Reinforced Carbon/Carbon Composites

Two-directional (2D) carbon/carbon composites made with woven cloth as reinforcement exhibit additional significant microstructural features due to the yarns running in two directions (warp and fill), passing over and under each other, and imparting a wavy nature to them. Although these composites are made by prepregging and compaction technique similar to for unidirectional (UD) composites, a significant portion of the composite remains as unreinforced matrix rich regions such as interyarn pockets, bundle cross-over regions, interply regions etc. The configurations of these regions will depend on the weave type (plain, twil, or satin weave), stacking, compaction techniques, etc. [22–25]. Figure 5.12 [1] shows, for example, a two-dimensional representation of the stacking patterns in plain-weave-reinforced laminates. In a single composite, a combination of these patterns may usually be seen. In satin weave laminates, the so-called stacking faults are reduced since in this case, using normal compaction techniques, warp yarns of one layer may touch the fill yarns of the neighboring layer. Accordingly, there will be more matrix regions in plain-weave-cloth-reinforced carbon/carbon composites than in satin weave laminates. Obviously, the former composites contain larger amounts of microcracks than the latter composites. However, in both composite types, transverse microcracks are observed in the intrabundle matrix (without fiber breakage).

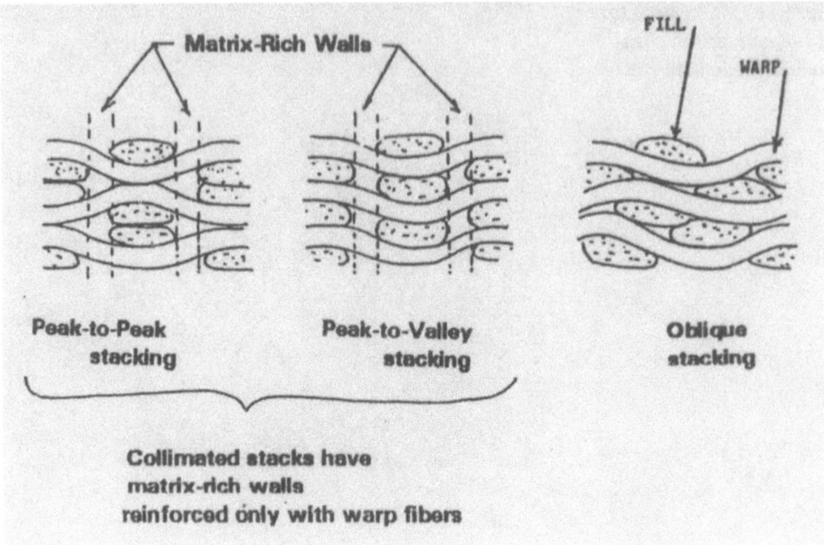

Fig. 5.12. Simplified two-dimensional description of stacking possibilities for plain weave laminates [1]

According to Jortner [1, 24, 25], in cloth-reinforced carbon/carbon composites made with thermosetting resins, the pockets of unreinforced phenolic matrix within the laminates are more resistant to microcracking during heat treatment than are the phenolic impregnated fiber bundles. Figures 5.13 and 5.14 present the different levels of cracking and microcracking observed in 2D carbon/carbon composites. Gao et al. [23] have classified the voids and cracks in 2D carbon/carbon composites according to their position, size, and orientation into four groups: coplanar and bundle cracks as those cracks aligned parallel and perpendicular to the cloth layers; interfiber cracks and interfiber pores as voids differing in cross-sectional area, present within the fiber bundles. All of these influence, in one way or another, the interlaminar shear strength, flexural strength, and transfer of the mechanical properties of the fibers to the composites.

In 2D composites, especially in thick ones, the heating rate has to be chosen carefully to avoid delamination of the composites during heat treatment [26]. Processing faults such as improper filtration of resins in the prepregs due to insufficient wetting of the fibers with the resin, incomplete compaction of the prepregs, or improper cure and carbonization cycles can result in dry zones in the polymer composites. These defects in the matrix

Fig. 5.13 a–c. Microcracking in 2D carbon/carbon composites

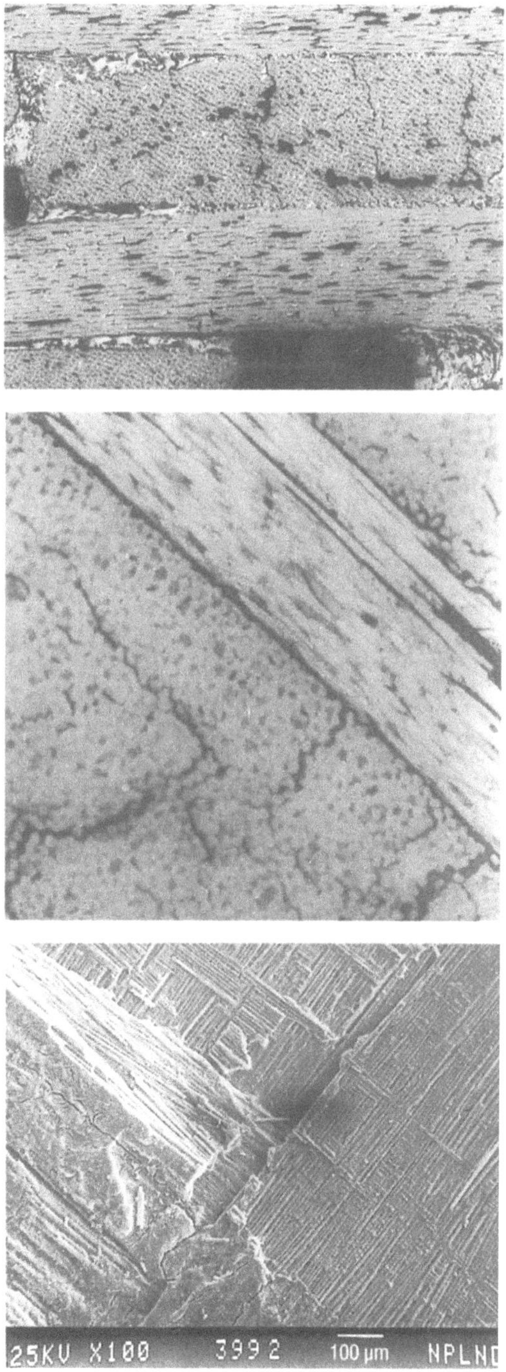

a

b

c

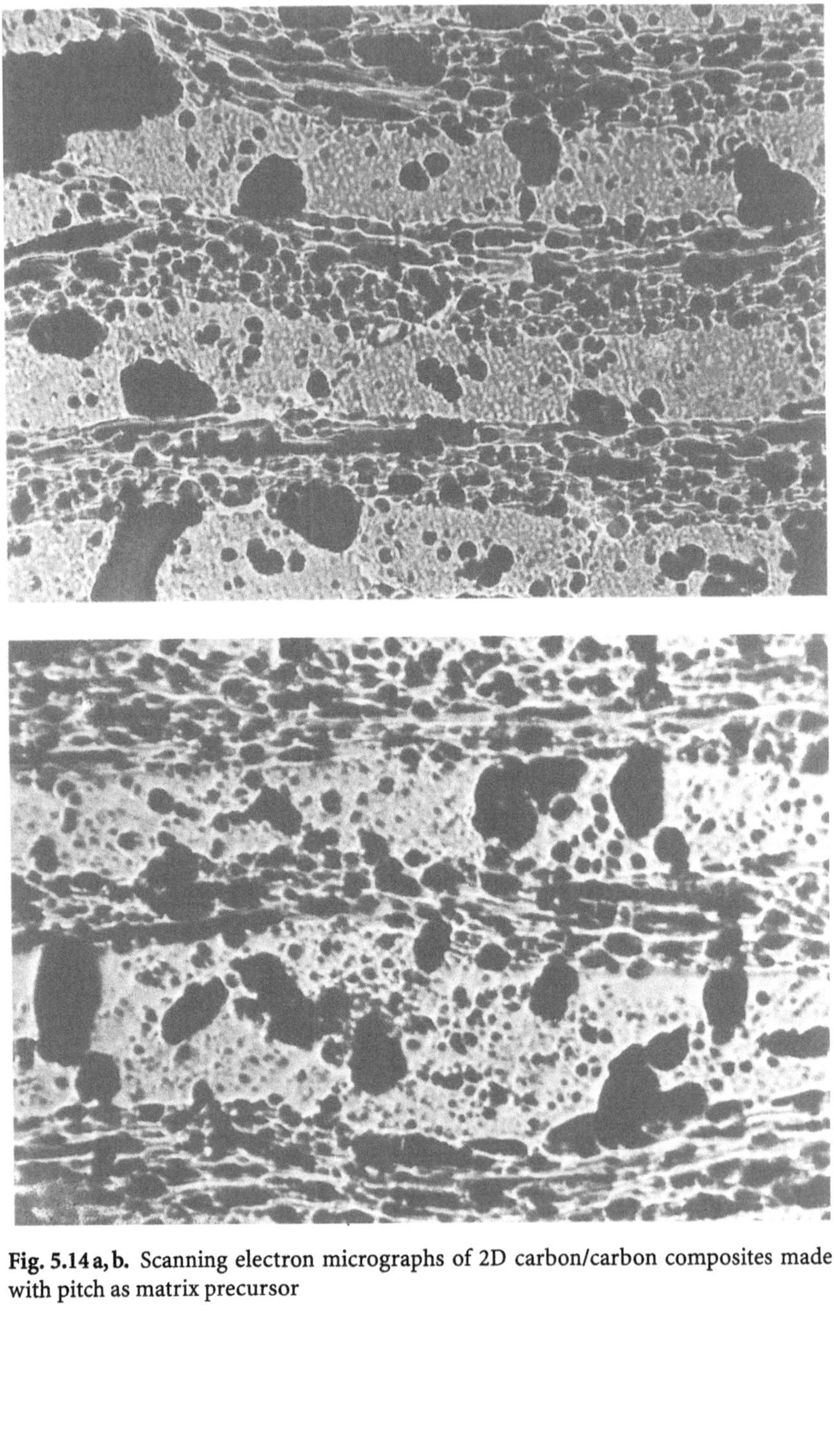

a

b

Fig. 5.14 a, b. Scanning electron micrographs of 2D carbon/carbon composites made with pitch as matrix precursor

are carried over to the carbonized state and are sometimes inaccessible to further densification. These dry zones can of course be avoided by proper fabrication or inspection techniques.

With thermosetting resin precursors, although the carbon yield is quite low, pyrolysis takes place in the solid-state form leading to volumetric shrinkage of the composites. However, in the case of pitch precursor composites, carbonization initiates in the liquid form during which most of the heterogenous reactions take place and volatiles are evolved before it finally solidifies. Evolution of these volatiles through the liquid matrix creates further pores and voids in the matrix-rich regions. Additionally, the pitch-based carbon matrix does not exhibit any shrinkage during carbonization, and, therefore, the defects created during carbonization of the composites such as voids, pores, or cracks get frozen into the macrostructure of the composites as shown in Fig. 5.15. These are generally well connected to each other and help in densification during subsequent impregnation cycles but still exhibit different structural domains in the ultimate composites.

5.2.1.3
Multidirectional Carbon/Carbon Composites

The situation becomes more complex in multidirectional (MD) carbon/carbon composites due to the rigid positions of the fiber bundles. In comparison with 2D composites, multidirectional composites contain more matrix-rich pockets, depending on the directional configurations; however, the yarn bundles are uncrimped. Like 2D composites, MD carbon/car-

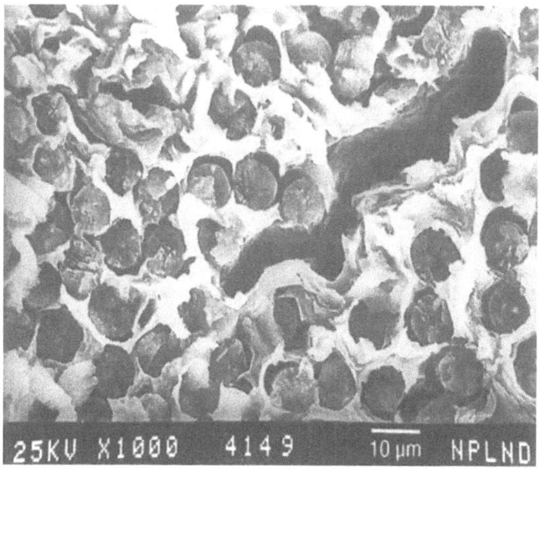

Fig. 5.15. Scanning electron micrograph of carbon/carbon composite made with phenolic as matrix precursor

bon composites also contain pores, voids, and microcracks [24, 27]. Figure 5.16 [24] shows a schematic representation of minimechanical features of 3D carbon/carbon composites. The various macro- and micro-defects shown in this figure are observed both in thermosetting resins as well as pitch-derived composites, although the extent of the defects may be different in the two cases. Composites infiltrated with pitch and processed by low-pressure pyrolysis exhibit more bubble pores than high-pressure pyrolyzed composites or those made with matrixes derived from thermosetting resins.

Similar to UD and 2D composites, MD composites also contain shrinkage cracks and thermal stress cracks. Since the latter composites are composed of a rigid fiber architecture which restrains the overall shrinkage, cracks are likely to be more pronounced in them than in UD and 2D composites. The void volume of thermal stress cracks depends on the maximum heat treatment temperature (MHTT) during processing and is found to be significantly larger if creep occurs at MHTT. All of these cracks are

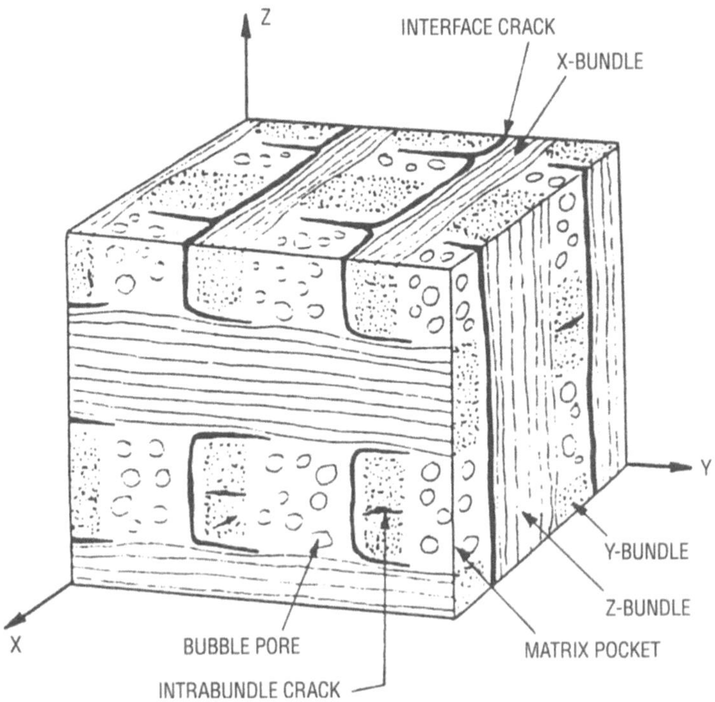

Fig. 5.16. Minimechanical features of 3D carbon/carbon composites [24]

periodic, occurring at regular intervals throughout the composite usually at similar locations within the composite unit cell as shown in Fig. 5.15. In finished composites, cracks formed by matrix shrinkage remain open during heating and cooling, whereas those formed by thermal stresses tend to close on heating and open on cooling. Another striking feature of the MD carbon/carbon structures is the billowing of the fiber bundles as shown in Fig. 5.17 [24, 27]. Although the fiber bundles are uncrimped and straight in the initial MD preform, up on densification with matrix precursors and subsequent heating cycles they become distinctively curved due to billowing of the bundles into the matrix pocket regions. This is attributes to creep or plasticity caused by compressive thermal stresses during heat treatment and tensile stresses during cooling accompanied by intrabundle cracking. Billowing is found to have a positive effect on the mechanical properties of MD carbon/carbon composites. A similar microcracking network has also been reported independently by Perry and Adams [28] and Greszezuk [29].

5.2.2 Matrix Microstructure in Thermosetting-Resin-Derived Carbon/Carbon Composites

Thermosetting polymer are usually highly cross-linked polymers characterized by random 3D intermolecular cross-links between polymeric

Fig. 5.17. Schematic representation of billowing [24]

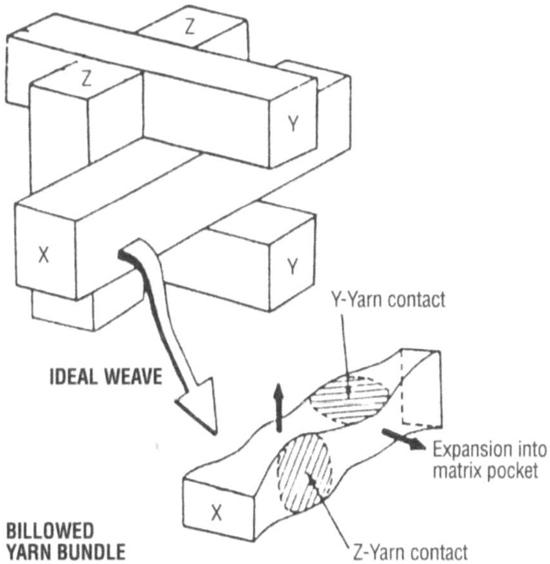

chains. These polymers, therefore, undergo solid-state pyrolysis to form isotropic carbon in bulk form with a morphology and structure that typically reflects that of the parent polymers. These carbons are nongraphitizable and comprise basic structural units (BSUs) of very small dimensions with no significant local molecular ordering. The density of these carbons is very low, in the range of 1.4–1.6 g/cc, with micropores in a closed form [30]. By high resolution transmisson electron microscopic examination, they are found to contain very small areas of short range, parallel graphitic layers which are strongly wrinkled and do not graphitize by thermal activation, not even at the highest heat treatment temperatures of 3000 °C [31]. Due to lack of the preferred orientation, these carbons are isotropic in nature and are optically inactive.

Ever since Franklins early observations on the effect of internal stresses on the graphitization process in various carbons, numerous experiments have been performed on the graphitization of glassy carbons under mechanical load as well as by the use of catalysts. The Japanese school led by Noda, Kamiya, and Inagaki [32, 33] performed early experiments on the application of hydrostatic pressure of the order of 1–10 kbars during heat treatment and found a dramatic accelerating effect on the graphitization of hard carbons. The results of Rieswig et al. [34] on glassy carbons as fillers and binders suggested that internal stresses generated during matrix shrinkage facilitate graphitization. Krosonov et al. [35] carried out a systematic study by applying a load over different temperature ranges during carbonization. They found that when the mechanical load was applied over a 400°–600 °C range, subsequent heat treatment of the resin-derived carbon to 2800 °C resulted in pronounced graphitization. This suggests that a stress-induced ordering effect, which occurs when the material is most deformable, is the principal factor responsible for the development of graphitic structure in hard carbons during subsequent high temperature heat treatment. However, the pressure has to be applied in a specific temperature range.

The widespread use of thermosetting resins as the matrix precursor in carbon/carbon composites has generated much interest in the studies on graphitization mechanisms in hard carbons in composites. Hishiyama et al. [36] were the first to report a detailed study on the graphitization behavior of a hard carbon matrix in a carbon/carbon composite. They observed that the anisotropic region in the matrix began at the interface and grew into the matrix. The optical activity of the matrix near the fiber surface was observed even in the carbonized stage (HTT 1000 °C). However, it became more predominant at higher temperatures. They attributed this to the large shrinkage stresses which developed at the fiber/matrix interface during carbonization of the composites causing stress graphitization of the matrix in the composites. The graphitic structure starts developing at a temperature of 2000 °C [37–39] (Fig. 5.18). The interbundle matrix and the matrix far away from the fibers remain isotropic in nature.

The type of carbon fibers and especially the surface characteristics of the carbon fibers controlling the fiber/matrix bonding at the polymer stage influence the graphitic structure in the matrix. A good correlation can be drawn between the surface energetics of carbon fibers, cross-sectional shrinkage during carbonization, and extent of graphitization in the thermosetting-resin-derived matrix [17–20]. Composites made with very strong fiber/matrix bonding, e.g., with surface-treated high-strength or high-modulus carbon fibers, exhibit a kinked lamellar structure with crystallographic tilt boundaries in the stress graphitized matrix around carbon fibers. In contrast, those made with non-surface-treated high-modulus carbon fibers exhibit long range lamellar orientation of the carbon matrix around the fibers (Fig. 5.19).

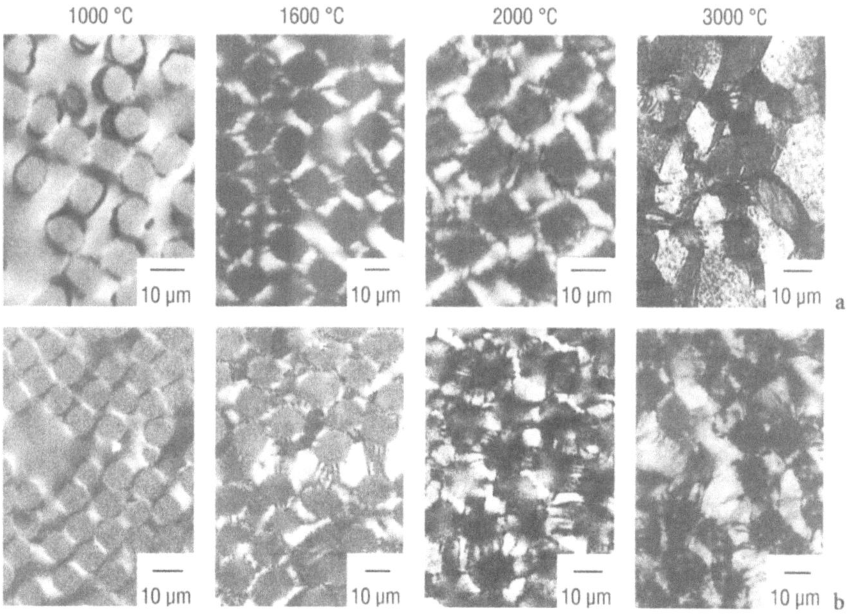

Fig. 5.18 a, b. Optical micrographs (under polarized light) of carbon/carbon composites heated to different temperatures. **a** Composites made with non-surface-treated carbon fibers. **b** Composites made with surface-treated carbon fibers [38]

a
b

Fig. 5.19 a, b. Optical micrographs (under polarized light) of heat-treated carbon/carbon composites made with surface-treated **a** and non-surface-treated **b** carbon fibers

The graphitization of the matrix in thermosetting resin precursor composites is found to be the same regardless of the polymer monomer, i.e., phenolic, polyfurfuryl alcohol, polyarylacetylene, etc. [40]. However, detailed X-ray studies on the composites made with the polymers indicate that the extent of graphitization of the phenolic-derived matrix is slightly higher than that of the polyfurfurylalcohol-derived matrix. The use of ion etching in conjuction with polarized light microscopy reveals the lamellar texture and optical activity simultaneously (Fig. 5.20). While the interfilament matrix in such composites exhibits highly aligned graphitic layers, the matrix in the regions far away from the filaments exhibits an isotropic texture (Fig. 5.21). Scanning electron microscopy and TEM examinations of these composites also reveal highly oriented lamellar regions near the fibers [41, 42].

PAN-derived carbon fibers subjected to a uniaxial tensile stress during oxidative stabilization and carbonization are known to be nongraphitic carbons. However, recent studies by Kyotani et al. [43] have revealed that thin films of PAN polymerized and carbonized between the layers of a

Fig. 5.20 a, b. Scanning electron micrographs of carbon/carbon composites showing alignment of interbundle matrix parallel to fiber axis

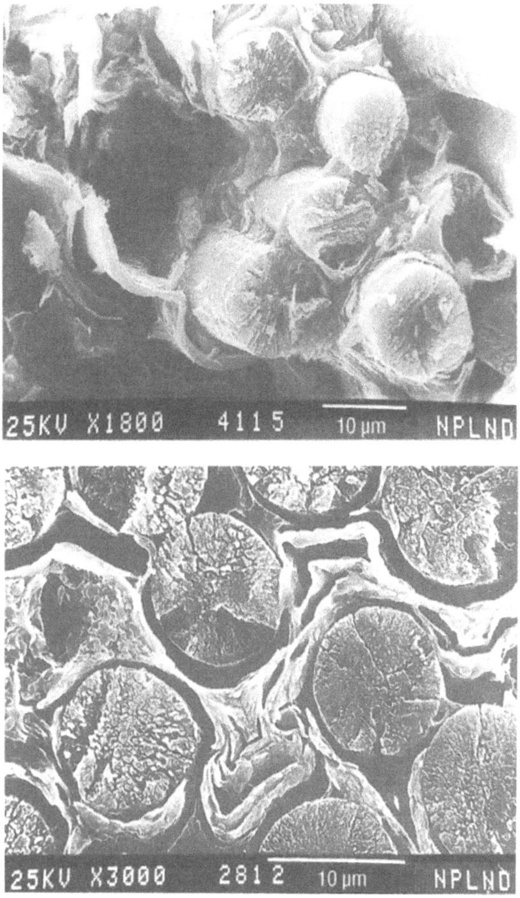

a

b

montmorillonite clay thus experiencing multiaxial stress, results in a crystalline graphitic structure. By analogy, considering thermosetting-resin-derived carbons in bulk form and as thin sheets in the interfilament regions of carbon/carbon composites, Rellick et al. [44] have proposed that lamellar formation in the resin-derived matrix of carbon/carbon composites may also be the result of predominantly multiaxial deformation stresses developed in the matrix resulting from restrained pyrolysis shrinkage during their conversion to carbon. They have proposed three types of stresses caused by shrinkage restraint during carbonization, which are shown schematically in Fig. 5.22 [44].

These are: (1) axial tensile stress, which is a consequence of the large matrix pyrolysis shrinkage and the combination of high axial modulus and

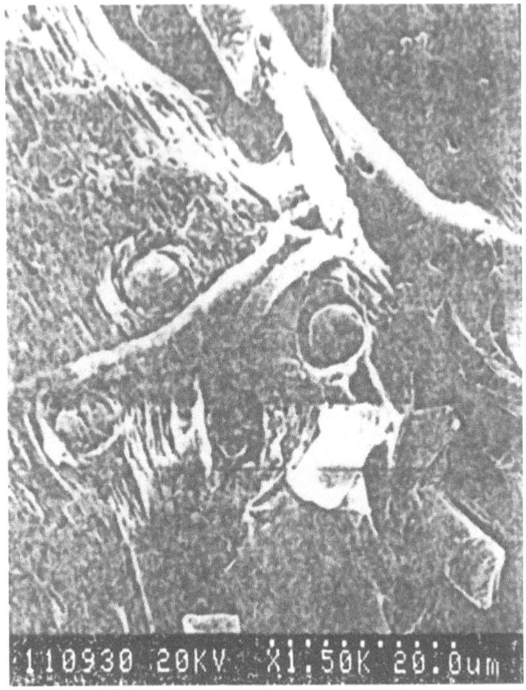

Fig. 5.21. Scanning electron micrograph of carbon/carbon composites with anisotropic and isotropic carbon matrixes

low linear expansion of the fibers; (2) compressive stress acting radially; and (3) tensile stress acting circumferentially.

The development of lamellar structure in the matrix is favored by (a) large amount of circumferential tensile stress in the cross-sectional plane, which when combined with the axial tensile stress, produces a state of biaxial tension in the matrix and (b) a small or negative (compressive) value

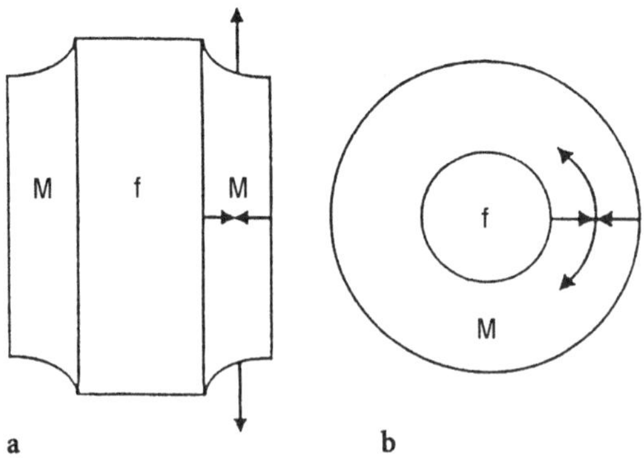

Fig. 5.22 a, b. Matrix stresses in a single fiber composite during pyrolysis. **a** Longitudinal section. **b** Transverse section [44]

of the radial stress in the same plane. The latter contributes significantly to the production of lamellar structure with kinks and crystallographic tilt boundaries. Accordingly, in the thermosetting resin precursor of carbon/carbon composites, the state of orientation of the matrix depends on the nature and extent of the stresses experienced by it during carbonization. The interfilament matrix experiencing biaxial stresses gives rise to a graphitizable lamellar matrix. In contrast, in the regions where the in-plane stresses are uniaxial (far away from the fibers, the bulk matrix or interbundle matrix pockets which are debonded from the fibers and contract in on itself during carbonization) nonlamellar, twisted ribbon-like carbon microstructure is observed.

Another prominant feature of the matrix microstructure, expecially in the lamellar regions, is the extensive kinking of the layer planes. Kinking is typically observed in the lamellar matrix around fibers with uneven surfaces with the kinks conforming to the depression and asperities on the fiber surface. Kinked matrixes are also observed in composites made with excessively surface-treated carbon fibers. These may be due to the excessive compressive stresses in the direction perpendicular to the fibers generated during pyrolysis of the polymer composites. Rellick and Adams [45] have also reported the observation of transversely oriented matrix immediately at the fiber surface with the c axis oriented tangential to the direction of the fiber axis and in some cases parallel to the direction of fiber axis, as reported by Evangalides [46] for pitch-based composites. They have attributed this to the deformation of the matrix during the critical carbon chain forming process that accompanies the matrix pyrolysis and in some cases to the formation of a carbon network of interpenetrating fibrils that effectively link fibers to the matrix.

The extent of the graphitizability of the matrix will obviously change, or to be more precise, will decrease from unidirectional to bidirectional and multidirectional composites owing to the nature and extent of the shrinkage stresses developed during pyrolysis of the composites. Most of the features observed in unidirectional composites are also seen in 2D and MD carbon/carbon composites, but the matrix continuity is not maintained.

Addition of fillers such as fine-grained graphite or coke powder to thermosetting resins also influences the matrix microstructure in carbon/carbon composites. These additions disturb both the hoop orientation and the longitudinal orientation of the graphitic structural units in a carbon/carbon matrix [47, 48] (Fig. 5.23).

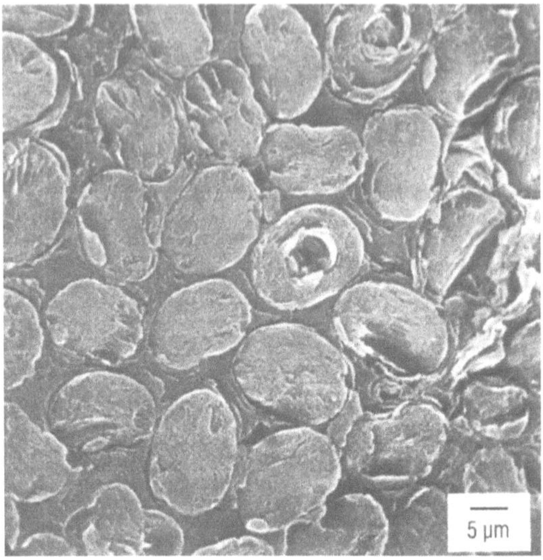

Fig. 5.23. Microstructure of a carbon/carbon composite with a phenol matrix to which graphite was added heated to 2800 °C

5.2.3 Matrix Microstructure in Pitch-Derived Carbon/Carbon Composites

In contrast to the microstructure of thermosetting-resin-derived carbon, that of thermoplastic pitch-derived carbon is controlled by nematic liquid crystal (Mesophase) formation during liquid-phase pyrolysis of the pitches. The flow of the oriented molecules of the liquid crystals during their formation and subsequent deformation results in the introduction of disclinations in the structure. Figure 5.24 [49] gives a schematic representation of the wedge disclinations identified in the carbonaceous mesophase. As polymerization continues and the carbonaceous mesophase hardens as coke, the disclinations are restrained. With further heat treatment, ordering of the layers to obtain a graphitic atomic structure occurs. Thus, the disclinations formed in the liquid crystal remain in the heat-treated solid matrix. These disclinations are not defects, cannot be annealed out, and contribute significantly to such properties as fracture, graphitizability, and thermal expansivity of the composites [50].

During pyrolysis of the pitch precursor in the fabrication of carbon/carbon composites, the disc-like molecules of the mesophase tend to align parallel to the surface of the carbon filament forming a circular sheath several microns thick around each filament. The parallel alignment of the filatments and mesophase sheath force the disclination lines to be parallel

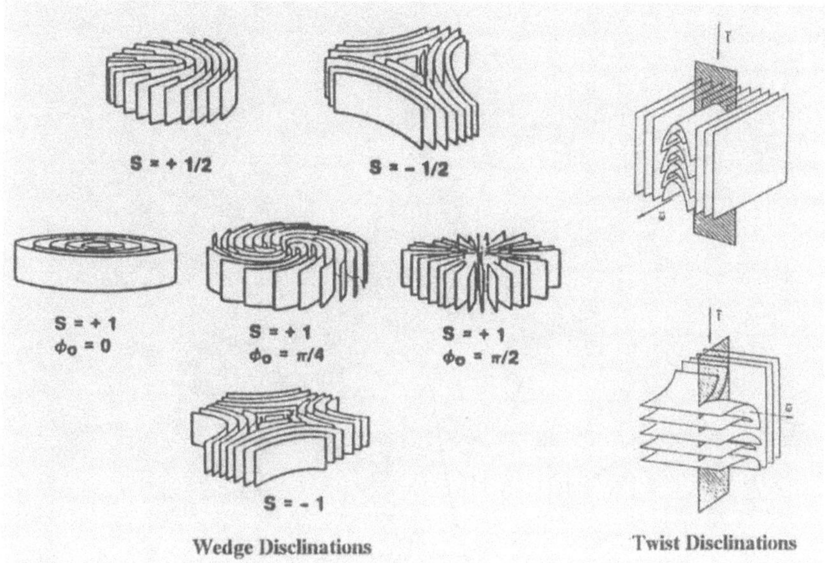

Fig. 5.24. Models for the wedge and twist disclinations of the carbonaceous mesophase [49]

to the fibers. However, the strength and type of disclination is determined by the local arrangement or pattern of the arrays of filaments. Different negative wedge disclinations among the filaments in a fiber bundle are shown in Fig. 5.25 [51].

The alignment of the mesophase on the surface of the carbon filaments in a bundle and, therefore, the disclination structure, may be influenced by the properties of the materials (type of pitch and carbon fibers and their configuration), processing conditions, etc. White and Shaeffer [51] reported that the wetting behavior of mesophase appears to be independent of fiber variety and weave, applied pressure, pitch variety, etc. The typical wetting behaviour and molecular alignment for the mesophase pitch on carbon fibers is shown in Fig. 5.26 [51]. The well-oriented sheath-like high-modulus skin around the carbon fibers, as shown in Fig. 5.27 [52], dominates the microstructure of carbon/carbon bundles. Even under the influence of magnetic fields sufficiently strong to reorient bulk mesophase, the mesophase layers within approximately 7μm of the fiber surface retain their sheath-like orientation [53, 54]. On subsequent heating, from the point of mesophase hardening (500 °C) to graphitization temperatures and cooling, extensive shrinkage cracking and fiber/matrix debonding is observed due

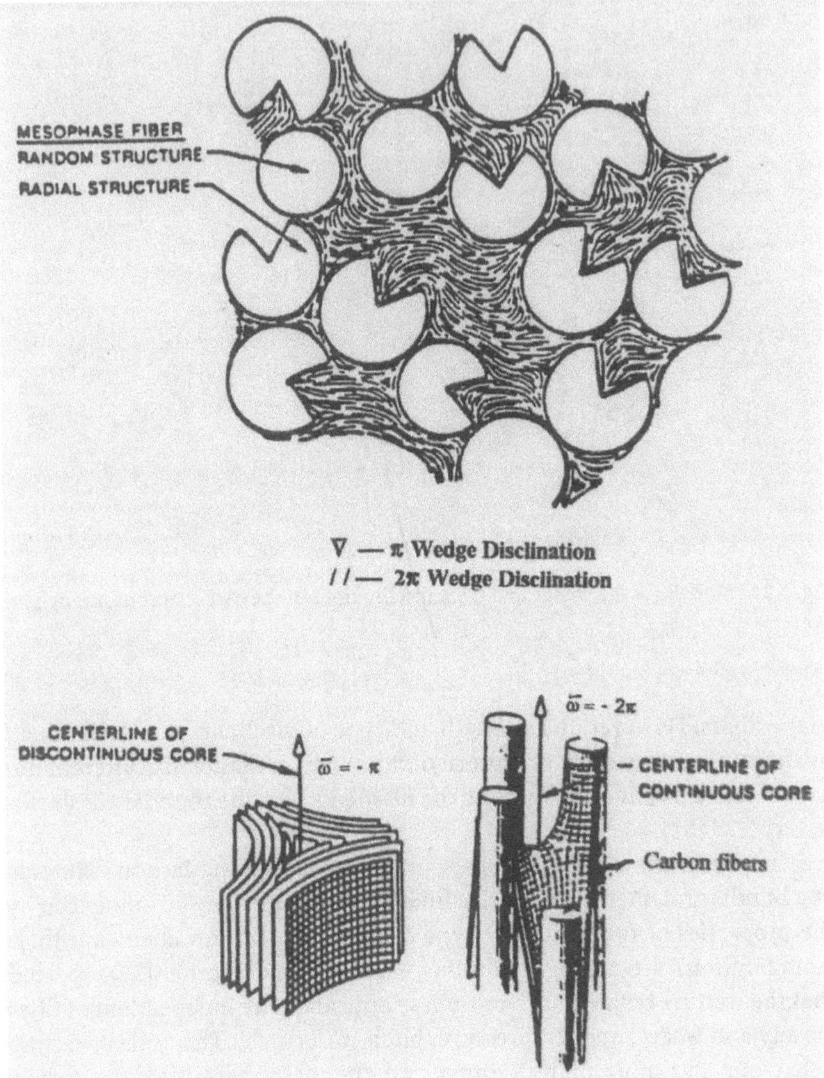

Fig. 5.25. Alignment of mesophase layers in a bundle of VSA-11 fibers [51]

to fold sharpening and local distortions. This results in a loose deformed sheath with little apparent bonding to the filaments [50].

Some variations in microstructure of the pitch precursor matrix between carbon filaments have been reported by Evangelides [54]. The lamellar structure aligned with the axis of the filaments is observed for low-

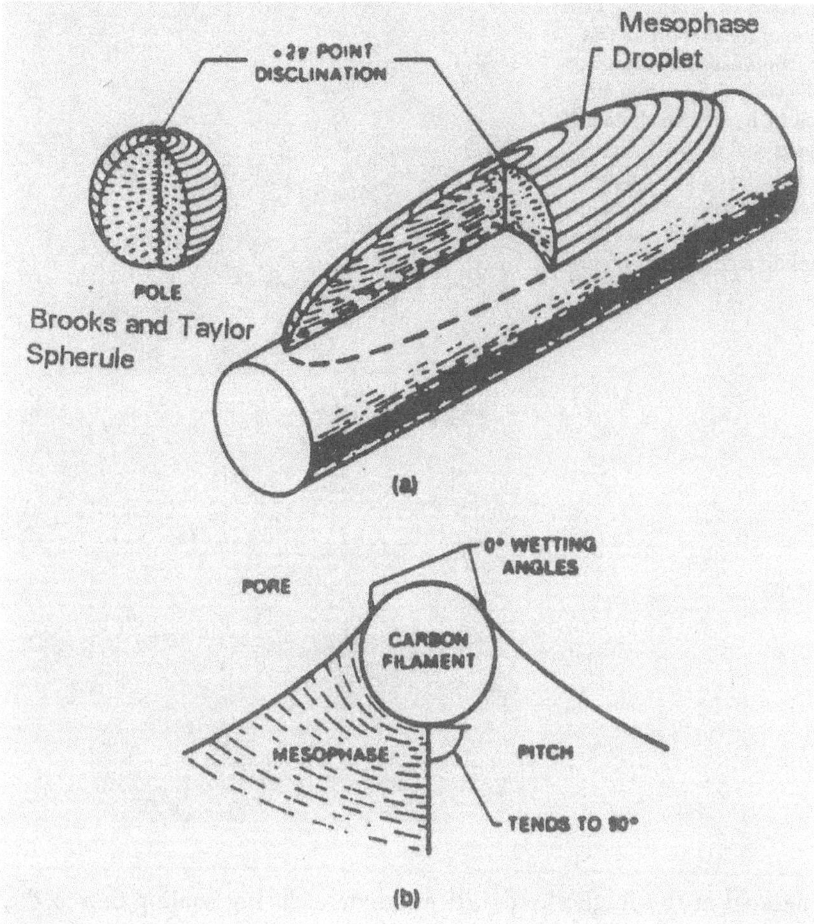

Fig. 5.26. Typical wetting behavior and molecular alignment for mesophase pitch on carbon fiber [51]

pressure processing of the composites, whereas a significantly different pitch microstructure is observed with high-pressure processing. Within a fiber bundle, the graphitic basal planes are no longer completely parallel to the filament surfaces but are transversely and randomly oriented (Fig. 5.28) [54, 55]. In addition, this structure does not posses the long range order of the low-pressure-processed composites. As the spacing between filaments increases, the alignment or sheath effect around the fibers diminishes and the matrix can contain various grain orientations. In the bulk matrix both within and between fiber bundles, shrinkage cracks are observed which are

Fig. 5.27 a–d. Preferred orientation of the carbon matrix in carbon/carbon composites made with QI-free coal tar pitch (**a, b**) or with chemically treated coal tar pitch (**c, d**). **a, c** Polarized light micrograph of cross section. **b, d** Scanning electron micrograph of etched cross section

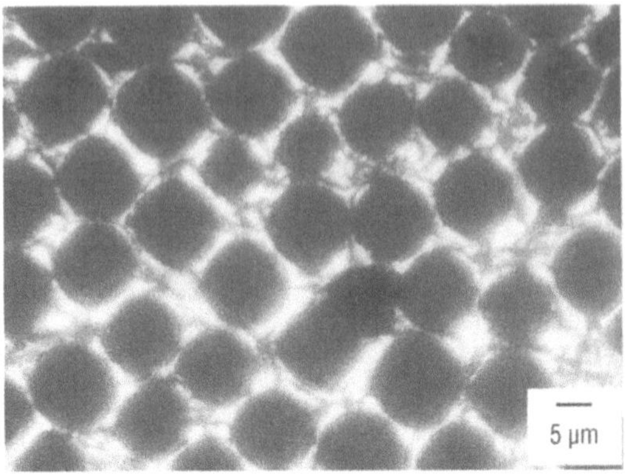

a

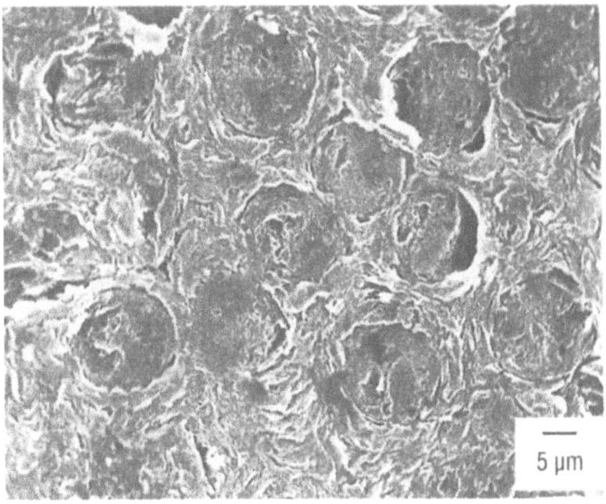

b

generated in the fragile hardened mesophase during cooling due to the mismatch in thermal expansion of the fibers and the mesophase carbon.

In order to achieve high densities in the finished carbon/carbon products, multiple impregnation/carbonization cycles have to be performed with possible intermediate heating to graphitization temperatures. These cyclic heating/cooling operations are likely to further affect the microstructure of the carbon matrix. Weisshaus et al. [56] have reported the transformation of continuous sheat-like morphology to leaf-like morphology of the matrix in carbon/carbon composites after four or five densification/graphitization cycles. The sheath-like orientation of the pitch carbon matrix has been observed in all types of carbon/carbon composites made with round carbon fibers irrespective of PAN or pitch (Fig. 5.29) [57]. However, due to the relatively less active surface in pitch carbon fibers, the matrix debonds from the fibers and the majority of shrinkage cracks are

Fig. 5.27 (continued)

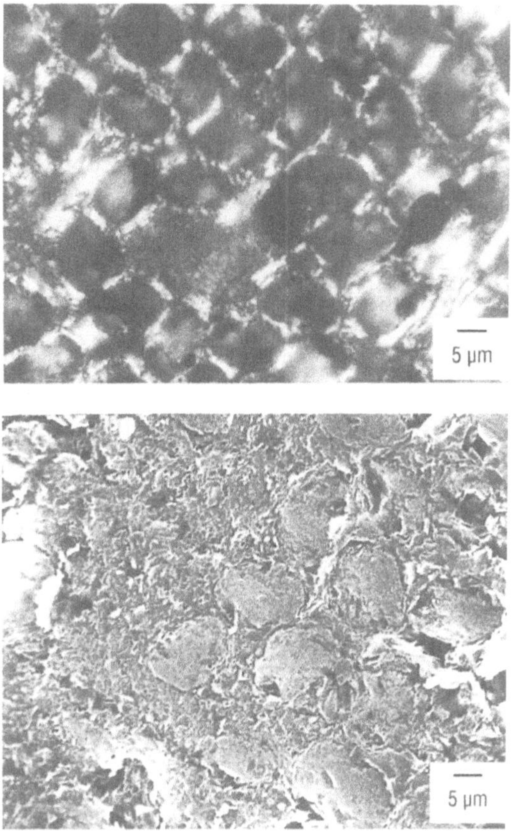

c

d

a

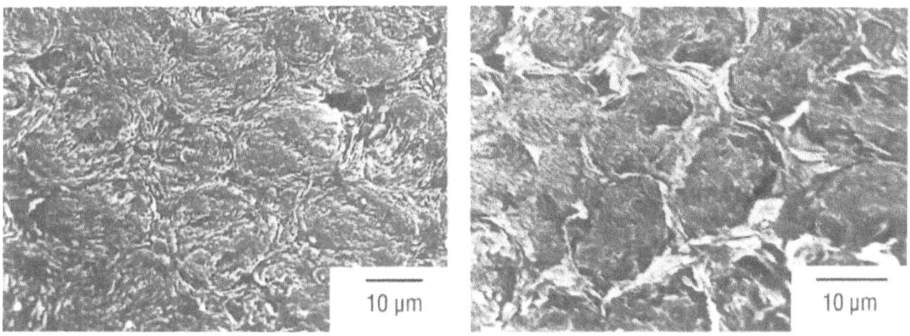

b

Fig. 5.28 a, b. Scanning electron micrographs of (**a**) sheat of aligned matrix produced by low-pressure impregnation and (**b**) transversely oriented matrix produced by high-pressure impregnation

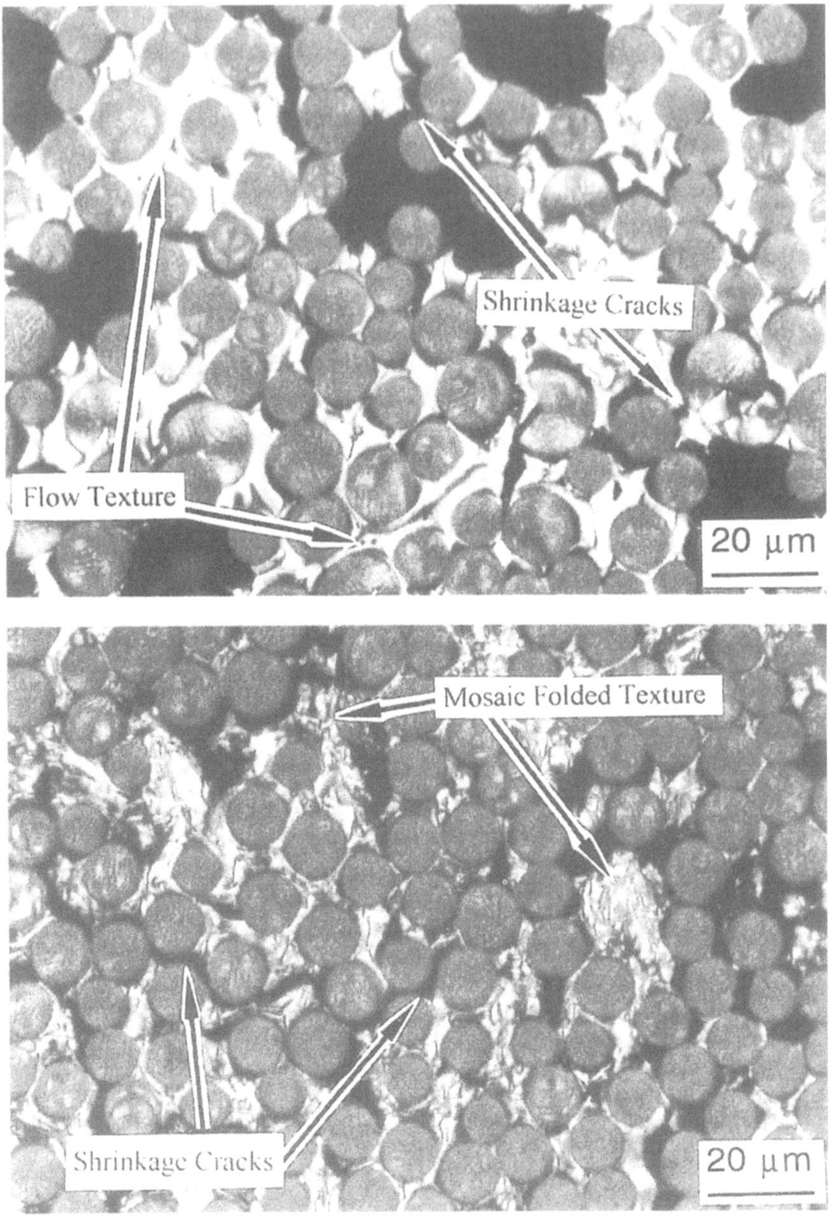

Fig. 5.29 a, b. Crossed polarized images of a cross section of a P55/AR120 carbon/carbon composite heat treated to (**a**) 1100 °C and (**b**) 2400 °C [57]

developed at the fiber/matrix interface, instead of in the bulk matrix. In the newly developed ribbon-fiber-reinforced carbon/carbon composites also, the carbon matrix orients parallel to the surface of the fibers. However, very few mosaic textures are observed in the graphitized ribbon fiber composites (Fig. 5.30) [57]. This indicates that the matrix structure is more linear than that observed in round carbon fiber composites (Fig. 5.31). The matrix, with a large flow-like domain structure, becomes oriented almost entirely in the long direction of the fibers and even in the spacing between the packed fibers.

The majority of the problems associated with pitch precursor carbon/carbon composites due to mesophase softening and pitch bloating during carbonization can be overcome by incorporating oxidation stabilization steps. This has been demonstrated by White and Shaeffer [51], Huttinger [58], and Mochida [59]. The oxidation stabilization step not only preserves the matrix microstructure established while the mesophase was fluid or in a deformable state but also increases the carbon yield with reduced microcracks and fiber/matrix debonding. The degree of anisotropy of the matrix can be altered by a combination of various matrix types of the same or different origins, the processing temperatures and soak times at the maximum temperatures.

Addition of thermosetting resins such as phenolic, PFA, or polymers such as plyethyleneterephthalate (PET) to pitches changes the microstructure of the resulting carbon from coarse mosaic to fine mosaic and isotropic [60–62]. When pitches are treated with dehydrogenating and oxidizing chemical additives, nongraphitizing residues are formed. The addition of sulfur to the pitch diminishes the sheath structure around the fibers. Carbon/carbon composites made from pitches with chemical additives exhibit an isotropic carbon structure near the filaments as well as in the bulk matrix regions [52]. These texture controlled matrixes have not yet been commercially exploited for carbon/carbon products.

5.3 Interfaces in Carbon/Carbon Composites

Interfaces and their properties critically influence the bulk properties of a material, static as well as under dynamic conditions of chemical, thermal and mechanical loading. Interfaces in carbon products have been studied specially w.r.t structure and surface porosity of the filler cokes. Although the processing of carbon artifacts and carbon/carbon composites is very similar, the interfaces in carbon/carbon composites are more complex due

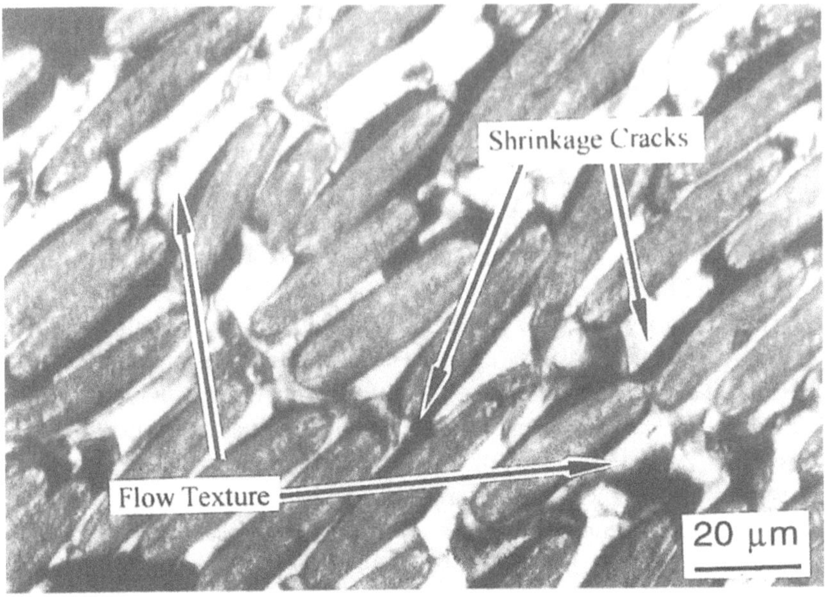

a

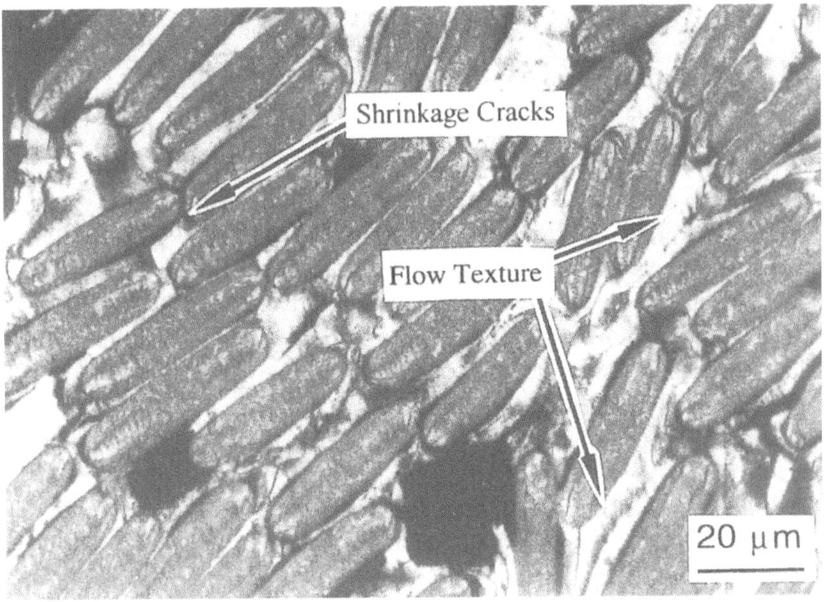

b

Fig. 5.30 a, b. Crossed polarized images of a cross section of ribbon/AR120 carbon/carbon composite heat treated to (**a**) 1100 °C and (**b**) 2400 °C [57]

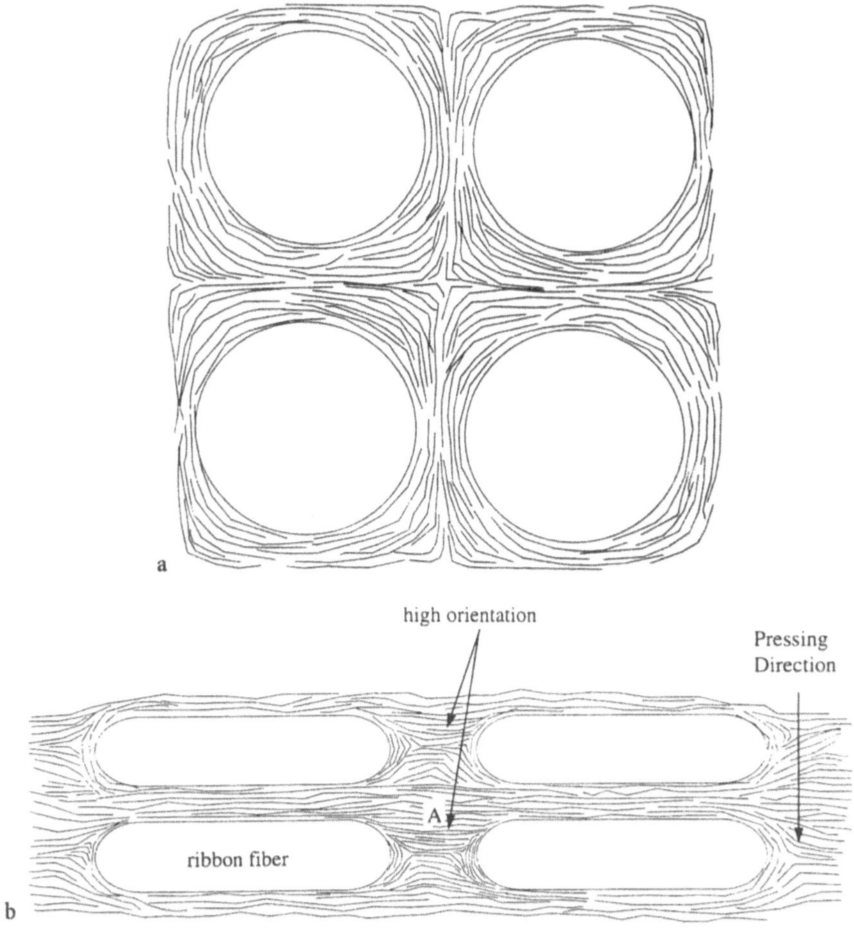

Fig. 5.31 a, b. Illustration of the sheat-like orientation observed in pitch matrix carbon/carbon composites reinforced with (**a**) round fibers and (**b**) ribbon fibers

to the presence of primary and secondary carbons derived from different precursors either individually or in combination. In general, the interfaces in carbon/carbon composites can be catagorized as (i) between filaments and matrix within yarn bundles; (ii) between fiber coating, if any, and matrix; (iii) between fiber yarns; (iv) between the microstructures that exist in the matrix; (v) between matrix and voids; and (vi) between different layers of matrix if processed by the multiple impregnation technique or if a combination of different matrix precursors has been used etc. Figure 5.32 shows a SEM micrograph of a carbon/carbon composite made by the mul-

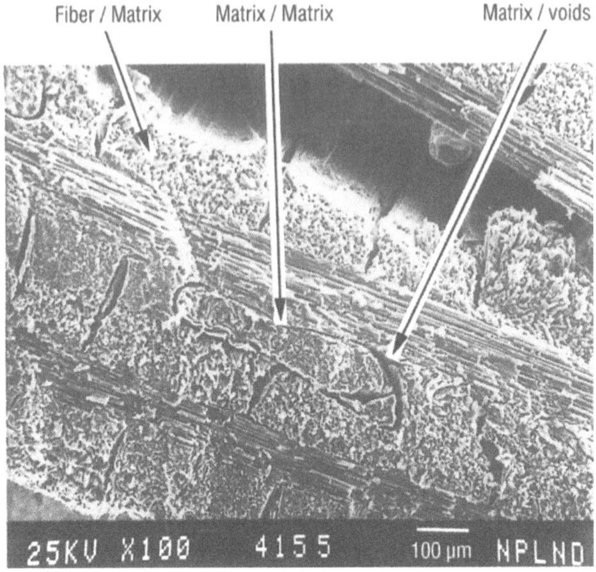

Fig. 5.32. Scanning elecron micrograph of a composite made by the multiple impregnation route

tiple impregnation technique, which exhibits various, typical interfaces. The interfaces formed depend on the constituents used such as type of fibers and matrix precursors, and on the processing conditions and may vary in thickness from a few tenths of a micron to a few microns.

Optical microscopy, SEM, TEM, and SAD have been used to directly characterize the interfaces in carbon/carbon composites [63, 64]. The concept of the interface and its role in the alignment of matrix around the reinforcing fibers was already introduced in the previous section in the context of the macrostructure and microstructure of different matrixes; some further details will be discussed in this section. The subject of chemical versus mechanical bonding between carbon fibers and the matrix in carbon/carbon composites has long been a subject of interest and controversy. It is still unknown whether the bonding between carbon fibers and matrix is of chemical, frictional, or mechanical nature. However, considering the spectrum of variations in the constituent properties, processing parameters etc., it is proposed that fiber/matrix bonding in carbon/carbon composites may be any of these types or a combination thereof, depending on the system. An interpenetrating network of tangled fibrils or ribbons extending between fibers and matrix is seen in some cases. This is due to surface microporosity of the fibers which could facilitate the interpenetration of polymeric chains during matrix pyrolysis. In certain cases, especially in

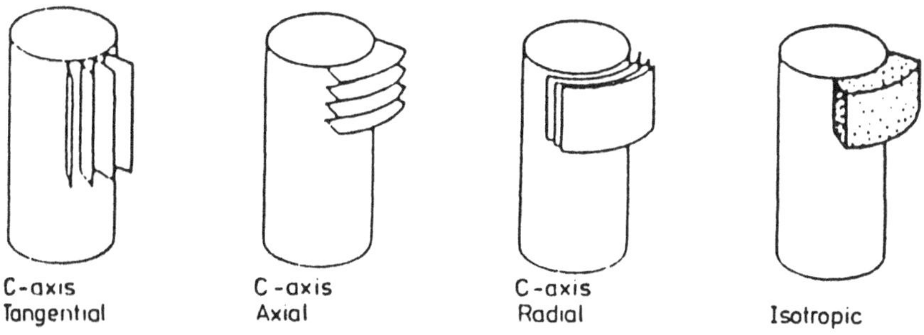

Fig. 5.33. Possible orientations of carbon matrix around a carbon fiber

mesophase pitch carbon fibers, a number of regions are seen where the matrix layer joins the fibers at oblique angles.

Considering the types of microstructures which develop in carbons derived from different precursors and the fiber/matrix interactions at the interfaces, the various possibilities for matrix orientation at the fiber/matrix interface are shown schematically in Fig. 5.33. These are (i) c axis tangential, (ii) c axis axial, (iii) c axis radial, and (iv) isotropic [54]. An excellent example of the effect of fiber/matrix interface on the matrix microstructure is the graphitization near the surface of the otherwise amorphous isotropic carbon derived from thermosetting resins [38, 65]. The chemical nature of the reinforcing fibers play a significant role in this case. The chemical bonding between fibers and thermosetting resins at the polymer stage induces different stresses at the fiber/matrix interface during carbonization. This results in stress graphitization of the matrix initiating at the interface and progressing into the bulk matrix. The graphitic matrix planes are oriented around the fibers with c axis radial which may exhibit a kinked lamellar structure if the fiber/matrix bonding is too strong as in case of carbon/carbon composites reinforced by surface-treated high-strength carbon fibers. However, when these composites are reimpregnated with thermosetting resins to fill up the voids in the bulk matrix or the interstices between fiber and matrix, different microstructures are found in these secondary carbons. Prior to reimpregnation, the composites are either isotropic or less graphitic due to the lack of intimate interface, whereas the secondary carbons have a long range graphitic structure [45, 66]. Formation of transverse oriented matrix is also seen in some cases and has been attributed to fiber/matrix bonding at the polymer stage [38].

Carbon/carbon composites made with a pitch matrix precursor, especially those made by the multiple impregnation technique, are liable to contain more interfaces than those made with a thermosetting resin precursor [67]. The flow characteristics of mesophase pitch in relation to the surfaces of the solid carbons, when used either as reinforcement or as a primary binder matrix, together with the processing conditions of temperature and pressure control the interfaces in these composites. According to Hay et al. [68], covalent bonds between carbon phases, when at least one phase has graphitic orientation, will form only if the graphitic lamellae are not parallel to the interface. However, the mesophase-pitch-derived matrix is found to form a sheath-like structure even around radial textured mesophase pitch fibers despite the presumed availability of edges with which mesophase might have formed covalent bonds [69]. This is in line with the observation of Cranmer et al. [70] that alignment of mesophase materials adjacent to the mesophase/substrate interface is primarily controlled by the flow motion of mesophase spherules.

TEM examination of PAN/pitch composite interfaces by Sohn et al. [71] revealed the presence of microfissures in the matrix along the fiber/matrix interfaces. SiC coating on carbon fibers results in the random orientation of the carbon matrix at the interface. The orientation in the carbon layer fabricated by CVD on the fibers are also found to vary depending on the surface characteristics of the carbon fibers. Although much work has been reported on the matrix microstructure at the fiber/matrix interface, there is hardly any conclusive report on the interfacial strength under different configurations of matrix orientations and fibers. Compared with PAN/pitch composites, the mesophase pitch fiber/resin plus CVI matrix carbon/carbon composites appear to have a significantly weaker fiber/matrix interface. Fig. 5.34 [72] is a brightfield TEM transmission electron micro-

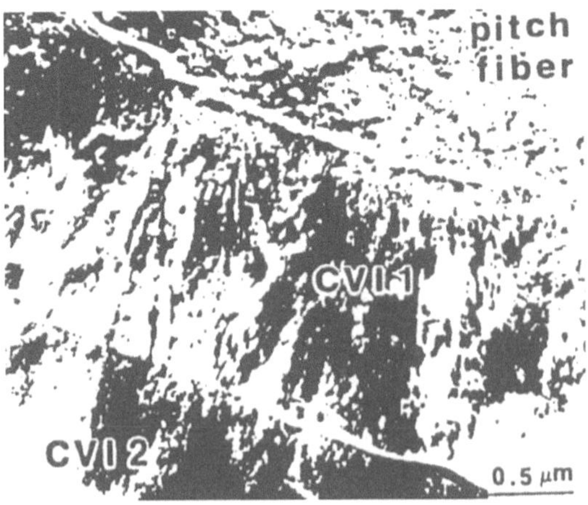

Fig. 5.34. Brightfield transmission electron micrograph of pitch fiber/resin plus CVI matrix carbon/carbon composite [72]

graph of the composite showing pitch fiber as well as the two CVI layers in the matrix. The interface between two CVI layers and between a CVI layer and resin matrix appears to be much stronger than that between the fiber and the first CVI layer which might have weakened due to graphitization of the composites before the second CVI.

5.4 X-ray Diffraction Studies on Carbon/Carbon Composites

X-ray diffraction (XRD) has been extensively used to obtain information about the structure, mainly the degree of graphitization, of carbon materials in powder form. Single crystal graphite exhibits sharp diffraction patterns characteristic of a crystalline material. The pattern is schematically shown in Fig. 5.35 [73]. The peaks {002}, {004}, {101}, and {110} indicate the crystalline order in carbon. The relative intensity and shape of these peaks are taken as a measure of crystalline content (degree of graphitization) in the material. Figure 5.36 [73] shows typical diffraction patterns of manufactured carbons. The parameters most often obtained from XRD on carbon materials are

d_{002} = interlayer spacing (c spacing)
L_a = stack width
L_c = stack height

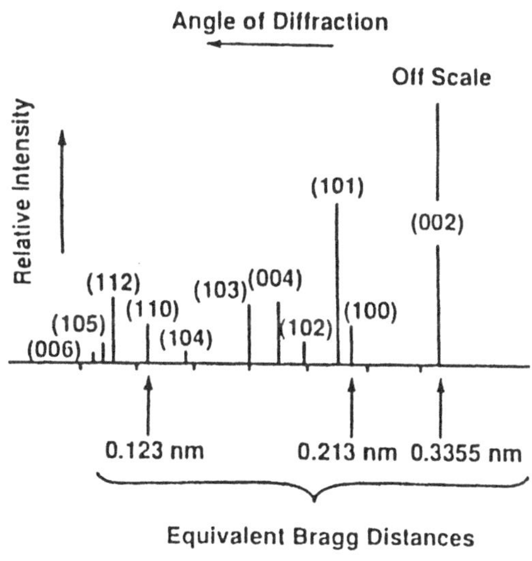

Fig. 5.35. Schematic representation of XRD lines in graphite [73]

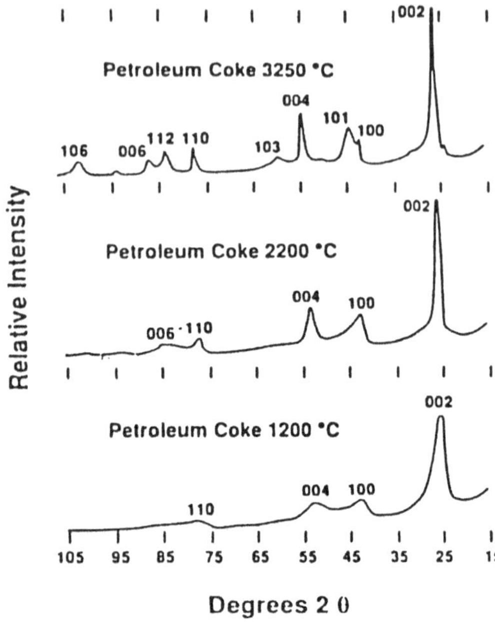

Fig. 5.36. XRD pattern of petroleum coke heated to various temperatures [73]

The standard Bragg equation is used to determine d_{002} spacing:

$$n\lambda = 2d \sin \theta$$

where n is the order of diffraction, λ is the wavelength of x-ray used, and 2θ is the angle of diffraction corresponding to the 002 line. The line width of d_{002} reflections give the c-axis crystallite size L_c. The in-plane crystallite size L_a is provided by study of the (100) diffraction line and its half width at half maximum intensity.

Although XRD has been extensively used in studying the structure of manufactured carbons and carbon fibers, this technique has not been fully exploited to study the structure of carbon/carbon composites, mainly because of the presence of two phases in these materials. The fibrous phase being more crystalline renders the overall XRD pattern of carbon/carbon composites somewhat complex (Fig. 5.37), although it looks similar to that of any carbon product heated to heat treatment temperature of the composites. However, by using high-resolution XRD and peak separation/integration techniques, the crystallinity in carbon/carbon composites, especially of the individual phases, can be estimated. The development of the 101 peak at $2\theta = 42.6$ in CVD carbon/carbon composites reflects the devel-

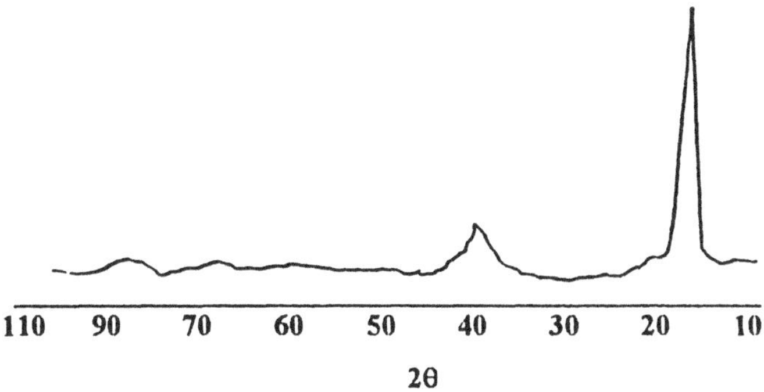

Fig. 5.37. XRD pattern of carbon/carbon composites

Table 5.2. Crystallite dimensions of some typical high performance carbon/carbon composites made by different routes

Material	d_{002} (nm)	L_c (nm)
PAN/CVD	0.337-0.34	28.5-9.0
PAN/resin	0.339-0.342	14.7-6.2
PAN/pitch	0.337-0.339	40.4-18.6
Pitch/pitch	0.337-0.338	48.2-23.4
VGCF/CVD	0.336-0.337	67.4-53.8

PAN, polyacrylonitrile; CVD, chemical vapor deposition; d_{002}, interlayer spacing; L_c, stack height; VGCF, vapor grown carbon fibers.

opment of three-dimensional ordering in the carbon matrix. The crystallite dimensions of typical carbon/carbon composites made through different routes are compiled in Table 2.4.2.

5.5 References

1. Jortener J (1990) MTC Annual Conf. Southern Illinois Uni, Carbondale, p 51
2. Pierson HO, Lieberman ML (1975) Carbon. 13, p 159
3. Bokros JC (1969) In: Walker Jr PL (ed) Chemistry & Physics of Carbon. Vol 6, p 1
4. Murdie N, Ju CP, Don J, Wright MA (1993) Carbon-carbon matrix materials. In: Buckley JD, Edie DD (ed) Carbon-carbon materials and composites. Noyes Publications, p 105
5. Gebhardt JJ, Stover ER, Yodsnukis JJ (1972) Proc 4th National SAMPE Tech Conf Polo Alto

6. Ehrburger P, Lahaye J, Bourgeois C (1981) Carbon. Vol 19, p 1
7. Bauer DW, Kotlensky WV (1972) SAMPE Quarterly. Vol 4, p 24
8. Granoff B, Pierson HO, Schuster DM (1973) Carbon. Vol 11, p 177
9. Delhaes P, Trinquecoste M, Pecault A, Gsma J, Oberlin A (1984) Chimie Physique 81, p 809
10. Granoff B (1974) Carbon. Vol 12
11. Kimura S, Yasuda E, Takase N, Kayusa S (1980) High Temp. High Pressure, Vol 13, p 193
12. Oh SM, Lee JY (1988) Carbon. Vol 26, p 763
13. Vaidyaraman S, Lackey WJ, Agarwal PK, Freeman GB (1995) Carbon. Vol 33, p 1211
14. Bouratt X, Dupel P, Pailler R, Naslain R (1995) 23rd Bien. Carbon Conf, p 84
15. Forrest MA, Marsh H (1983) J Materials Science. Vol 18, p 973
16. Oberlien A (1984) Carbon. Vol 22, p 521
17. Manocha LM (1988) Composites. Vol 19, p 311
18. Fitzer E, Geigl KH, Huettener W (1980) Carbon. Vol 18, p 265
19. Manocha LM, Yasuda E, Tanabe Y, Kimura S (1988) Carbon. Vol 26, p 333
20. Manocha LM, Bahl OP, Singh YK (1989) Carbon. Vol 27, p 381
21. Appleyard S, Rand B, Carbon 92, Essen, p 718
22. Manocha LM, Bahl OP, Singh YK (1989) In: Jones FR (ed) Interfacial aspecta in composite materials, Butterworth (Pub), p 310
23. Gao F, Patrick JW, Walker A (1993) Carbon. Vol 31, pp 103
24. Jortener J (1976) Proc AMMRC MS 7602, p 81
25. Jortener J (1990) Carbon. Vol 30, pp 153
26. Savage G (1992) Carbon-carbon composites, Chapman & Hall (Pub)
27. Jortener J (1986) Carbon 24, pp 603
28. Perry JL, Adams DF (1976) Carbon 14, p 61
29. Groszezuk LB (1977) 13th Biennial Carbon Conf, p 70
30. McEnaney B, Mays T (1994) In: Thomas CR (ed) Essentials of Carbon-carbon composites. Royal Society of Chemistry publication, p 143
31. Fishbach DB (1971) In: Walker Jr PL (ed) Chemistry & Physics of carbon. Vol 7, 1
32. Kamiya K, Inagaki M, Mizutani M, Noda T (1968) Bull Chem Soc Japan 41, 2169
33. Inagaki M, Oberlin A, de Fonton S (1977) High Temp.-High Press 9, p 453
34. Reswig RD, Levinson LS, O'Rourke JA (1968) Carbon 6, 124
35. Krosonov A et al. (1969) Dok Akad, Nauk SSSP 185, 1316
36. Hishiyama Y, Ingaki M, Kimura S, Yamada S (1974) Carbon 12, 249
37. Yasuda E et al. (1995) In: Kyoto, Composites 95 Proc Japan US CCM-VII, p 159
38. Manocha LM (1994) Carbon. Vol 32, p 213
39. Murdie N et al. (1990) 13th Conf Metal, Carbon & Cer Matrix comp NASA CP 3054, p 349
40. Zaldivar RJ, Rellick GS (1991) Carbon. Vol 29, p 1155
41. Murdie N et al. (1991) Seventh MTC conf, p 81
42. Ju CP, Don J (1990) Material Characterisation. Vol 24, p 77
43. Kyotoni T, Sonobe N, Tomila A (1988) Nature 331, 331
44. Rellick GS, Chang DT, Zaldivar RJ (1992) J Mat Res 7, 2798
45. Rellick GS, Adams PF (1994) Carbon. Vol 32, p 127
46. Evangalides JS, Meyer RA, White JL, Zimmer JE (Sept 1976) Carbon Nose Tip Materials Testing - Microstructure & Fracture of Carbon Materials Rep TOR -0076 (6726-02) 6, Vol 1

47. Yasuda E, Tanabe Y, Manocha LM, Kimura S (1988) Carbon. Vol 26, p 225
48. Yasuda E, Ariyoshi J, Akatsu T, Tanabe Y (1995) Carbon 95, Japan-US CCM VII
49. Zimmer JE, White JL (1982) Adv in Liquid Crystal 5, 157
50. Zimmer JE, Waitz RL (1987) Disclinations and Fracture. Fourth Annual Conf Materials Technology Center, p 143
51. White JL, Shaeffer PM (1989) Carbon. Vol 27, p 697
52. Fitzer E, Huettener W, Manocha LM (1980) Carbon. Vol 18, p 291
53. Zimmer JE, Weitz RL (1988) Carbon 26, 579
54. Evangalides JS (1974) Microstructure & Fracture of C/C Composites TOR - 0075 (5626)-2. The Seripan Conf El Segu, California
55. Meyer RA, Gyetvay SR (1976) Carbon Carbon Composites matrix microstructure and its possible influence on physical properties. In: Petoleum Derivad carbon, p 381
56. Weisshaus H et al. (1993) Carbon 28, p 125
57. Klett JW, Edie DD (1995) Carbon. Vol 33, p 1485
58. Christ K, Huttinger KJ (1993) Carbon 31, p 731
59. Mochida I, Fujiura R, Korai Y (1992) Tanso No 155, p 398
60. Ogawa I, Sakai M, Inagaki M (1985) J materials Science. Vol 20, p 17
61. Inagaki M, Washiyama M, Goto K (1987) J Mat Sc. Vol 22, p 3315
62. Dillon F, Thomas KM, Marsh H (1993) Carbon. Vol 31, p 1337
63. Weizhou P, Tianyou P et al. (1988) Interfaces in Polyner, ceramic and metal matrix composites, Ishida H (ed) Elsevier Pub, p 399
64. Ju CP et al. (1991) J Mat Sc. Vol 26, p 6753
65. Fitzer E, Huttener W (1981) J Phy D Appl Physics 14, p 347
66. Pleger R, Brane W, Meyer RA (1993) Proc Biennial Carbon Conf, p 398
67. Band B (1993) in: Thomas CR (ed) Essentals of carbon-carbon composites
68. Hays D, Patrick JE, Walker A (1983) Fuel 62, 1145
69. Peebles Jr LH, Meyer RA, Jortener J (1988) Interfaces in Carbon - Carbon Composites. In: Ishida H (ed) Interfaces in polymer ceramic & metal matrix composites, p1
70. Cranmer JH, Plotsker TG, Peebles Jr LH, Uhlmann DR (1983) Carbon 21, p 201
71. Sohn KY et al. (1988) Carbon 26, p 157
72. Don J (1991) Seventh MTC conf, p 169
73. Pierson HO (1993) Carbon, Graphite, Diamond and Fullerenes. Noys Publication

6 Mechanical Properties of Carbon/Carbon Composites

6.1 Introduction

The vast interest in carbon/carbon composites is based on the unique mechanical and thermal properties of these materials, which in turn reflect the basic characteristics of carbon. Graphite, the crystalline form of carbon, can assume a variety of structures – fine to coarse mozaic, oriented, lamellar and columnar – with varying degrees of anisotropy. These structures can be recognized in bulk carbon and graphite materials even though reinforcing fibers also exhibit some of these structures. Since each structure is associated with characteristic properties [1], carbon/carbon composites with desired properties can be fabricated by choosing appropriate reinforcing fibers and controlling the microstructure of the matrix [2]. The available degree of freedom with respect to the fiber type and distribution and matrix microstructure enables the designer to select and obtain the proper type of materials that will withstand adverse environments.

Carbon/carbon composites, in the true sense, cover a very large range of materials. Therefore it is important from design point of view to know the behavior of several different systems under different test and operating conditions before selecting the materials for particular applications. Furthermore, it is necessary to insure a high degree of reproducibility. This is particularly essential for materials which are to be used under severe operating conditions. The properties which are often addressed in the literature are strength and stiffness, fracture toughness, fatigue and frictional properties, thermal conductivity, thermal expansion and resistance to ablation. The operating mechanisms for these properties are quite different. The mechanical bond strength and crack propagation mechanism determine the mechanical properties, whereas the thermal properties are governed by thermodynamic and thermal transport phenomena [1, 2]. Moreover, the properties of multiphase materials such as carbon/carbon composites are determined by the properties of their constituents, and their interrelationship and processing procedures. More so, because the constituents, both reinforcement and matrix are likely to undergo a change in properties during processing such as large shrinkage or expansion, processing-induced thermal stresses due to repeated thermal cycling, change in crystallinity etc. These secondary factors influence the composite properties in different ways. Therefore, in the following sections various properties of carbon/carbon composites are discussed individually.

The mechanical properties of advanced composites are generally related to the properties, volume fraction and direction of the fibers relative to the direction of the applied stress. Considering these factors, the mechanical properties of polymer matrix composites or in a broader sense, of ductile matrix composites are calculated by simple rule of mixture. However, in carbon/carbon composites, the matrix has a failure strain of less than 0.5 %, which is lower than that of the fibers. Consequently, theories of reinforcements which have been developed for commercial composites with high-strain matrix systems and strong fiber/matrix bonding do not apply to carbon/carbon composites. Moreover, matrix structure (both at mini and micro scales) and fiber/matrix bonding in these composites differ depending on the type of reinforcements and processing routes. Accordingly, different principles will prevail for the mechanical properties of carbon/carbon composites processed through various routes. A complete description of the scenario is beyond the current level of understanding due to lack of data required for it. Furthermore, the data concerning mechanical properties which are available in the literature are based on the constituents and the processing conditions adopted by individuals. Based on the of data available, some trends can be drawn. These are reviewed here in an attempt to illustrate the principles and factors which influence the mechanical properties of carbon/carbon composites and to optimize these in order to get high-performance composites.

6.2 Basic Concepts Governing Mechanical Properties of Carbon/Carbon Composites

6.2.1 The Young's Modulus of the Composites

Young's modulus of carbon/carbon bundles could be much greater than that calculated from the stiffness of the virgin fibers and their volume fractions [3]. This might, to some extent, be attributed to the change in crystallinity of the fibers due to process-induced thermal stresses and repeated thermal cycling [4]. However, the type of changes to the filament properties depend on the initial elastic modulus of the fibers and on the extent of interactions with the matrix phase. Low-modulus carbon fibers with a well-packed graphitizable matrix are expected to exhibit more changes than high-modulus carbon fibers [5]. The properties of various carbon matrix precursors also undergo substantial changes during initial thermal treatments, resulting in carbon matrixes with a significantly different microstructure which varies from amorphous to graphitic with random or parallel alignment [6].

Because of the great difference between Young's modulus of baked carbons (produced from the same matrix precursors as those used for making carbon/carbon composites) and that of the reinforcing carbon fibers, the contribution of the matrix to Young's modulus of carbon/carbon composites is usually neglected. However, a significant contribution of the matrix, especially of the interfilament region, to the modulus of the composites can be expected if the matrix is highly crystalline and well packed around the fibers with graphite planes oriented parallel to the fiber direction. This matrix structure can be achieved comparatively more easily by using pitch as matrix precursor and choosing the proper processing conditions (normal pressure carbonization) [4]. Matrixes in the interfilament region, with possibly large variance in microstructure, possess a range of mechanical properties. A tensile modulus of 3–40 GPa has been reported for structures having an amorphous or graphitic order with random alignment and 400–700 GPa for a highly developed sheath structure [6]. In a thermosetting-resin-derived matrix, graphitization of the matrix with lamellar-type microstructure oriented parallel to the fibers is achieved through stress graphitization, making additive contribution to Young's modulus of the composite. Figure 6.1 shows a comparision between Young's modulus of composites made with different matrix precursors after various final heat treatment temperatures [7]. As is evident from the figure the matrixes of all of the composites contribute considerably towards Young's modulus of the composites. Using a mesophase pitch and/or the intermediate stress graphitization technique, a value twice as high as that calculated from a rule of mixtures can be obtained [4, 6]. This preferred orientation of the matrix permits full utilization of the bonds in the matrix, which could never be achieved without the influence of the fibers [4]. In this case the matrix has an extremely well-aligned microstructure initiated by the fiber skeleton. The effects of the same matrix precursor can be further enhanced by using carbon fibers with a highly aligned graphitic structure (high-modulus

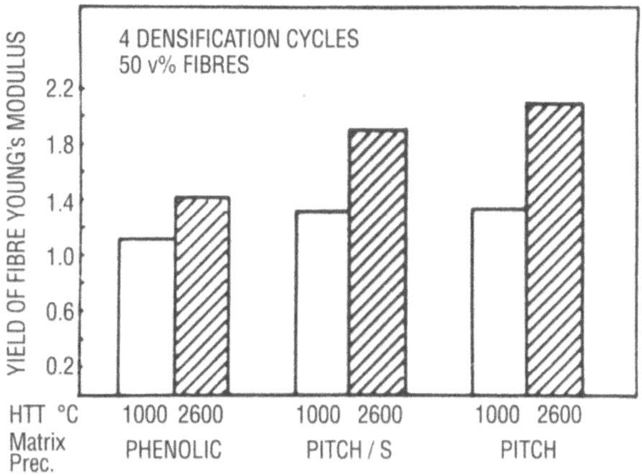

Fig. 6.1. Yield of fiber Young's modulus in UD carbon/carbon composites made with different matrix precursors and heat treated to 1000 and 2600 °C [7]

mesophase pitch carbon fibers). Although a stiffening factor of 1.5–2.0 can be achieved by controlling the microstructure of the matrix, even then the theoretical Young's modulus of carbon is utilized only by 30%–50%. Transversely oriented graphite structure (TOG), as discussed by Evangalides [8] is unlikely to contribute to the modulus of the composites.

6.2.2 Strength of the Composites

The prediction of the strength of brittle matrix composites is rather complicated since it is governed by the mechanisms responsible for crack initiation and crack propagation. Figure 6.2 [3] shows the theoretical stress/strain diagrams of fiber reinforced composites made from brittle fibers and different matrixes (ductile and brittle). In the case of ductile matrix composites, the crack is initiated from the fibers, obviously at very high stress levels, and propagates towards the matrix, which acts as a bridging medium. The strength of these composites is totally governed by the strength of the reinforcing fibers and their volume fraction in the direction of application of load. However, in brittle matrix composites, with the matrix having a lower failure strain than the fibers, crack initiation takes place in the matrix at much lower stress levels and fiber strength is not fully utilized. Therefore, in order to have maximum utilization of the fiber strength, the failure strain of the matrix must be increased. Alternatively, the decisive influence of the matrix on the fracture of the composite must be diminished, e.g., by modifying adhesion between the fibers and the matrix and/or by introducing energy-absorbing microcracks in the matrix.

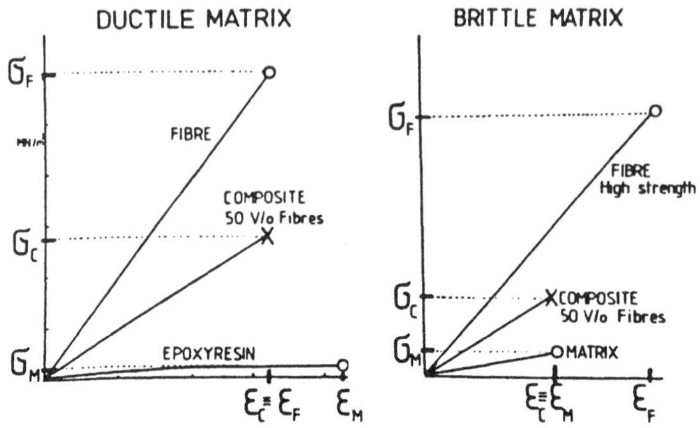

Fig. 6.2. Theoretical stress/strain diagrams of composites made with brittle fibers and different types of matrixes (ductile and brittle) [3]

The aim is to provide alternative mechanisms for energy absorption during fracture propagation. The fracture and strength of carbon/carbon composites is governed by the Cook–Gorden theory for strengthening of brittle solids [9] which states that if the ratio of the adhesive strength of the interface to the overall cohesive strength of the solid is in the right range, a large increase in the strength and toughness of otherwise brittle solids may result. In addition to this, there are many process related factors which may adversely effect fiber strength utilization in carbon/carbon composites [2–5, 10–12]. These include (i) process induced changes in the mechanical properties of the fibers either due to chemical attack on the fibers by matrix pyrolysis gases or due to high heat treatment temperatures; (ii) residual stresses in the fibers or matrix induced during processing due to thermal expansion mismatch between fibers and matrix; (iii) matrix microcracking either due to carbonization shrinkage or during cooling, giving rise to centers of stress concentration; (iv) stacking faults and matrix voids resulting in delaminations; (v) fiber/matrix debonding resulting in poor stress transfer capabilities.

Extensive research has been carried out to understand the mechanism of influence of these parameters on the strength of carbon/carbon composites with a view to develop technology for making carbon/carbon composites with maximum possible strength. Like with any composite material with a polymer matrix, in carbon/carbon composites too, the fiber layup architecture and size of the carbon fibers control the ultimate strength of

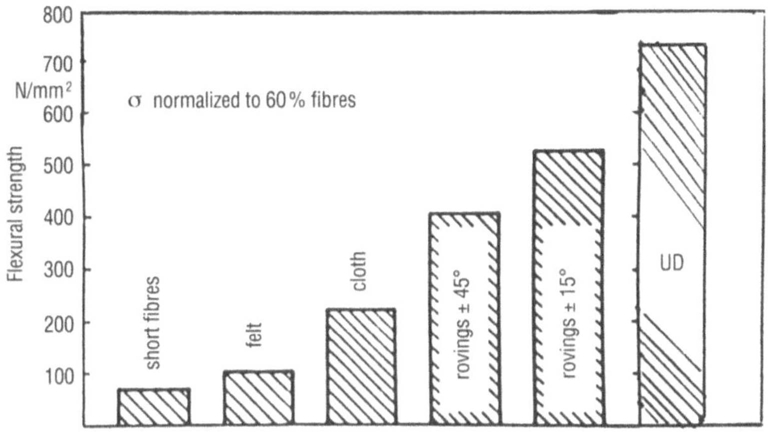

Fig. 6.3. Comparison of flexural strength of carbon/carbon composites with fiber architecture

the composites. This is seen in Fig. 6.3. Although various types of architecture alongwith multidirectional (MD) structures are used for making carbon/carbon products, unidirectional (UD) composites are often made for research purposes simply because it is easy to make them and to understand and correlate different mechanisms operating during fabrication and evaluation. Therefore, reports in the literature mainly deal with UD or two-directional (2D) composites.

6.3 Mechanical Properties of Composites with Pyrolytic Carbon Matrix

The effect of fiber content on the mechanical properties of UD carbon/carbon composites derived by chemical vapor deposition (CVD) is shown in Fig. 6.4. As is evident from this figure the strength of the composites decreases with a fiber volume content over 60%. This agrees well with the theoretical calculations based on the optimum interfilament distance and the matrix thickness between the filaments [6].

In carbon/carbon composites with pyrolytic carbon, tensile strengths in the fiber direction of 1000–1400 MN/m^2 have been reported which correspond to 80%–90% of the fiber properties' translation [13–15]. At lower heat treatment temperatures (HTT < 1500 °C), composites made with high-strength carbon fibers have been found to exhibit better strength than those made with high-modulus fibers while the opposite trend is observed

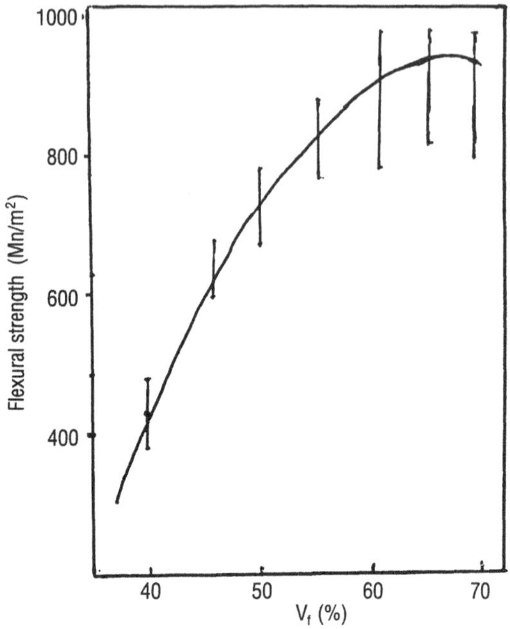

Fig. 6.4. Variation in flexural strength of CVD-derived UD carbon/carbon composites with fiber volume V_f contents

at HTT > 1500 °C. Any kind of chemical surface treatment of the fiber results in 10%–20% lower ultimate strength of the composite than those made with non-surface-treated fibers. This could be due to surface damage of the fibers caused by evolution of surface groups during heating of the fibers at infiltration temperature. On heat treatment of the composites to graphitization temperatures, composites made with high-modulus carbon fibers exhibit a significant increase in strength while those made with high-strength fibers loose strength.

The type of CVD technique chosen for composite fabrication has also been found to influence the mechanical properties. Composites with 50% fiber volume contents and made by the isothermal process are reported to exhibit much higher strengths than those made with the temperature or pressure gradient technique [14]. However, some authors prefer the latter for making high-performance carbon/carbon composites. This reveals the importance of matrix structure and fiber/matrix interactions for fracture and mechanical properties of carbon/carbon composites. In the absence of pores and macrovoids, smooth laminar (SL) carbons although brittle, exhibit higher compressive strengths than isotropic (ISO) or rough laminar (RL) carbons [16]. The fracture and ultimate mechanical properties are governed, in addition, by the fiber/matrix interface and the orientation of matrix microcracks. In composites having a relatively weak fiber/matrix interface, the crack propagates through the matrix and is deflected at the fiber/matrix interface, causing debonding between fiber and matrix. Such a situation causes the crack to deviate from the original paths. The propagation of the crack gets further modified by the presence of matrix microcracks. The matrix microcracks perpendicular to the crack propagation direction improve the properties, especially the strength and fracture toughness, whereas, those parallel to it serve as paths of rapid crack propagation.

CVD-derived ISO and SL carbon matrixes are strongly bonded to the reinforcing carbon fibers. As-deposited RL and SL carbons do not contain microcracks. However, after heat treatment of the composites, circumferential microcracks develop in the SL matrix as well as at the interface between RL and SL due to anisotropic thermal contraction [17]. Both the strong fiber/matrix bonding and microcracks in the matrix result in low-strength carbon/carbon composites. Because of this RL structures are preferred for making high-performance carbon/carbon composites. As-deposited RL composites possess a similar or somewhat higher flexural strength and modulus than SL composites. Upon heat treatment to graphitization temperature, SL matrixes exhibit a higher degree of graphitization with the others following in the order RL > ISO. With heat treatment temperature, there is only marginal change in properties of the latter composites. SL matrix composites exhibit a significant drop in strength and modulus, whereas ISO matrix composites show an increase in modulus [16–19]. The decrease in strength of CVD carbon/carbon composites following high temperature heat treatment is due to the formation of cracks and the annealing of matrix defects such as cross-links and misaligned layer planes which permit shear and basal plane slip [20]. Circumferential microcracks due to anisotropic thermal contraction of pyrolytic carbon deflect the path of the advancing crack parallel to the fibers and result in a terraced fracture in the matrix [19]. In all CVD carbon/carbon composites the matrix is always found to possess a combination of microstructures (SL+RL+ISO), the relative amount of each one depending on the processing conditions. Some authors have also observed columnar type microstructure with a similar effect as laminar structures [21]. A fracture in multiphase matrix systems will obviously be the combined effect of the constituents and interactions among these. Kimura et al. [20, 22] have studied the fracture behavior of CVD multiphase carbon/carbon composites. As seen in Fig. 6.5, composites with ISO + RC matrix fracture predominantly in tensile mode without fiber pullout. A crack initiated in RC carbon matrix propa-

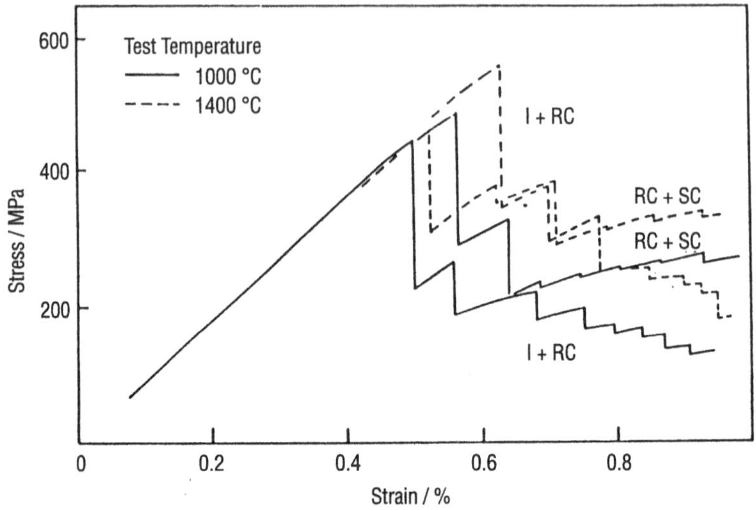

Fig. 6.5. Stress–strain curves of CVD-derived multiphase matrix carbon/carbon composites heat treated to 2000 °C [20]

gates perpendicular to the fiber axis without any deflection owing to the strong interface between RC and ISO structures as well as between the carbon fiber and ISO matrix. These composites also exhibit a high compressive strength and modulus. In contrast in case of RC + SC matrix composites, the crack encounters multiple deflections due to microcracks in the SC matrix. These composites possess a somewhat lower interlaminar shear strength but higher fracture energy.

The conclusions derived from unidirectional composites have been found to hold true also for CVD felt, 2D, and MD carbon/carbon composites [19, 23–27]. In MD composites, the relative fiber orientations in different directions also influence the properties of the composites in those directions. Process macrodefects, such as voids, closed pores etc. also contribute to the ultimate mechanical properties of the carbon/carbon composites. CVD composite materials are stronger (36 % in tension and 85 % in compression) than composites made with phenolic-based carbon matrixes. Table 6.1 summarizes the properties of some of these composites as described in the literature.

Table 6.1. Mechanical properties of various carbon/carbon composites

Composite system	Density (g/cc)	Volume fraction (%)	Flexural modulus (GPa)	Flexural strength (MPa)	Tensile modulus (GPa)	Tensile strength (MPa)
HM/resin	1.79	NA	NA	NA	106.0	349.0
T300/resin	NA	50.0	NA	200.0	NA	NA
Sigrafil/coal tar	1.4	NA	NS	180.0	NA	NA
Felt/coal tar	NA	NA	14.0	25.0	NA	NA
UD	1.45	NA	NA	NA	175.0	1350.0
MPF3/coal tar	NA	NA	80.0	450.0	NA	430.0
T300/PAA	1.623	8 0.0	58.0	145.0	NA	NA
8HSG/CVD	1.63	34.0	23.0	172.0	NA	NA
8HS PAN/CVD	1.55	34.0	10.0	104.0	NA	NA
8HS/pitch	1.77	35.0	15.0	160.0	NA	NA
8HS/phenol	1.50	35.0	52.0	240.0	NA	NA
8HS/HA43	1.56	35.0	72.0	300.0	NA	NA
3D CVD	1.52	44.0	NA	XY 96 Z 148	NA	NA
3D CVD/pitch	1.88	NA	XY 70 Z 148	XY 100 Z 330	NA	NA
3 D Resin/pitch	1.9	55.0	XY 70 Z 148	XY 192 Z 280	NA	NA
3D Resin/pitch	1.84	41.0	XY 35 Z 68	XY 55 Z 150	NA	NA
Felt/CVD/1100	1.8	16	NA	227	NA	NA
Felt/CVD/ 3000	1.88	16	16	88	NA	NA

MPF, mesophan pitch fibers; HS, high strength; HM, high modulus; UD, unidirectional; PAA, Polyarylacetylene; HSG; 8HSG, 8H satin graphitized; CVD, chemical vapor deposition; PAN, polyacrylonitrile; 3D, three-directional; NA, data not available.

6.4 Carbon/Carbon Composites Fabricated by the Liquid Infilteration Technique

6.4.1 Thermosetting-Resin-Derived Carbon/Carbon Composites

Thermosetting resins such as phenolic and polyfurfuryl alcohol are the most often used precursors for making carbon/carbon products for various structural applications. These resins are capable of making strong chemical bonds with the surface of carbon fiber, depending on their surface functionality. This has been dealt with to some extent in Chap. 4 and 5, wherein it was observed that shrinkage during carbonization of the polymer matrix composites is influenced considerably by the fiber/matrix bonding. Figure 6.6 [5] shows the flexural strength of UD composites made with different types of carbon fibers at different stages, from polymer to the final carbon/carbon composite. It can be seen from the figure that though the strength of polymer matrix composites made with high-strength carbon fibers is higher than of those made with high-modulus carbon fibers, opposite is true for carbon/carbon composites. Moreover, the strength of the composites made with high-modulus non-surface-treated carbon fibers increases with further densification of the composites, but it decreases for those made with high-strength carbon fibers. Similar effects are observed when carbon fibers of the same high modulus but with different surface functionalities are used as reinforcement [28–30]. This is evident from Fig. 6.7.

Contrasting differences are observed also in their fracture behaviors. Composites made with strong fiber/matrix bonding fail catastrophically with no fiber pullout (Figs. 6.8 and 6.9) while others fail in mixed tensile cum shear mode with fiber pullout depending on the extent of bonding [30–32]. However, upon further heating of the composites to graphitization temperature the strength of the two composites exhibit different

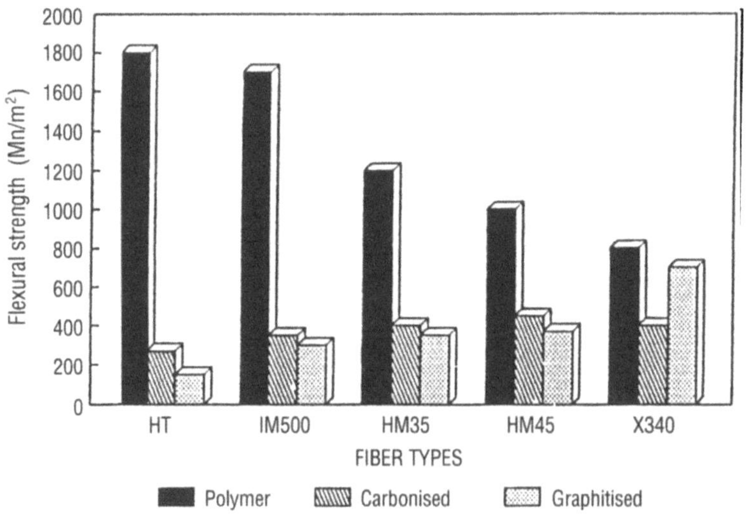

Fig. 6.6. Flexural strength of UD composites at different stages of processing (polymer to carbon) with different types of carbon fibers [5]

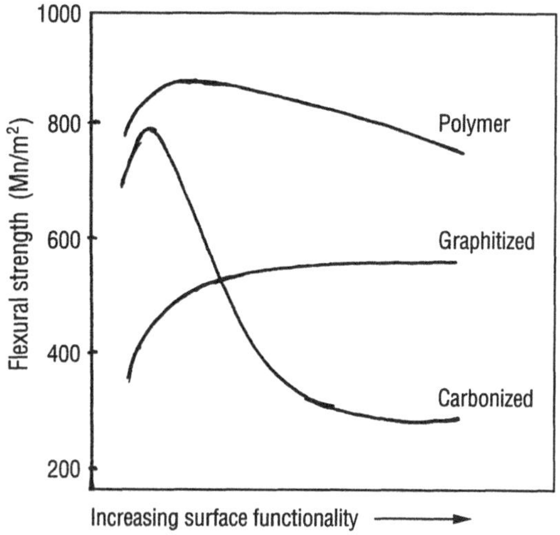

Fig. 6.7. Flexural strength of UD composites (polymeric to graphitic) made with high modulus carbon fibers of varying surface functionality

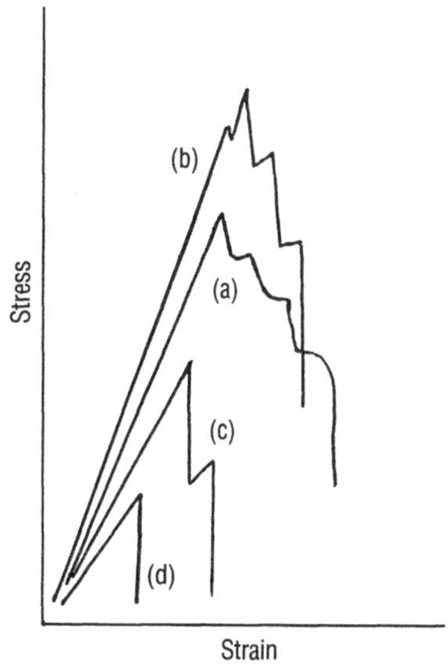

Fig. 6.8. Stress–strain curves of carbon/carbon composites (heat treatment temperature of 1000 °C) made with HM carbon fibers surface treated for (*a*) 0 min, (*b*) 15 min, (*c*) 60 min, and (*d*) 200 min

trends. Composites possessing strong fiber/matrix bonding and low strength at the carbonized stage exhibit higher strengths at the graphitized stage and vice versa. Based on the experience with different fiber/matrix systems, a general trend can be derivd as shown in Fig. 6.10 [32]. The fracture behavior of the composites also changes with heat treatment. This is evident from stress–strain diagrams (Fig. 6.11) and scanning electron micrographs of the fractured composites (Figs. 6.12 and 6.13). The change

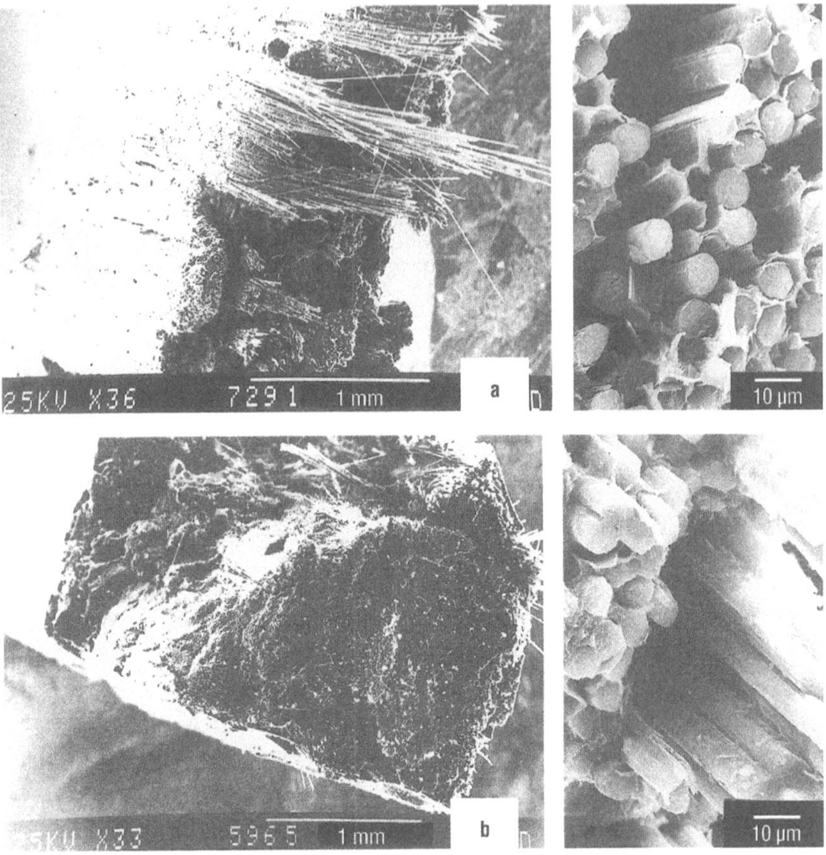

Fig. 6.9 a–d. Scanning electron micrographs of fractured carbon/carbon composites made with HM carbon fibers surface treated for **(a)** 0 min, **(b)** 15 min

in the strength pattern on heat treatment is a consequence of the change in failure modes [33, 34]. The failure modes prevailing in different temperature regions are shown schematically in Fig. 6.14 [33]. In carbonized composites, the matrix dominated-failure prevails. The shrinkage stresses produced in the carbonized composites due to inhibition of shrinkage of the precursor along the fiber axis result in numerous pyrolysis cracks in the composites. The size, shape, location, and geometry of these cracks depend on the fiber/matrix bonding at the polymer stage. Moreover, hindrance of the well-bonded matrix shrinkage in the direction of the fibers also leads to the build up of a high degree of compressive prestresses in the fibers and tensile stresses in the matrix. Such composites can be described as

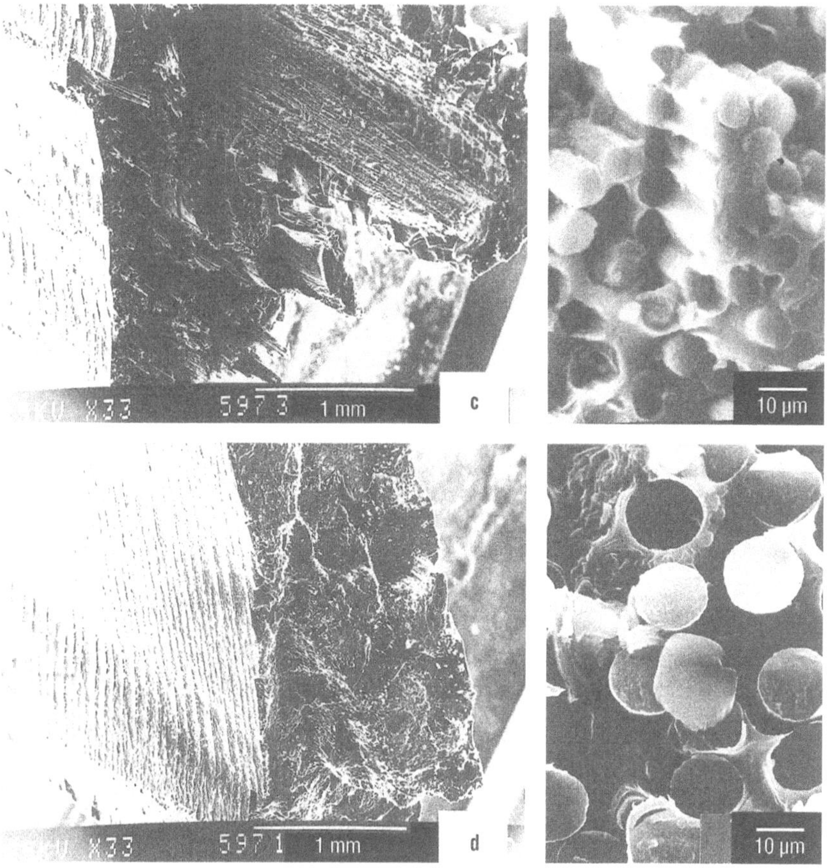

Fig. 6.9 (continued) (**c**) 60 min and (**d**) 200 min

precracked and prestressed material. On stressing the composites during testing, the preexisting matrix cracks, which are usually normal to the reinforcement direction, propagate towards the fiber/matrix interface. In the case of type I composites (weak fiber/matrix interface) the advancing crack is deflected at the interface and thus absorbs the fracture energy. If the matrix content is insufficient and the fiber/matrix bonding is too weak, the stress distribution amongst the fibers through shear stresses at the interface will not take place. Such composites will fail at low stress levels with long fiber pullout. This is the case with carbonized composites made with non-surface-treated high-modulus carbon fibers. In the case of composites with high density and intermediate bonding, the stress distribution at the

Fig. 6.10. Influence of heat treatment temperature on flexural strength of carbon/carbon composites having (*A*) weak fiber/matrix bonding and (*B*) strong fiber/matrix bonding at the polymer stage [32]

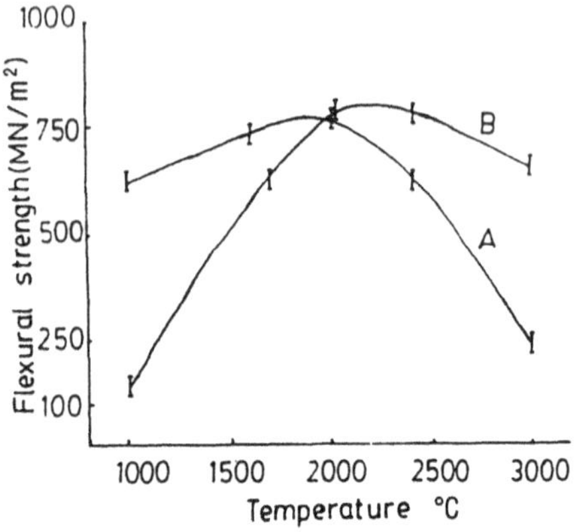

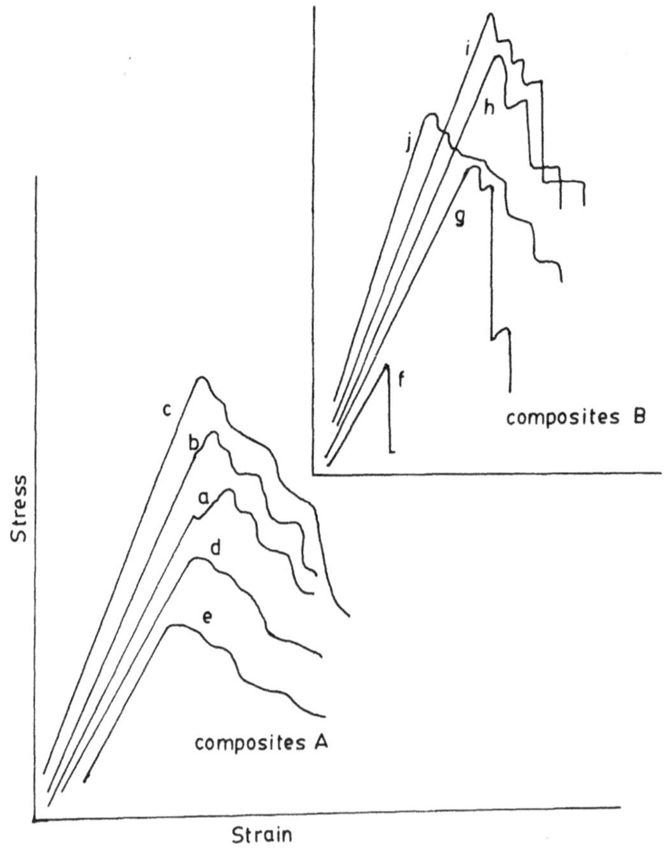

Fig. 6.11. Stress–strain curves of carbon/carbon composites with (**A**) weak and (**B**) strong fiber/matrix interactions at the polymer stage, heat treated at different temperatures: *a* and *f*, 1000 °C; *b* and *g*, 1600 °C; *c* and *h*, 2000 °C; *d* and *i*, 2400 °C; *e* and *j*, 3000 °C [32]

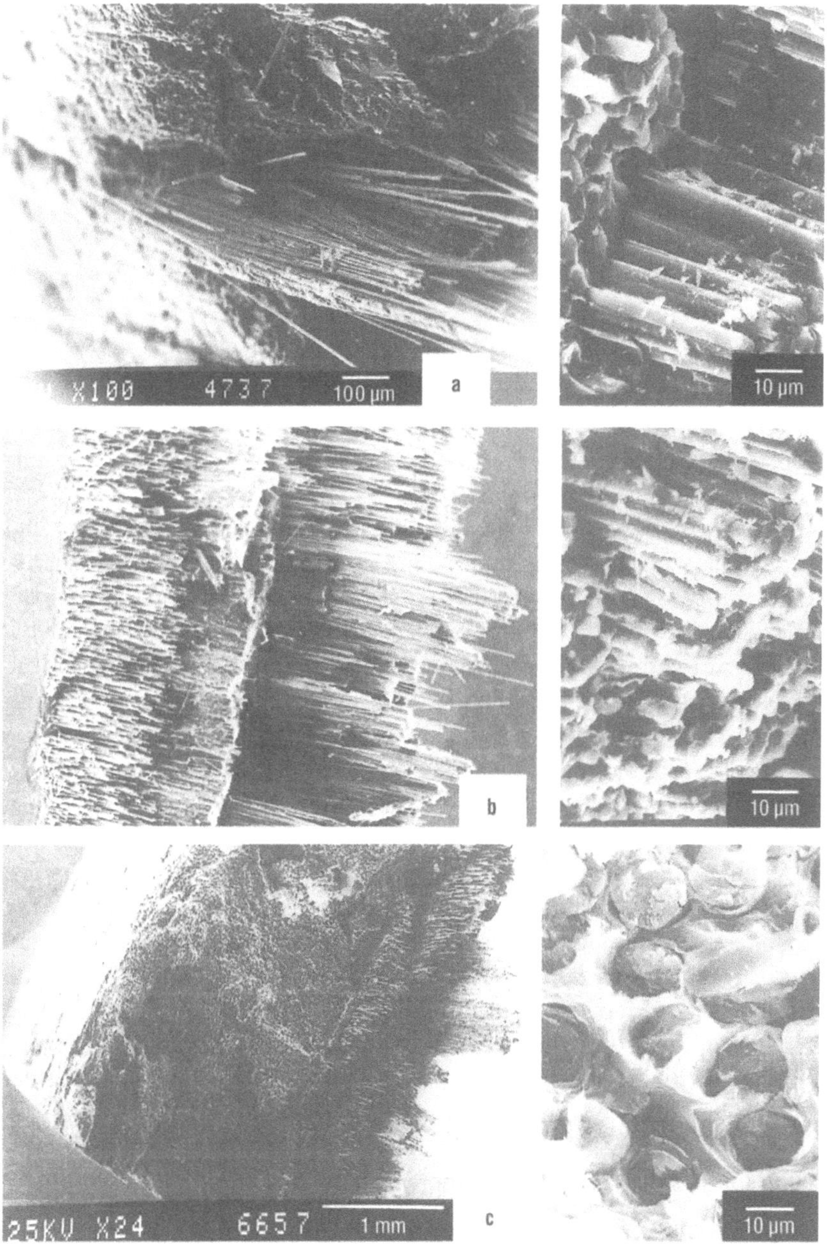

Fig. 6.12 a–d. Scanning electron micrographs of fractured carbon/carbon composites (A in Fig. 6.11) heat treated to different temperatures: (**a**) – (**d**) long fiber pullout with unorganized matrix around carbon fibers changing to mixed made failure with oriented matrix closely packet around the fibers

Fig. 6.12 (continued)

interface takes places through shear stresses. Such composites fail at high stress levels with short fiber pullout and exhibit a mixed mode failure. Densified composites made with non-surface-treated high-modulus carbon fibers and those made with low surface energetic carbon fibers behave in this way. In composites with very strong fiber/matrix bonding, no fracture energy is absorbed at all and the intramatrix cracks perpendicular to the fiber axis advance through the fibers. This results in catastrophic failure of the composites at a strain which is characteristic of the matrix and is far below the ultimate strain of the fibers. The fibers do not share the load [32–36].

In the temperature range of 1200 °C–2200 °C, mixed mode failure prevails. The heat treatment in this temperature range weakens the fiber/matrix interface resulting in crack blunting. This continues until an optimum level is reached where the interfacial bond strength is weak enough for deflection of the crack and strong enough to effectively utilize the stress transfer capabilities of the matrix. As a net result, the composites fail in mixed mode with low fiber pullout. On further heating to 2500 °C and above, the graphitization of the interfilament matrix increases. The graphitized matrix has a low shear strength. The composites fail in a jagged fashion. As the graphitization of the matrix increases, the intramatrix cohesive energy decreases. The fiber/matrix interfacial strength also becomes very low. A crack initiating from the matrix propagates fast up to the fiber in a jagged manner where it is deflected along the fiber direction, decoupling the matrix from the fiber. The failure is dominated by the fiber failure. Therefore, the matrix contributes least to the composite strength. The failure of the

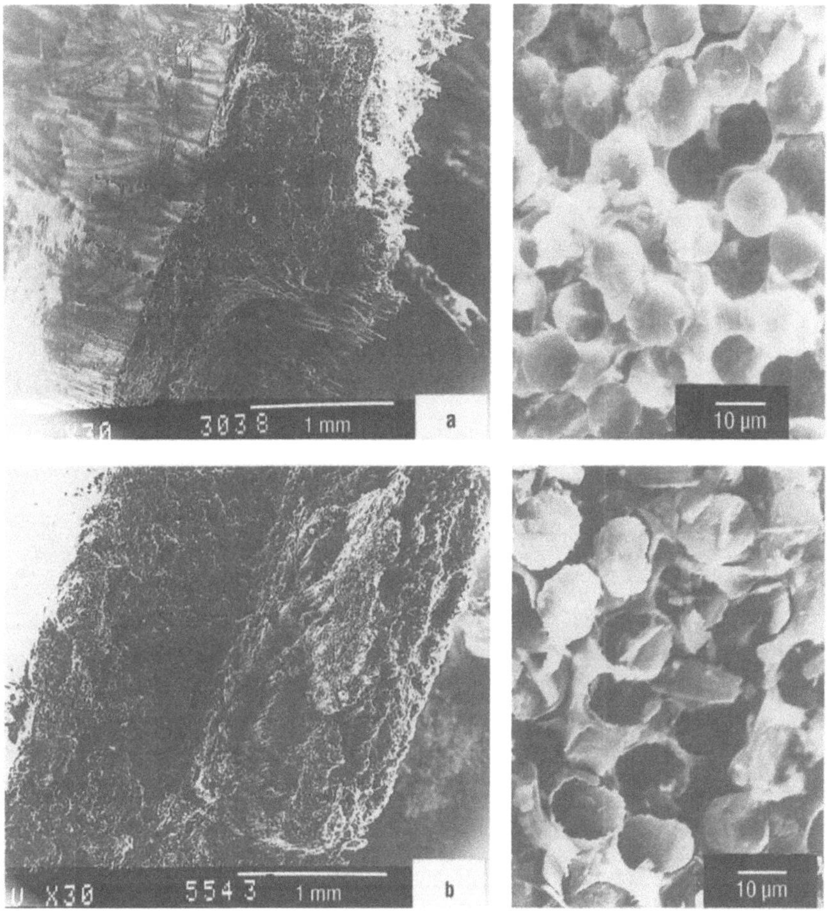

Fig. 6.13 a–d. Scanning electron micrographs of fractured carbon/carbon composites (**B** in Fig. 6.11) heat treated to different temperatures: (**a**) – (**d**) Brittle fracture of the composites with strong fiber matrix bonding changing to shear mode failure, oriented matrix closely packet around the fibers

composites is dominated once again by the failure mode of the matrix, i.e., predominantly by delamination [35]. The decisive influence of fiber/matrix bonding on fracture behavior of carbon/carbon composites not only prevails for PAN-based carbon fibers but is also reported for pitch based carbon fibers [37]. However, in pitch-carbon-fiber-based composites, the fracture behavior of the carbon fibers, which is characteristic of the graphitic plane orientation and hence the modulus of the fiber, also contributes to the overall fracture of the composites [38].

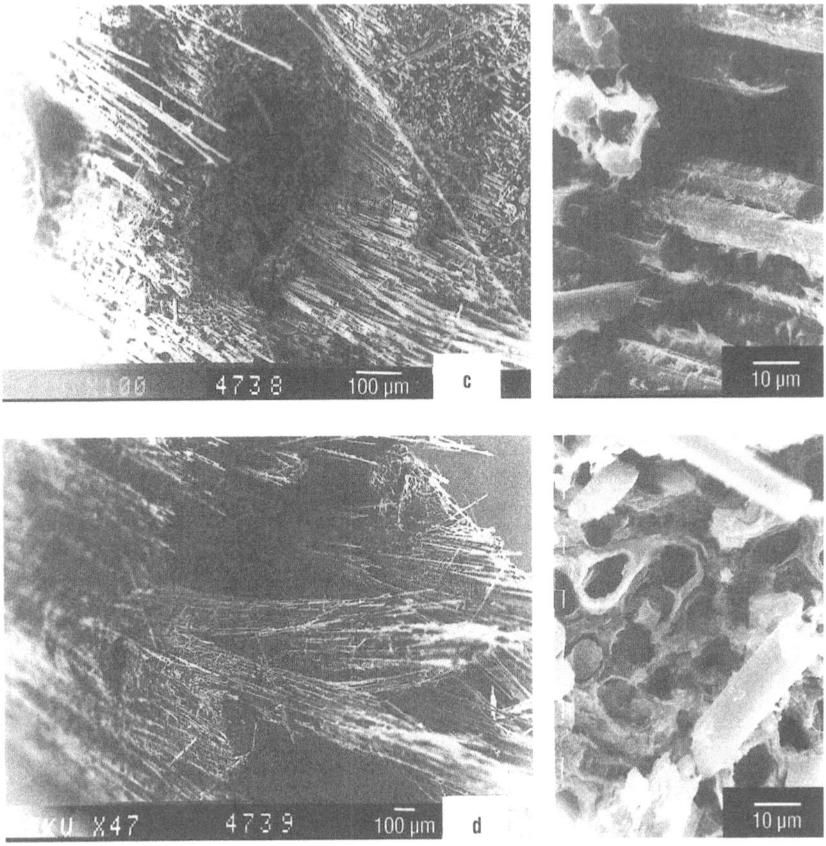

Fig. 6.13 (continued)

The shrinkage of the matrix and, hence, the concentration of cracks can be diminished and controlled by the addition of fine powder (1 – 2 µm) fillers, such as cokes or graphite powder, to the resin. Addition of about 10% graphite powder significantly decreases the cracks and results in an increase in the ultimate strength of the composites (Fig. 6.15) [39]. In addition to decreasing in the number of pyrolysis cracks in the matrix, the graphite filler also increases the failure strain as well as the fracture toughness of the composites. However, one must ensure uniform distribution of the filler in the matrix in interfilament and interbundle regions so that the binder bridges formed by the matrix are not weakened.

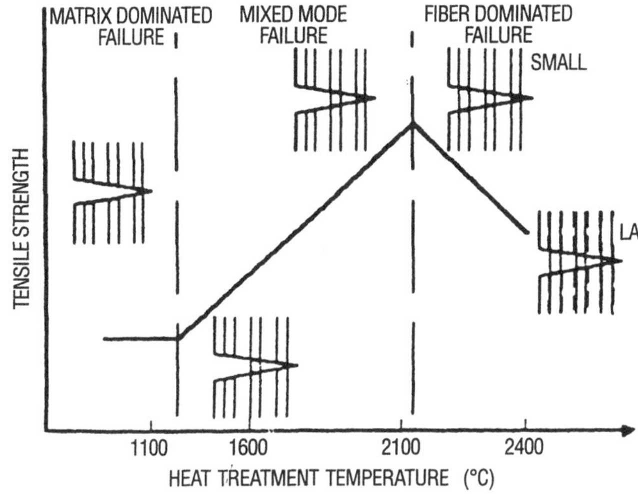

Fig. 6.14. Fracture modes vs heat treatment temperature [33]

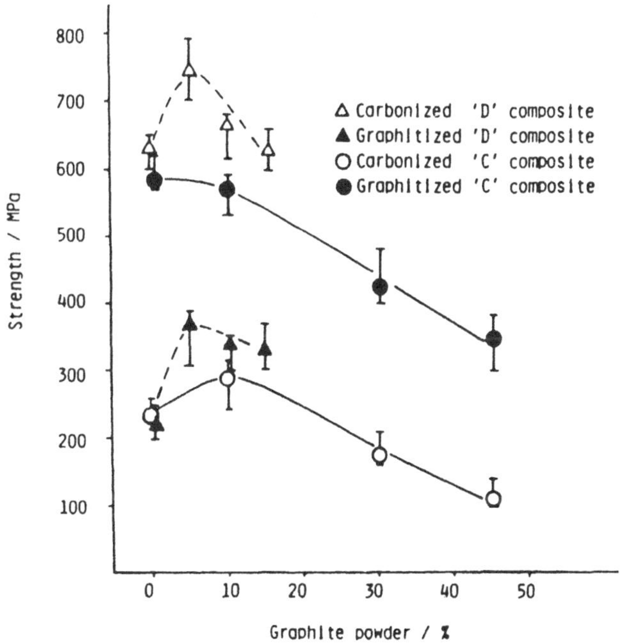

Fig. 6.15. Increase in strength of resin-derived carbon/carbon composites by graphite powder addition; *C* surface treated and *D* non-surface-treated HM carbon fibers [39]

6.4.2 Pitch-Derived Carbon/Carbon Composites

In the case of pitch-precursor-based carbon/carbon composites, chemical bonding between the carbon fiber surface and matrix plays a similar role as in the case of thermosetting-resin-derived composites [28]. In addition, the porosity, which usually develops during carbonization of thermoplastic pitches, is the major strength-controlling factor in these composites. Therefore, all of the processing variables, such as heating rate, gas pressure during baking, sulfur or graphite powder addition, densification through reimpregnation/carbonization cycles, which lead to high density carbon/carbon composites also result in an improvement in flexural strength. Another important feature of pitch as matrix precursor for carbon/carbon composites is the development of connected pores during the first carbonization, which can be easily filled up on subsequent impregnations. As with all polycrystalline graphites, densification cycles of pitch-bonded carbon/carbon composites result in a rapid decrease in porosity. The decrease in porosity results in an increase in flexural strength as well as shear strength of the composites. A three- to fourfold improvement in flexural strength of pitch-based carbon/carbon composites has been observed after three to four densification cycles, i. e., with a decrease in porosity from 25 % to 8 % (Fig. 6.16). However, the maximum impact strength is found in composites with a porosity of about 20 %.

As discussed in Chap. 4 the processing time for pitch-based carbon/carbon composites can be decreased by using thermally stabilized pitch. This method not only reduces the processing time but also has a significant effect on the strength and fracture behavior of carbon/carbon composites. Christ and Huttinger [40] studied the effect of fiber surface treatment on the strength and fracture of carbon/carbon composites made with stabilized pitch. They also observed that composites made with non-surface-treated fiber generally fail by delamination. The use of stabilized matrix results in an enhancement of the strength. The opposite results are obtained with composites made with strongly surface treated fiber and a stabilized matrix. In this case, the composites fail by pure brittle fracture at an extremely low strength. These results are similar to those obtained with thermosetting resins as matrix precursor. On heating to a temperature of greater than 2000 °C, the composites made with surface-treated fibers and stabilized pitch exhibit a much higher strength (almost twice) than to those made with pitch without stabilization (Fig. 6.17) [40]. It is well known in graphite technology that densification by impregnation is more effective

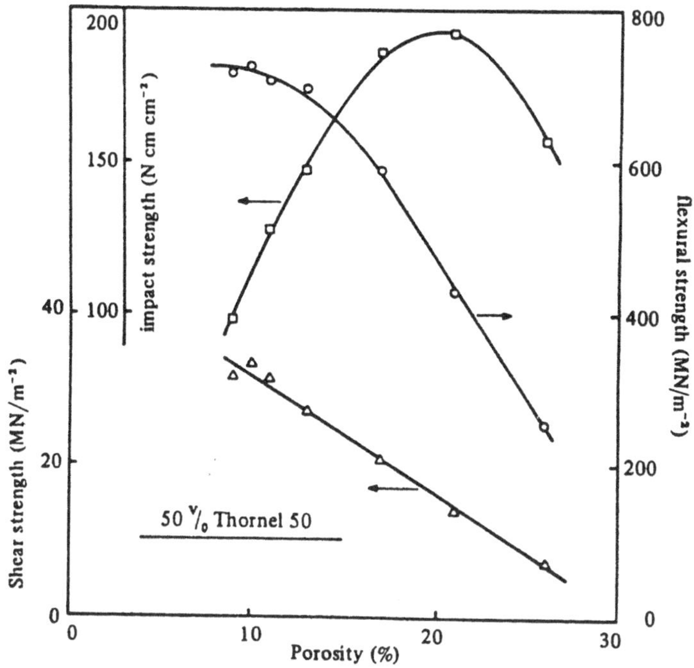

Fig. 6.16. Mechanical properties of pitch bonded carbon/carbon composites as a function of porosity

after intermediate graphitization. Therefore, if carbon/carbon composites are subjected to graphitization temperature in between the impregnation cycles, not only the density but also the flexural strength of the resulting composites is improved. Figure 6.18 [4] compares the density and flexural strength of composties which have undergone a single graphitization after repeated densification with that of composites which have undergone repeated intermediate graphitization after each impregnation.

Young's modulus of carbon fibers is fully utilized in pitch-derived carbon/carbon composites. A well-oriented matrix contributes as much as the fibers towards the modulus of the composites. Therefore, one observes a large stiffening factor of 1.5 to 2.00 (Fig. 6.19) [4] in pitch derived carbon/carbon composites. Even then, only 30%–50% of the theoretical Young's modulus of carbon is utilized in the carbon/carbon composites. Although the carbon matrix does not contribute to an improvement in the strength of carbon/carbon composites, it does control the fracture behavior decisively and enhances the final fiber strengthening factor. A 100% fiber-strengthening factor can be achieved if the failure strain of the composite

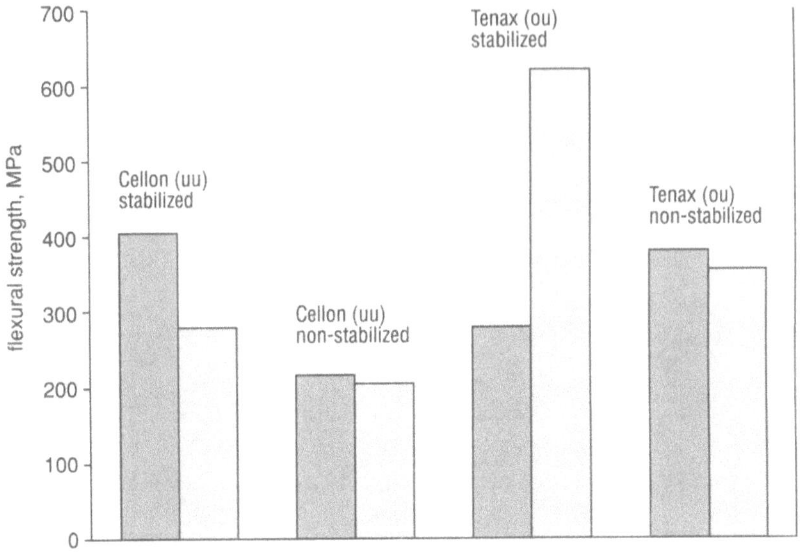

Fig. 6.17. Comparison of the flexural strength of carbon/carbon composites made with different carbon fibers and pitch without and after stabilization. The composites were heat treated to 1000 °C (*criss-cross columns*) and 2000 °C (*hatched columns*) [40]

is increased to that of the fiber. A further criterion for achieving the best mechanical properties is the selection of the most suitable precursor materials and fabrication processes. In order to improve mechanical interlocking by pore filling carbon, the precursor must transform into a carbon with accessible pores and should also have sufficient yield and strength.

General conclusions regarding the influence of fiber/matrix adhesion on fracture behavior of the composites drawn from the unidirectionally reinforced model carbon/carbon composites are also applicable to two directionally and multidirectionally reinforced carbon/carbon composites [41–43]. However, with respect to carbonization shrinkage and the stresses which it generates, the additional fiber orientation directions change these characteristics from anisotropic to quasitropic and isotropic. In carbon-cloth-reinforced composites the weave pattern, fiber crimp angle, broken filaments, and the stacking sequences introduce additional parameters. Crimping is also a source of weakness.

In woven cloth composites, stresses applied in the plane of the laminate will induce stress component perpendicular to the plane of the fiber axis. Crimping of the fibers coupled with poor transverse properties of the fibers

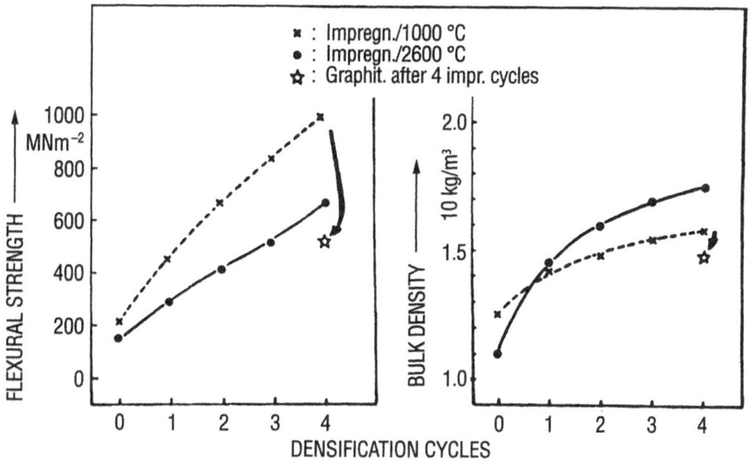

Fig. 6.18. Variation of bulk density and flexural strength of carbon/carbon composites with densification cycles. The composites were subjected to graphitization heat treatment either after each densification cycle or after complete densification [4]

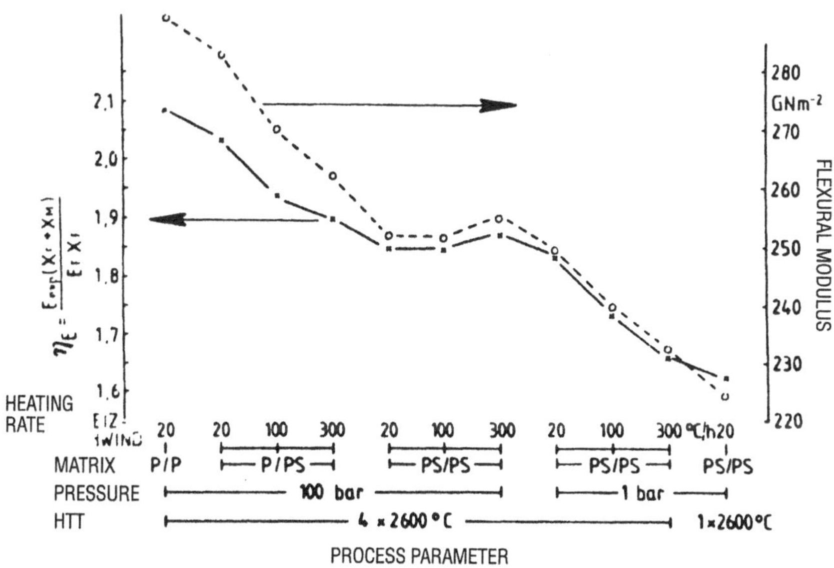

Fig. 6.19. Young's modulus and apparent translation of Young's modulus as influenced by process conditions of pitch bonded carbon/carbon composites

are, therefore, additional parameters influencing crack propapation and mechanical properties of 2D carbon/carbon composites. Two main aspects of crimping are (i) the frequency of crimping, i.e, the number of crossovers, which depends on the type of weave pattern and (ii) the crimp angle, which additionally depends on the tow size [42–44]. Figure 6.20 shows a typical example of the effect of the weave pattern on the flexural properties of 2D carbon/carbon composites [42]. Composites made with 8H satin cloth exhibit high flexural strength with mixed mode fracture, whereas those made with plain weave cloth possess low strength and brittle fracture (Figs. 6.21 and 6.22). This is obvious since the former weave pattern contains less fiber crossovers, low crimp angles, and more void pockets underneath the bundle crossovers than plain weave carbon cloth. Additionally, the layup sequence may introduce additional porosity.

The process variables and the resulting additional interfaces and microstructural defects and voids influence the mechanical properties of the ultimate 2D and MD carbon/carbon composites. If the layup is not a balanced weave, the testing in warp and fill direction also exhibit a wide difference

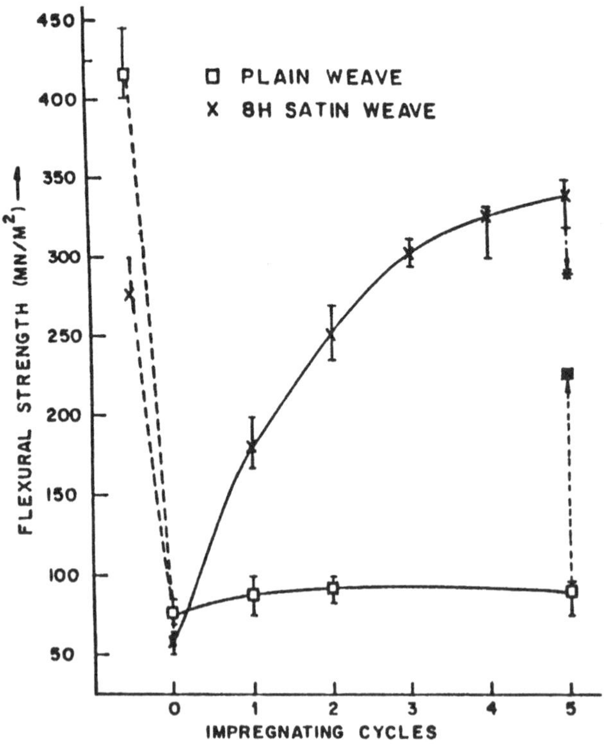

Fig. 6.20. Variation of flexural strength with the number of impregnation cycles of carbon/carbon composites made with different PAN based graphitized cloths

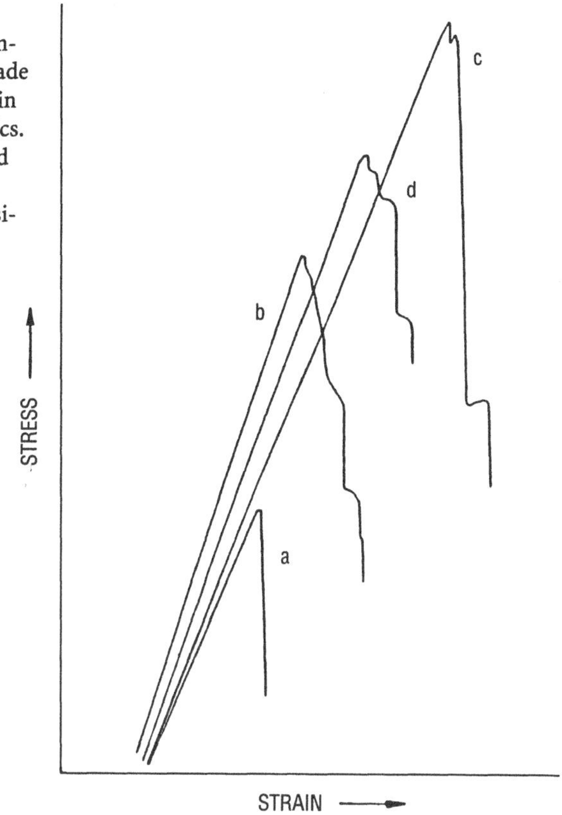

Fig. 6.21. Stress–strain diagrams of carbon-carbon composites at different stages made with plain weave and 8H satin PAN based graphitized fabrics. Densified carbonized (*a*) and graphitized (*b*) composites made with plain fabric; densified carbonized (*c*) and graphitized (*d*) composites made with 8H satin fabric

in mechanical properties (Fig. 6.23) [6, 45, 46]. The influence is expected to be greatest on the fatigue and creep properties. Studies by Anand and Gupta [47] on failure mechanism in 2D composites under compressive and shear loading also show that crimp angles associated with the weave patterns are significant in defining the plane of shear failure. However, they did not observe any deformation in the initial microcracks or voids adjacent to the terminal shear crack formed at the peak stress. The initial crack density is found to be important for compressive strength. Under uniaxial compressive testing, a shear band is formed across the specimen diagonal in low-crack-density (LCD) composites whereas in high-crack-density (HCD) materials the failure is found to be predominantly through interply delamination. In general, good fiber/matrix bonding is desirable in 2D carbon/carbon composites which will obviously have a positive influence on interply bonding and mechanical performance of the composites under

Fig. 6.22 a–f. Scanning electron micrographs of fractured 2D carbon/carbon composites made with plain fabric reinforcement (**a, c, e**) and 8H satin reinforcement (**b, d, f**). The composites were either pyrolyzed (**a, b**), densified (**c, d**), or graphitized (**e, f**)

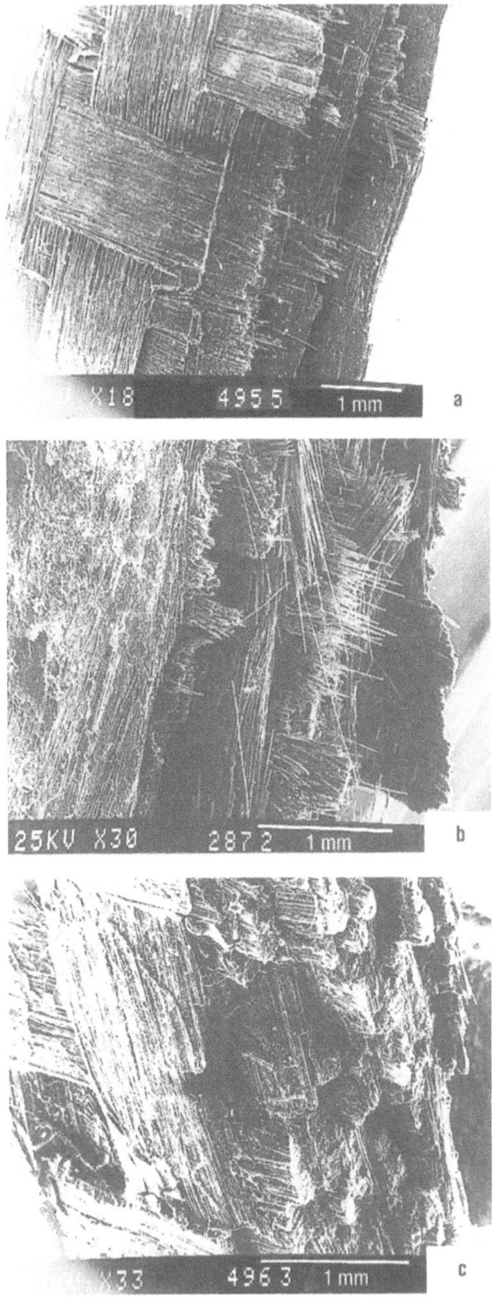

Fig. 6.22 d–f (continued)

Fig. 6.23 a–c. Flexural strength of 2D carbon/carbon composites parallel and perpendicular to lamination as a function of fiber content (*a*) s parallel, (*b*) o perpendicular to lamination

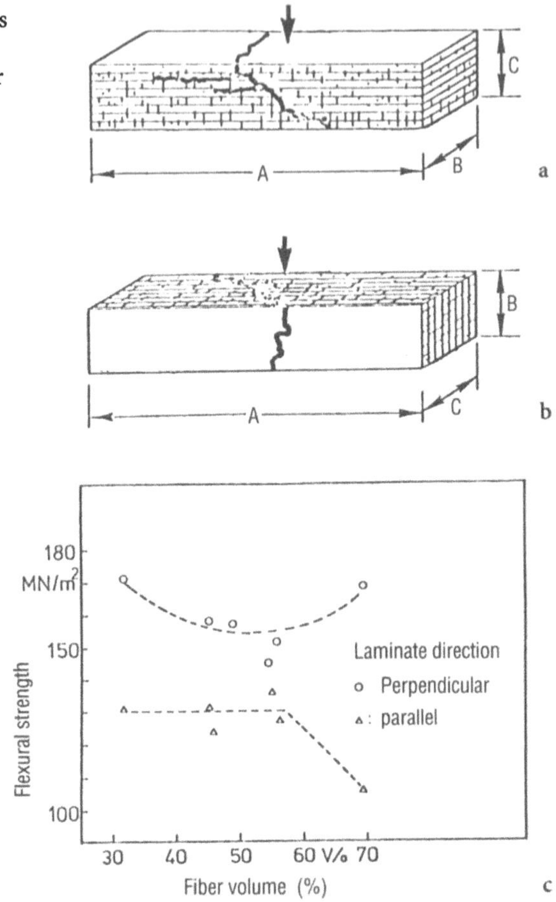

tensile, compressive, and shear loading. Three-directional carbon/carbon composites contain a repeating volume of elements consisting of straight fiber bundles in three directions and matrix pockets. The failure of a fiber bundle involves a process of debonding, breaking, and pullout. The fracture of a 3D carbon/carbon specimen is the result of creation, extension, and cumulation of fiber bundle failure under external load [48]. As can be seen from the stress–strain diagram in Fig. 6.24.

Multidirectional carbon/carbon composites exhibit pseudoplastic fracture behavior at room temperature with strains up to 5% which is characteristically different to that of UD and 2D reinforced carbon/carbon composites. This difference is attributed to the presence of reinforcing fibers in different directions and the continuous crack structure of the matrix whose

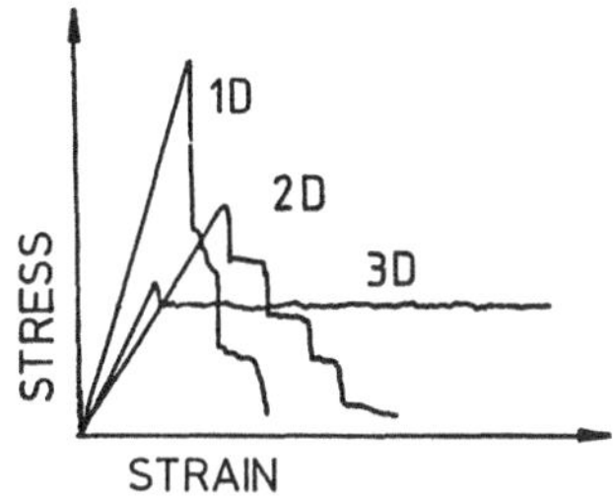

Fig. 6.24. Stress–strain curves of UD, 2D, and 3D carbon/carbon composites [2]

geometry is dependent on the fiber array. However, this also limits the ultimate mechanical properties of the composites in a particular direction due to restricted fiber volume fraction [25, 49, 50]. According to reports in the literature the scatter in mechanical properties of carbon/carbon composites is quite large.

6.4.3 Angle-Plied Composites

Angle-plied composites are formulated to compensate for the anisotropic matrix shrinkage and thermal expansion of composites. Figure 6.25 shows the effect of fiber orientation angle on longitudinal and transverse strength of angle plied carbon/carbon composites for resin- and CVD-derived carbon matrixes. The results agree well with Tsai's curve. The transverse strength of the composites increases with $\pm 10^{\circ}$ fiber orientation; however, the increase is very small compared with the decrease in longitudinal strength. It is important to consider these results when designing cylindrically woven carbon/carbon composites.

6.4.4 Sandwich Composites

Sandwich composites are mix form composites made of predetermined alternating layers of unidirectional and felt- or cloth-based prepregs. Figure 6.26 shows examples of such sandwich composites. The purpose of making sandwich composites is to increase the transverse strength of carbon/carbon composites while retaining as much axial strength as possible. An example of this is shown in Fig. 6.27. The exact strength values, however, will depend on the type of reinforcement used and the ordering adopted for making the sandwich.

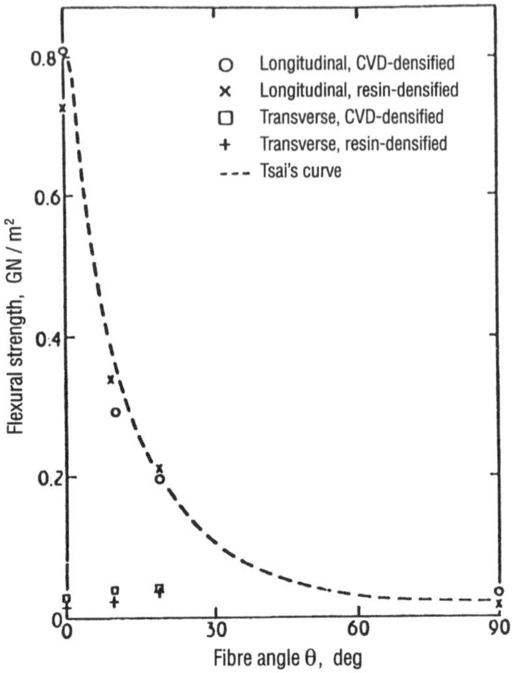

Fig. 6.25. Fiber orientation dependence of angle plied carbon/carbon composites

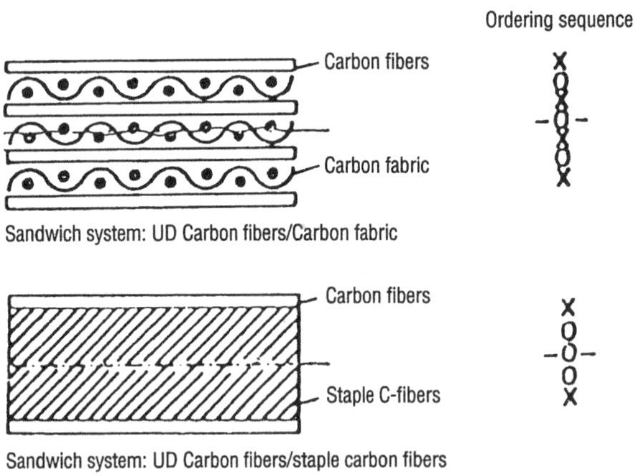

Fig. 6.26. Fiber ordering in sandwiched composites

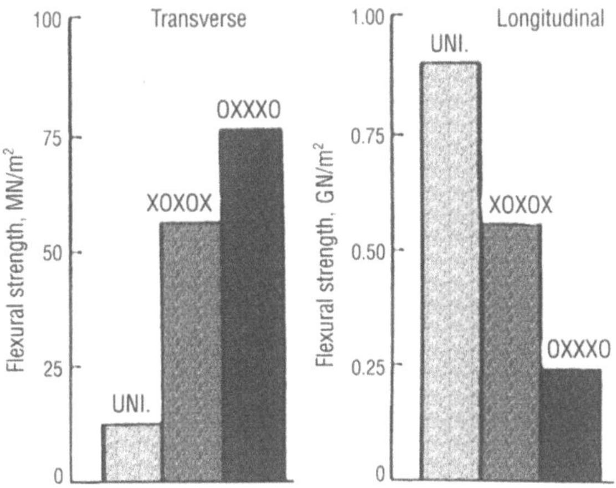

Fig. 6.27. Flexural strength of sandwiched composites in two directions

6.5
Fracture Toughness of Carbon/Carbon Composites

In addition to strength and stiffness, another property of considerable importance from a structural point of view is toughness, i.e., resistance to crack propagation. Carbon as such, both in bulk as well as in fibrous form, is brittle with a fracture toughness (energy required to propagate a crack) in the range of 0–200 J/m^2. However, if carbon fibers and carbon matrix are combined to form composite materials with optimum fiber/matrix bonding, the composites will exhibit a higher toughness and work of fracture than carbon in monolithic form. The pseudoplastic fracture behavior of these materials is demonstrated by hammering a nail through the composite, as shown in Fig. 6.28. The work of fracture varies widely and depends on the fiber layup, matrix precursor, and fiber/matrix bonding.

Table 6.2. Fracture toughness of some carbon/carbon composites

Materialsystem	V_f %	Sample size (mm)	K (Nm m$^{3/2}$)
Felt/pitch	25	1.8 × 5.0	83 ± 8
Felt/pitch	33	1.8 × 5.0	100 ± 8
Felt/CVD	9	9.5 × 9.5	34 ± 4
UD/pitch	50	4.8 × 4.8	15 ± 2
Fabric/pitch	35	2 × 10	15 ± 1
Angle ply/CVD	NA	NA	20 ± 3

V_f, fiber volume; K_{ic}, fracture toughness; see Table 6.1 for other abbreviations.

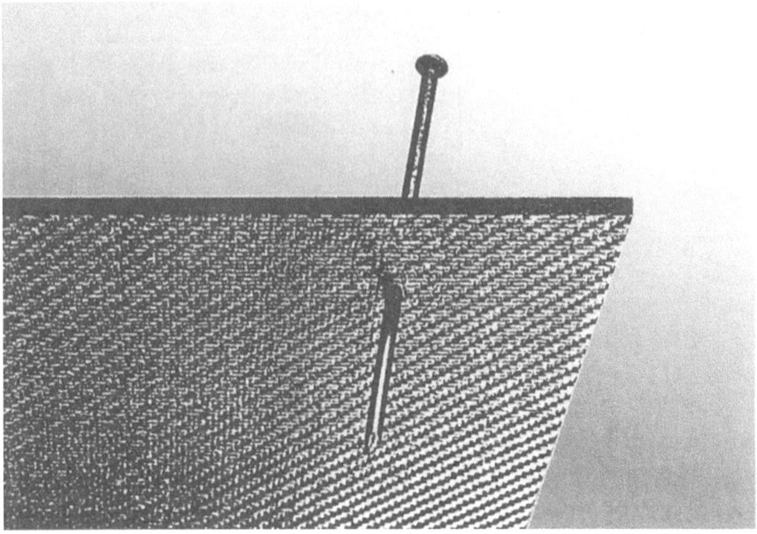

Fig. 6.28. Example of excellent fracture toughness of carbon/carbon composites

The micromechanical processes which impart toughness to carbon/carbon composites include matrix cracking, fiber/matrix debonding, fiber fracture, stress shielding, frictional forces at fiber/matrix interface, bridging effects, and fiber pullout. In fact, the factors discussed above which control strength and crack propagation in composites have a similar bearing on the fracture toughness of composites. Therefore, the 3D orthogonal carbon fiber woven composites exhibit the highest work of fracture. Carbon/carbon composites made using carbon felt exhibit the lowest values for work of fracture, while those made of random fibers or 2D cloth exhibit intermediate values [51–59]. Table 6.2 gives representative values for the work of fracture of different carbon/carbon composites. These are about three orders of magnitude greater than those of glass, glassy carbon, and graphite.

6.6 Fatigue and Creep Properties

Carbon/carbon composites exhibit good fatigue and creep resistance. Figure 6.29 [7] presents the results of torsional and flexural tests in which the fatigue properties of carbon-fiber-reinforced carbon are compared with those of carbon-fiber-reinforced polymers.

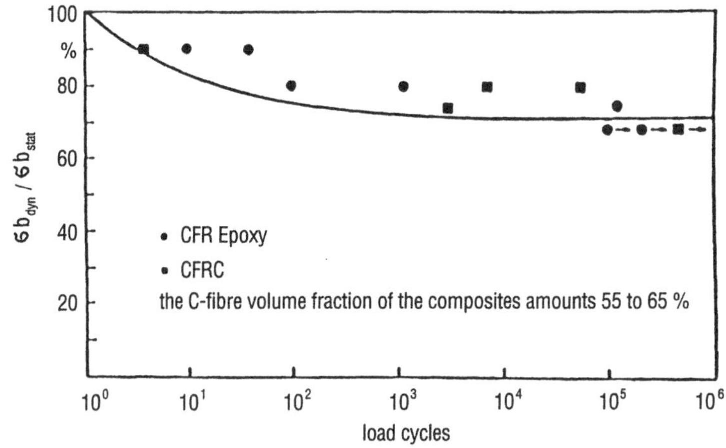

Fig. 6.29. Fatigue characteristics of carbon/carbon composites (dynamic/static loading) [7]

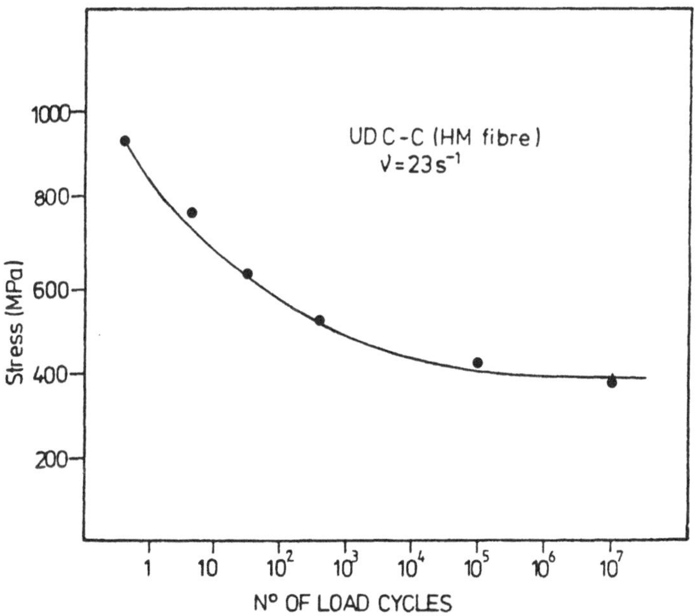

Fig. 6.30. Fatigue characteristic of carbon/carbon composites (alternating flexural loading) [57]

The fatigue loading of carbon/carbon composites is accompanied by opening/closing of slit pores [54]. In fact, slit pore volume is increased during extended cycling and it is probable that in local areas the fracture strength for the matrix will be surpassed. The consequence is the formation of small particles. With the proceeding of the local fracturing process, the matrix can be lost as dust from carbon/carbon composites [2]. Figure 6.30 [57] shows a Wohler curve for unidirectional carbon/carbon composites for sinus type alternating bending tests. It is evident that at a strain of 400 N/mm^2, i. e., 40 % of the static flexural strength, the material can withstand more than 10^7 stress cycles. This amount of stress is more or less in agreement with the critical stress, determined in a static flexural test, above which cracking of the matrix will occur.

6.7 Mechanical Properties at High Temperature

Carbon in its graphite form is structurally stable even at temperatures above 3000 °C in a non-oxidizing atmosphere. This material maintains its mechanical properties at much higher temperatures than any other ceramic material. A slight plastic deformation may be observed at approx. 2500 °C. This characteristic of graphite is reflected in the mechanical properties of carbon/carbon composites at high temperatures. In the absence of air, the room temperature mechanical properties of carbon/carbon composites are not only maintained at high temperatures, but rather a slight improvement in mechanical properties is observed at a temperature of around 2000 °C due to annealing of defects [55, 56].

However, the exact behavior depends on the initial processing parameters of the composites, i. e., fiber type, fiber orientation, matrix precursor, heat treatment temperature, matrix microstructure, etc. From a thermodynamical point of view the composites must be exposed to a much higher temperature than the temperature of their application. Though the literature data will much depend on the fiber, matrix type and processing conditions of particular carbon/carbon composites, Figs. 6.31 to 6.35 show typical cases of mechanical properties of various carbon/carbon composites at elevated test temperatures. In all of the composites, the strength as well as the modulus at elevated temperatures is higher than that at room temperature. At high temperatures, several characteristic phenomena responsible for the mechanical properties of the composites take place simultaneously. Annealing of gross defects result in an improvement of mechanical properties [55], whereas the closure of stress distributed

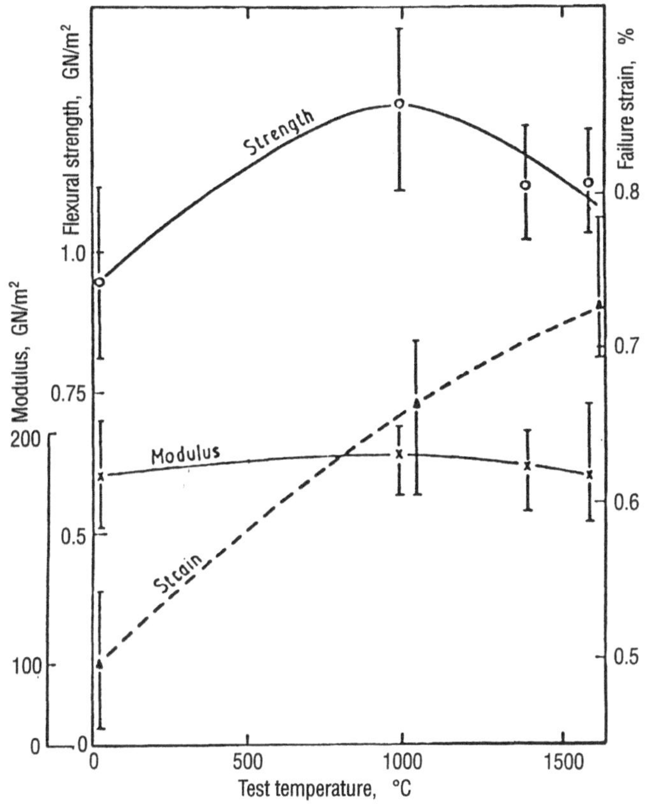

Fig. 6.31. Mechanical properties of UD carbon/carbon composites as a function of measurement temperature

microcracks due to thermal expansion of fibers and matrix change the fracture behavior of the composites [56–57]. As a combined result of all of these phenomena, the carbon/carbon composites exhibit an increase in strength and stiffness by 10%–60% above 1000 °C [55–60]. The compressive strength has been observed to increase by as much as 100%. The tensile and compressive modulus of the composites decreases with temperatures above 1600°–2000 °C. The fall in the modulus can be explained on the basis of the symmetric potential curve of graphite crystals. With an increase in temperature, the average atomic distance is increased, resulting in a decreases in the modulus. Like strength, the work of fracture of carbon/carbon composites also increases with test temperatures up to 1800°–2000 °C. As mentioned above, the structural changes at high temperature influence the stress transfer mechanism in the composites. This

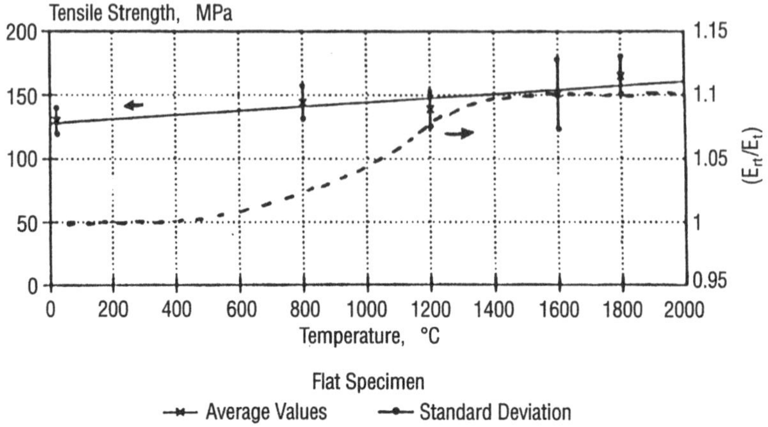

Fig. 6.32. Tensile strength of 2D carbon/carbon composites as a function of measurement temperature. E_{rt} – room temperature modulus, E_t – testing temperature modulus

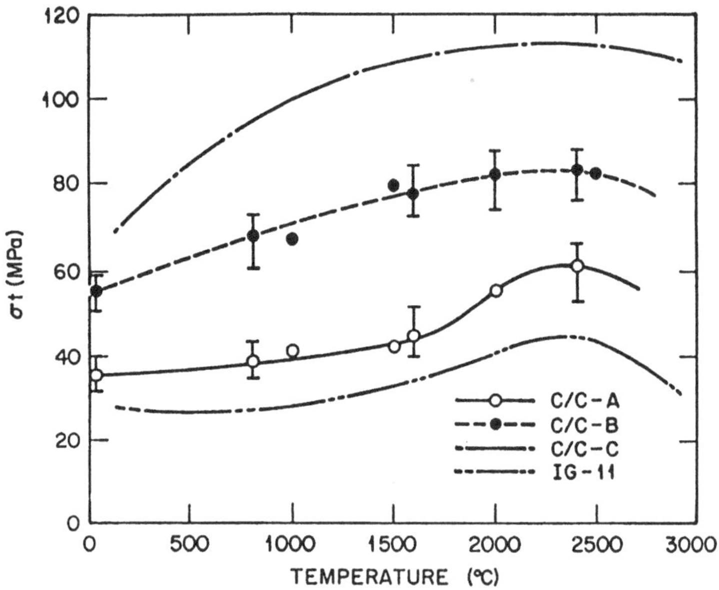

Fig. 6.33. Tensile strength of carbon/carbon felt composites and graphite as a function of measurement temperature: *A* Pitch; *B* PAN-based carbon-fibers

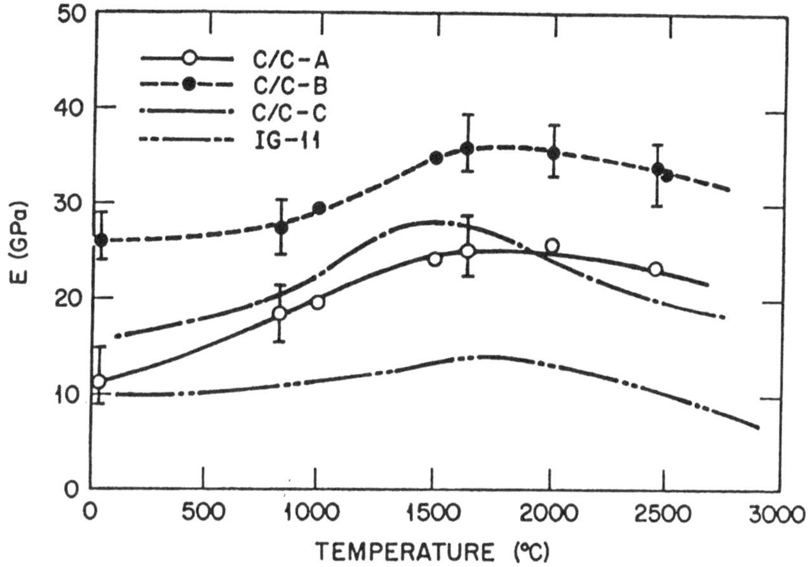

Fig. 6.34. Young's modulus (E)of carbon/carbon felt composites and graphite as a function of measurement temperature: *A* Pitch; *B* PAN-based carbon-fibers

Fig. 6.35. Tensile strength of 3D carbon/carbon composites in X and Z directions as a function of measurement temperature

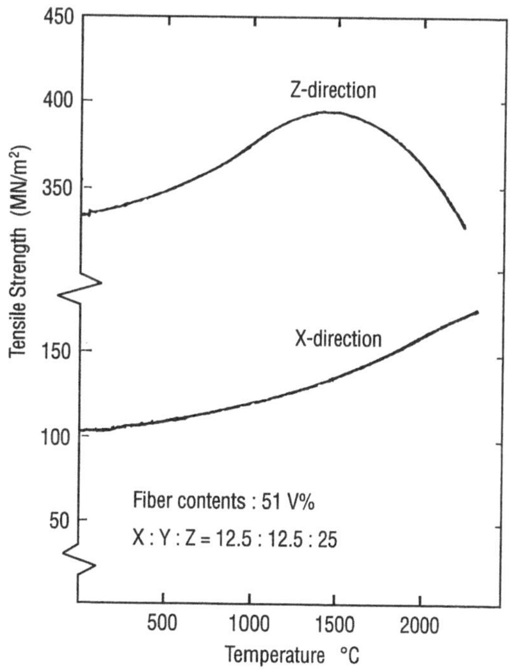

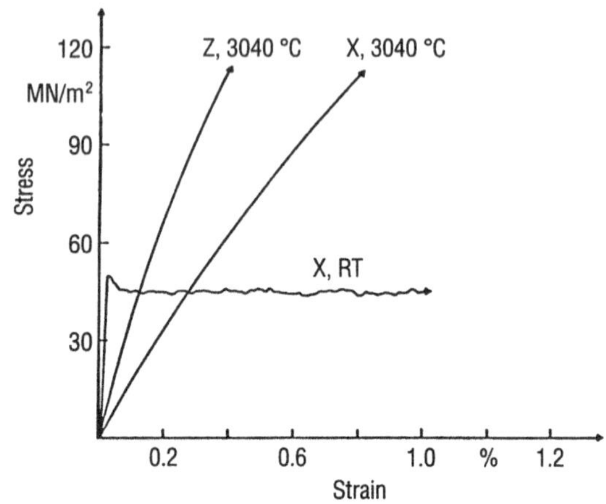

Fig. 6.36. Stress–strain curves of 3D carbon/carbon composites at room temperature and elevated temperature

results in change in the fracture behaviour of the composites. Figure 6.36 shows a typical example of the change in fracture behavior of a 3D carbon/carbon composite with elevated test temperatures. This change in 3D composites is also attributed to the closure of the well-connected cracks in 3D composites due to thermal expansion of the fibers and matrix.

Senet et al. [61] have studied the fracture behavior of 2D discontinuous carbon-fiber-felt-reinforced phenolic resin and pitch precursor carbon/carbon composites at elevated temperatures using both chevron-notch and straight-notch single-edge beam specimens. They observed that K_{ic} values of the chevron-notch specimens increased slightly with the test temperature through 1650 °C, while the straight-notch values remained almost constant throughout the temperature range up to 1650 °C. They suggest that fiber debonding, matrix damage, and fractional sliding during pullout are the critical constituents of an effective wake zone process responsible for the toughening of such composites. Figure 6.37 shows typical cases of an increase in fracture toughness of different types of carbon/carbon composites with increasing test temperatures. Composites made with PAN carbon fibers exhibit a much higher increase in fracture toughness than other composites [62].

The mechanical behavior of carbon/carbon composites at high temperature discussed above holds true for unidirectional, bidirectional and multidirectional reinforced carbon/carbon composites made with continuous fibers or felt and liquid- as well as gas-phase-derived carbon matrix [60]. The absolute strength values may vary depending upon the configuration. In comparison with other structural materials, carbon/carbon composites exhibit exceptionally high specific strengths at all temperatures (Fig. 6.38).

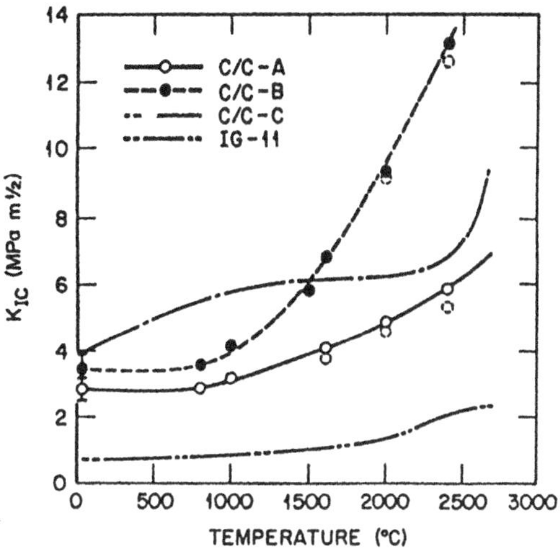

Fig. 6.37. Temperature dependence of fracture toughness of carbon/carbon composites and graphite

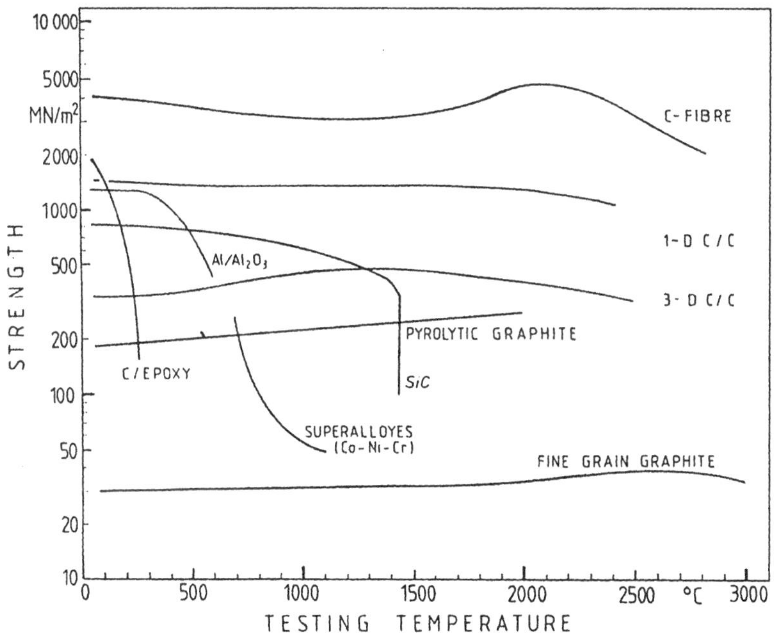

Fig. 6.38. Specific strength of carbon/carbon composites and other engineering materials at elevated temperatures

6.8
Effect of Oxidation on Mechanical Properties of Carbon/Carbon Composites

Carbon/carbon composites are susceptible to oxidation at high temperatures if not protected. Therefore, for the reliable performance of a carbon/carbon component, it is essential to know the effect of oxidation on both the mechanical and thermal properties of the composite. Although this is an important aspect, not much information is available in the literature. Oxygen attacks preferentially the less ordered carbon at matrix-rich regions, bundle/bundle interfaces, or fiber/matrix interfaces. Accordingly, the fracture behavior and the mechanical and thermal properties will be altered. Thermal conductivity will be reduced due to the porosity generated in the composites.

The change in mechanical properties are governed by the preferential attack by oxygen as stated above [63]. This will have a severe effect if the weight loss is due to attack on the fiber/matrix interface and is more than 5% [64, 65], which will result in a drastic decrease in the strength of the composites. If oxidation takes place at low temperatures (400° – 500 °C) and is less than 2%, it may alter the fiber/matrix or matrix/matrix interfaces

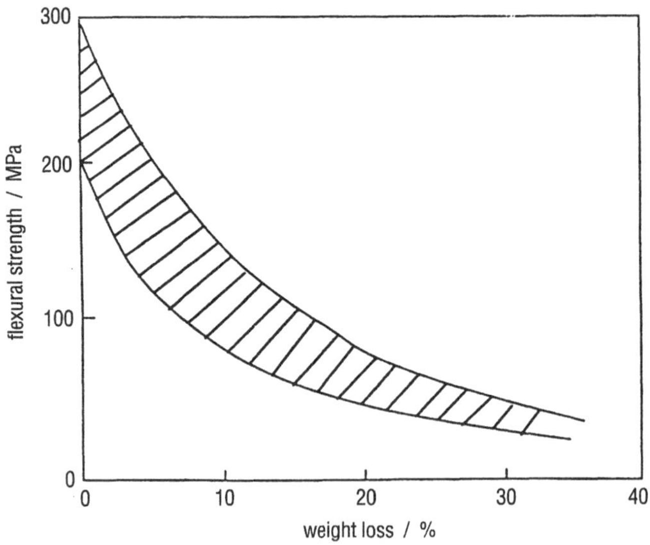

Fig. 6.39. Effect of oxidation in air at 600 °C on the flexural strength of 2D carbon/carbon composites [66]

resulting in change of fracture behaviour to mixed mode type and, hence, an improvement of mechanical properties [65]. However, in practice the application temperatures are much higher than 500 °C and the oxidation weight loss is also higher than 5 %. A general trend of the effect of air oxidation on the mechanical properties of carbon/carbon composites is shown in Fig. 6.39 [66]. Similar trends are observed for CVI based carbon/carbon composites.

6.9 Effect of Neutron Irradiation on Properties of Carbon/Carbon Composites

Nuclear graphite, as the name suggests, is widely used as a moderator. Low atomic mass with a high thermal stability are the main attractions of graphite. However, the graphite needs to be of high purity. Extensive work has been carried out on the effect of neutron irradiation on the thermophysical and mechanical properties of graphite [67, 68]. Carbon/carbon composites, being stronger and tougher than graphite, are widely used as plasma facing surfaces in Tokamak devices, especially for high heat flux surfaces, i.e., divertors [68–71]. Heat flux on these devices typically ranges from 5 to 30 MW/m and is of sufficient duration (hundreds of seconds).

In order to obtain materials with a reliable performance it is important to evaluate the behavior of carbon/carbon composites under normal and disruptive conditions of plasma in fusion devices. Like in graphite crystal, neutron iiradiation causes dimensional changes also in carbon/carbon composites. The nature of the dimensional changes in the composites will depend upon both the irradiation conditions (temperature, fluence) and the microstructure of the carbon fiber, i.e., their architecture and matrix structure. Figure 6.40 [58, 71] shows irradiation-induced dimensional changes (diametral and longitudinal) of a UD, 2D and 3D carbon/carbon composite irradiated at 600 °C. Composites made from more crystalline pitch carbon fibers exhibit less dimensional changes at a given fluence than carbon/carbon composites manufactured from less crystalline PAN carbon fibers (Fig. 6.41). Dimensional changes can be expected to be greater at elevated temperatures due to the closure of the pores and voids and reduced accomodation available within the composite structure.

Fast neutrons interact with graphite ejecting atoms from graphene planes into interstitial positions. At low doses, this results in the strengthening of graphite. At high doses the ejection of atoms amounts to structural disorder, which is evident from the increase in d_{002} spacing. This results in

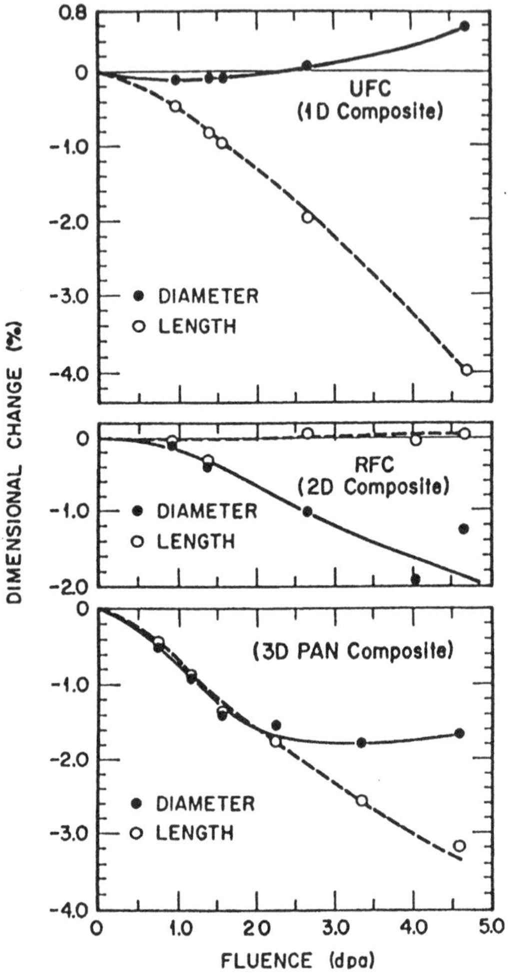

Fig. 6.40. Neutron-irradiation-induced dimensional changes of several carbon/carbon composites irradiated at 600 °C [71]

decrease in mechanical properties. Similar effects are observed with carbon/carbon composites. Hamada et al. [71] reported the effects of neutron irradiation at a low fluence level on the mechanical properties of carbon/carbon composites (Fig. 6.42). Up to the level of fluence used by them, they have observed an increase in strength. Sato et al. [72] have also observed similar effects (at a low fluence level of 10^{21} n/cm^2). Table 6.3 shows the mechanical properties of different carbon/carbon composites after low dose neutron irradiation. Young's modulus of the composites increases by more than 30 %. Strength as well as fracture toughness increases by 20 %–30 %. The increases are observed to be higher for pitch-based,

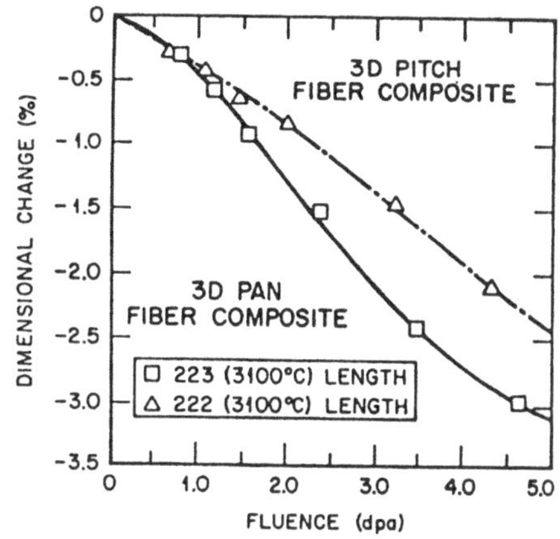

Fig. 6.41. Neutron-irradiation-induced dimensional changes in 3D carbon/carbon composites made with PAN- and pitch-based carbon fibers [58]

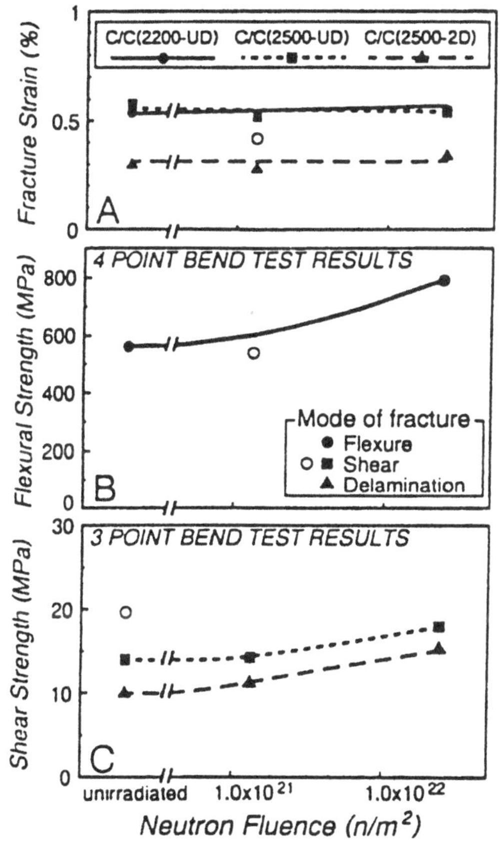

Fig. 6.42 A–C. Effect of neutron irradiation on (**A**) fracture strain, (**B**) flexural strength, and (**C**) shear strength of carbon/carbon composites [71]

Table 6.3. Mechanical properties of some carbon/carbon composites before and after irradiation

Material state	C/C felt A		C/C felt B		2D C/C	
	Before	After	Before	After	Before	After
Young's modulus E (GPa)	12.8	18.6	26.0	33.8	23.2	31.0
Compressive strength (MPa)	45.2	58.1	66.4	82.1	58.2	76.4
Bending strength (MPa)	66.3	82.4	96.7	120.6	114.8	148.0
Tensile strength (MPa)	34.8	44.2	55.4	64.9	88.6	114.2
Critical stress intensity Factor ($MPa/m^{1/2}$)	2.9	3.56	3.42	4.14	5.23	5.64

2D, two-directional.

highly crystalline composites and 2D carbon/carbon composites. However, at high doses, the mechanical properties of the composites tend to decrease. The fracture behavior also changes to brittle type resulting in a reduced fracture toughness.

Neutron irradiation produces defects in graphite basal planes which act as phonon scattering centers, resulting in an increase in thermal resistance and, hence, a decrease in thermal conductivity [73, 74]. Figure 6.43 shows

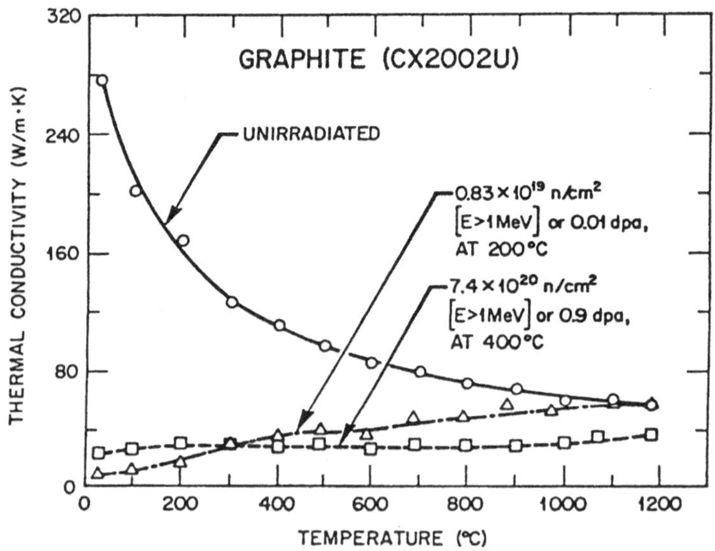

Fig. 6.43. Effect of neutron irradiation on thermal conductivity of carbon/carbon composite measured at elevated temperatures [58]

the decrease in thermal conductivity of carbon/carbon composites upon neutron irradiation. As can be seen from the figure, even at low fluences, the thermal conductivity is reduced by 30% or more. However, on thermal annealing at high temperatures, the damage is repaired substantially.

6.10 References

1. Pierson HO (1993) Hand book of Carbon, Graphite, Diamond and Fullerenes. Noyes Publications
2. Fitzer E (1987) Carbon. Vol 25, p 163
3. Fitzer E, Huttener W (1981) J Physics D Applied Physics. Vol 19, p 347
4. Fitzer E, Huttener W, Manocha LM (1980) Carbon. Vol 18, p 291
5. Jain PK, Bahl OP, Manocha LM (1992) SAMPE Q 23, p 43
6. Huttener W (1980) PhD Thesis, University of Karlsruhe
7. Huttener W (1990) Potential of carbon/carbon composites as structural materials. In: Figueireds JL et al. (eds) Carbon fiber, filaments & Comp Kluwar Pub, p 275
8. Evangalides JS (1977) 13th Bien Carbon Conf Irvine, p 7
9. Cook J, Gorden JE (1964) Proc Roy Soc London, A 282, 508
10. Jortener J (1985) Effect of Weak interfaces on thermostructural behaviour of C/C Composites. Office of Noval Research, Ann Report, Contrect N00014-82-C-0405, p 19
11. McEnaney B, Mays T (1993) Relationship between Microstructure and Mechanical Properties. In: Thomes CR (ed) Essential of carbon/carbon Composites, Royal Social of Chemistry (Pb), p 143
12. Chiopek J, Biazewicz S (1991) Carbon. Vol 29, p 127
13. Walker EJ, Hall J (1975) 12th Bien Carbon. Conf, p 299
14. Bojna L, Cook D, Hall J, Turner K (1974) 4th Lond Carbon. Con, p 50
15. Oh SM, Lee JY (1988) Carbon. Vol 26, p 769
16. Pierson HO, Northrop DA (1975) J Comp Materials. Vol 9, p 118
17. Granoff B, Pierson HO, Schuster DM (1973) Carbon. Vol 11, p 177
18. Delhaes P et al. (1984) J de chem Phys 81, 809
19. Sawage G (1992) carbon/carbon composites. Chapman and Hall Pub
20. Kimura S, Yasuda E, Takara N, Kasuya J (1981) HTHP. Vol 13, p 193
21. Bauer DW, Kotlensky WV SAMPE (Q) Vol 4, 10
22. Kimura S, Yasuda K,Yasuda E, Tanabe Y (1987) Tanso 128, 30
23. Buckley JD Ceramic Bulletin. Vol 67, p 364
24. Adams DF (1974) J Comp Materials 8, p 320
25. McAllister LE, Lachman WL (1983) Multidirectionally carbon/carbon composites. In: Kelly A, Milleko ST (ed) Handbook of Composites. Vol 4, Elsevier Pub, p 109
26. Lieberman ML, Pierson HO (1974) Carbon. Vol 12, p 233
27. Gebhardt JJ, Yodsnukis JJ, Stover ER (1972) SAMPE, p 507
28. Manocha LM, Bahl OP, Singh YK (1991) Carbon. Vol 29, p 357
29. Fitzer E, Geigl KH, Manocha LM (1978) 5th London and Carbon Conf, p 504
30. Manocha LM, Yasuda E, Tanabe Y, Kimura S (1988) Carbon. Vol 26, p 333
31. Fitzer E, Huttener W, Weiss R (1982) Carbon. Vol 20
32. Manocha LM (1994) Carbon. Vol 32, p 213

33. Zaldivar RJ, Rellick GS, Yang JM (1995) 22nd Bien. carbon conf, p 52
34. Tanaka H, Yasuda E, Kimura Y (1978) Trans JSCM. Vol 4, p 37
35. Appleyard S, Rand B (1992) Int Carbon conf Essen, p 718
36. Pardini LC, McEnaney B (1994) Carbon. Vol 32, p 815.
37. Zaldivar RJ, Rellick GS, Yang JM (1993) J Mat Res, Vol 8(3), p 501
38. Kowbel W, Shan CH (1990) Carbon. Vol 28, p 287
39. Yasuda E, et al (1988) Carbon. Vol 26, p 225
40. Christ K, Huttinger KJ (1993) Carbon. Vol 31, p 731
41. Dillon F, Thomas KM, Marsh H (1993) Carbon. Vol 31, p 1337
42. Manocha LM, Bahl OP (1988) Carbon. Vol 26, p 13
43. Aglan MA (1993) Carbon. Vol 31, p 1121
44. Jortener J (1990) MTC Annual Conference, SIU, Carbondale, p 51
45. Pollock PB (1990) Carbon. Vol 28, p 717
46. Fitzer E, W.Huttener W, Manocha LM (1978) 5th London Carbon Conference
47. Anand K, Gupta V (1993) Carbon. Vol 33, p 739
48. Dong-Huck J, Min-Jing L (1985) 17th Bienal Conf, p 280
49. Muller CK, Roy PJ. (1972) Fabrication & properties of AVCO 3D C/C materials; SAMPE
50. Adams DF (1976) Mat Sc Eng. Vol 23, p 55
51. Bouquet M et al. (1990) Composite Science & Technology. Vol 37, p 223
52. Zao JX, Bradt RC, Walker PL (1981) 15th Biemal Carbon Conf, p 274
53. Kawada H et al. (1993) 21st carbon conf, p 140
54. Lamicq PJ (1984) Int Carbon Conf Bordeux, France
55. Hill J, Thomas CR, Walker EJ (1974) 2nd Int Carbon Fiber Conf London, p 1
56. Fitzer E, Terwiesch B (1972) High Temperature, High Pressure 4, p 359
57. Karlisch K (1977) PhDThesis University of Karlsruhe
58. Burchell TD, Oku T. Materials Property data for fusion reactor plasma facing carbon/carbon composites preprint
59. Huettener W, Weiss R, Dietrich G, Meistring R (1990) Proc ESA Symp ESTEC, Noordwijk (NL) (ESA S-303-1990) p 91
60. Lieberman ML, Curlee RM, Braten FH, Noles GT (1975) J Comp Mat 9, p 337
61. Senet S, Grimes RE, Hunn DL, White KW (1991) Carbon. Vol 29, p 1039
62. Sato A, Karumada A, Iwaki H, Komatzu Y (1989) Carbon. Vol 27, p 791
63. Zhao JX, Bradt RC, Walkar PL Jr (1985) Carbon. Vol 23, p 9
64. Pickup IM, McEnaney B, Cook RG (1986) Carbon. Vol 24, p 535
65. Ahearn C, Rand B (1996) Carbon. Vol 34, p 239
66. Crocker P, McEnaney B (1991) Carbon. Vol 29, p 881
67. Kelly BT (1981) The Physics of Graphite applied Science Publishers, London
68. Burchell TD et al. (1992) J Nuc Mater 191, p 295
69. Ando T et al. (1992) J Nucl Mater 191, p 1423
70. Tanabe Y et al. (1991) Carbon. Vol 29, p 905
71. Hamada K, Sato S, Kohama A (1994) J Nuc Mater 212, p 1228
72. Sato S, Kurumada A, Kawamate K, Ishde R (1990) Fusion Engg and Design. Vol 13, p 159
73. Burchell TD (1994) Proc 39th Int SAMPE symposium. Vol 39, p 2423
74. Maruyama T, Harayama M (1992) J Nuc Mat. Vol 195, p 44

7 Thermal Properties of Carbon/Carbon Composites

7.1 Thermal Transport Phenomena

Thermal transport properties of a solid material are generally controlled by two mechanisms: (i) electron charge cloud drift and (ii) lattice vibrations, known as phonons. In the latter case, thermal transport properties such as thermal conductivity or thermal diffusivity are directly proportional to the mean free path of phonons [1]. In a perfect crystal, phonon scattering takes place predominantly by four mechanisms; the crystal boundaries, the natural isotopic composition, the inharmonic interaction with other phonons, and the conduction of electrons [2]. Figure 7.1 illustrates the contri-

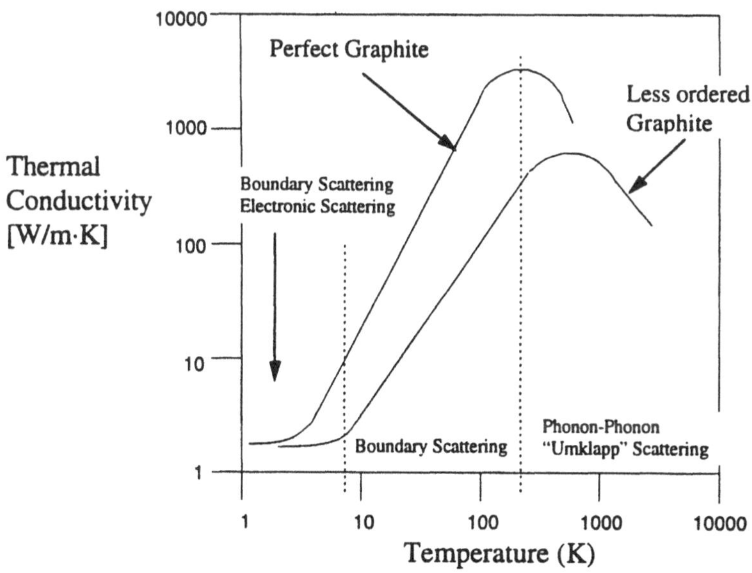

Fig. 7.1. Scattering mechanisms contributing to thermal conductivity of graphite in different temperature domains [5]

bution of different scattering mechanisms to the thermal conductivity of a graphite crystal over different temperature ranges. As can be seen from this figure, the conduction of electrons is not significant above 10 K while phonon–phonon interaction (or Umklapp scattering) is effective at high temperatures. Therefore, at intermediate temperatures, phonon scattering at crystal boundaries or defects is the dominant mechanism [3]. Moreover, graphites are anisotropic materials with varying degrees of anisotropy from fine-grained isotropic graphite to highly anisotropic pyrolytic graphite. The thermal conductivity in graphite is dominated by the layer plane conduction. The absolute magnitude of the thermal conductivity is determined by the crystallite size and the orientation of the layer planes. For highly ordered graphite, the thermal conductivities parallel to basal planes may be two orders of magnitude higher than those in the perpendicular direction. The ratio changes with structural order in graphite. As shown in Fig. 7.1, different phonon scattering mechanisms operate in different temperature regions. At low temperatures, the phonon concentration is small and the mean free path is roughly equal to the crystallite size l_a of the crystal. As the temperature is raised, the thermal conductivity increases with square of the temperature and reaches a maximum at a temperature near the dielectric maxima [4]. This temperature is approximately 180 K for pure graphite and higher for less ordered graphite. Above this temperature, phonon–phonon or Umklapp scattering dominates, the mean free path decreases, and the thermal conductivity decreases. At high temperatures, thermal conductivity values tend to come together. Therefore, thermal conductivity values for graphite based on different mechanisms vary quite widely, both as a function of the degree of graphitization (anisotropy) and as a function of temperature (Fig. 7.2.) [1].

In carbon/carbon composites, the fibers and matrix are both composed of graphite. Therefore the mechanisms discussed above are also relevant for the thermal transport properties of carbon/carbon composites. However, these composites have a heterogeneous structure consisting of fibers, matrix, and pores. The presence of multiple components increases the phonon scattering, affecting the thermal conductivity. As discussed in Chap. 1 and 5 carbon fibers and the matrix may exhibit a variety of microstructures depending on the precursor and processing conditions. Obviously, these microstructures will posses varying thermal properties. This is evident from the thermal conductivity of different carbon fibers shown in Fig. 7.3 [5]. The thermal conductivity of various carbon fibers is compard with that of some selected metals in Figs. 7.3 and 7.4 [6]. The shape, size, and extent of porosity may also vary, depending on the constituents and processing

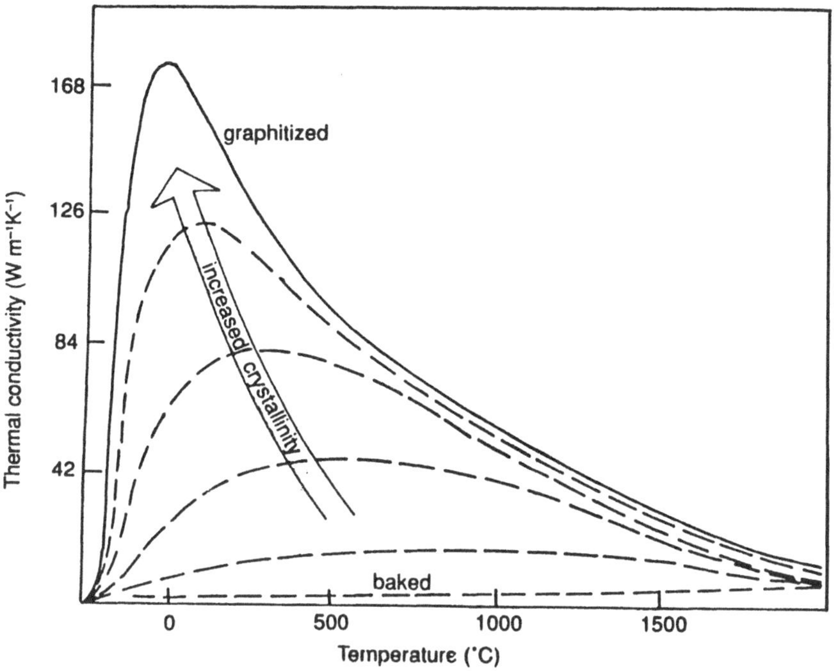

Fig. 7.2. Thermal conductivity of graphite as a function of temperature and heat treatment temperature [1]

conditions. Therefore, although the thermal transport properties of carbon/carbon composites are complex, a material with tailor made thermal properties can be fabricated by the proper choice of constituents, their configuration, and processing conditions.

In general, in carbon/carbon composites processed by polymer pyrolysis or chemical vapor deposition (CVD), the graphite matrix orients around the fibers with the graphite planes oriented in the fiber direction. Therefore, the thermal conductivity (K_c) of carbon/carbon composites in the direction parallel to the fibers can be reasonably estimated by the modified rule of mixtures [5, 7]

$$K_c = K_f V_f + K_m (1 - V_f - V_p)$$

Where K_f and K_m are the thermal conductivities of fibers and matrix, respectively in the ab direction and V_f and V_p are the volume fractions of the fibers and pores, respectively.

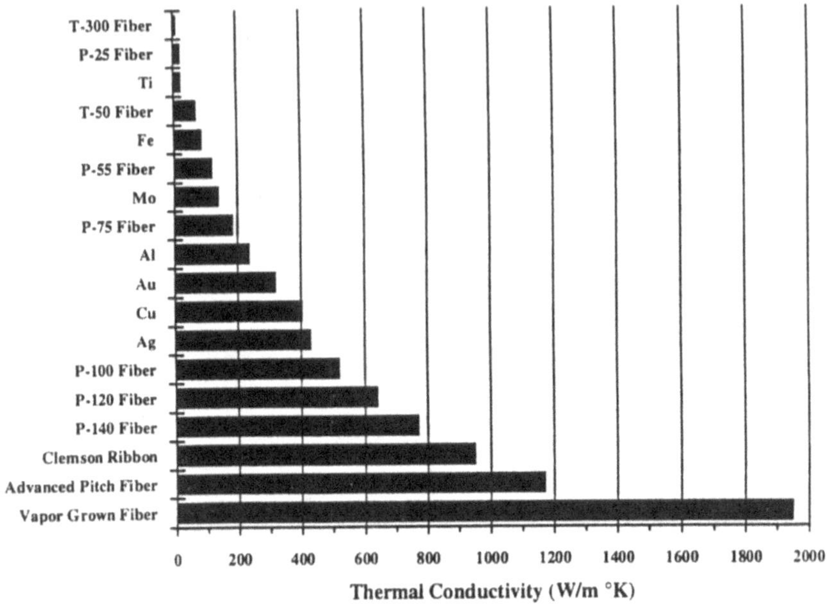

Fig. 7.3. Thermal conductivity of different carbon fibers [5]

Fig. 7.4. Room temperature thermal conductivity and electrical resistivity of carbon fibers and selected metals

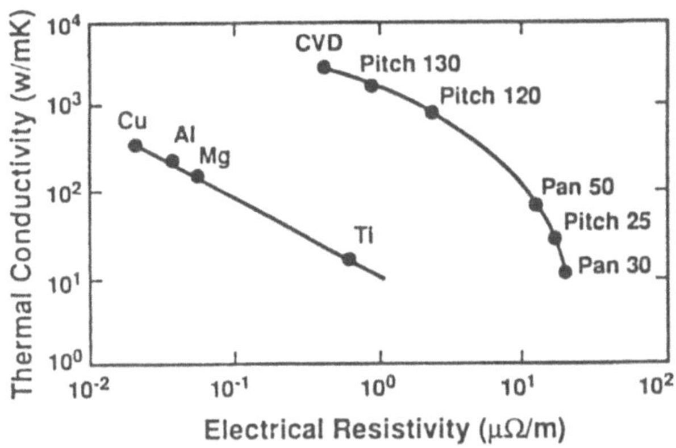

The thermal transport properties in the direction perpendicular to the fibers are difficult to predict. This is because the heat flow does not follow a simple path. Both the fibers and the pores serve to disrupt heat flow, thus decreasing thermal conductivity. Moreover, PAN- as well as pitch-based carbon fibers possess different heterogeneous microstructures along the cross section which introduces an additional parameter. A number of mathematical models have been developed to predict the transverse thermal conductivity of two phase materials with polymers [8, 9] as well as carbon matrixes [5, 7]. These models, however, cannot be generalized for different fiber and matrix types or for different fiber configurations, especially for random fiber orientations or multidirectional reinforcements.

Thermal conductivity of carbon/carbon composites has been measured mostly by two techniques (i) comparative method using steady state thermal gradient technique and (ii) nonequilibrium laser flash method. In the former method, the test piece is placed directly adjacent to a standard reference piece of known comparable thermal conductivity [10]. A temperature gradient is maintained across the samples and the sides are insulated. Once the sample reaches steady state, the heat flux is measured and thermal conductivity is calculated. This technique is quite simple to perform but considerable time is required to reach steady state and the utmost care must be taken to make good contacts between the test sample and the reference. The thermal conductivity of the reference sample should be comparable to the thermal conductivity of test piece. This requires the use of different reference samples for different types of composites at different temperatures. The sample size is also normally large (25–50 mm in diameter and 20–50 mm in thickness).

The nonequilibrium or flash method is employed to measure the thermal diffusivity of the materials (normally 10 mm in diameter and 5–10 mm in thickness) and is based on the one dimensional transient heat flow equation [11, 12]. The front face of the sample held at ambient temperature of measurement receives a pulse of energy from the laser which shortly after raises the back face temperature by a degree or two. This is measured with an infrared pyrometer and recorded with a computer. The data is plotted in a dimensionless form, compared with the theoretical solution curve, and the thermal difusivity is calculated. The thermal conductivity (K) of the material can be calculated using the formula

$$K = \alpha . \varrho . C_p$$

Where α, ϱ, and C_p are the diffusivity, density, and specific heat of the samples, respectively.

Although this method requires small samples and is quickly performed, it is based on one-dimensional heat flow and therefore much care needs to be taken, especially for carbon/carbon composites with a high thermal conductivity where the conduction in the other direction may also take place through the fibers. However, in most of the recent literature, the reported thermal diffusivity values are usually based on the laser flash measurement technique [11, 13–14].

As discussed above, the thermal conductivity of carbon/carbon composites is dependent on the volume fraction and organization of its constituents, i.e., the fibers, matrix, and various defects. The absolute values are governed by the thermal transport properties of the constituents. Although thermal properties of carbon/carbon composites govern their application, there are not many reports dealing with the systematic correlation of these properties with various factors. The thermal conductivity data available in the literature generally pertains to products made by the manufacturers. Very few groups have carried out a systematic evaluation of the thermal conductivity of carbon/carbon composites.

7.2 Carbon/Carbon Composites with Thermosetting Resin-Derived Carbon Matrixes

Fitzer et al. [15] have studied the thermal conductivity of unidirectional composites made with thermosetting-resin-derived carbon matrixes. Figure 7.5 shows the change in thermal conductivity of the carbon/carbon

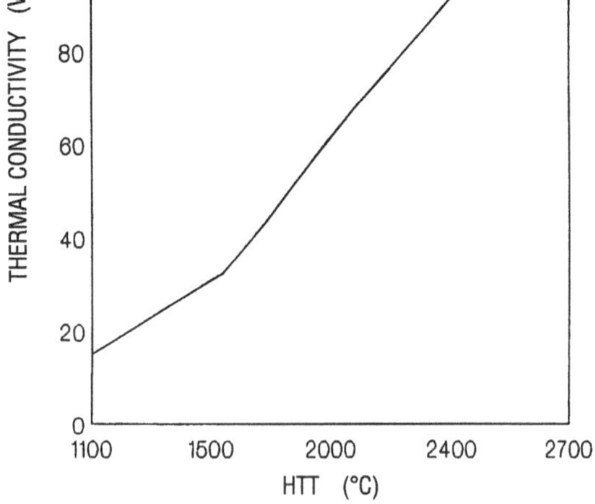

Fig. 7.5. Effect of heat treatment temperature (HTT) on thermal conductivity of carbon/carbon composites made with thermosetting resin

composites with heat treatment temperature. The steep increase is due to stress graphitization of the matrix. A similar effect has also been observed by Kimura et al. [16]. The low values of thermal conductivities in Fig. 7.5 are due to the low thermal conductivity of the reinforcing fibers as well as the network of cracks present in the matrix formed during carbonization of the thermosetting resin. If the formation of cracks in the matrix could be controlled, the absolute value of the conductivity could be increased by six-fold. The thermal conductivity of two-directional (2D) as well as multidirectional (MD) composites made with thermosetting-resin-derived carbon matrix are compiled in Table 7.1.

7.3 Composites with Pitch-Derived Matrixes

In contrast to thermosetting-resin-derived carbon/carbon composites, pitch-precursor-derived carbon/carbon composites exhibit very high thermal conductivity. This is evident from Fig. 7.6. The conductivity in the fiber direction increases markedly with the number of densification cycles. The absolute value is further increased on graphitization of the composites at

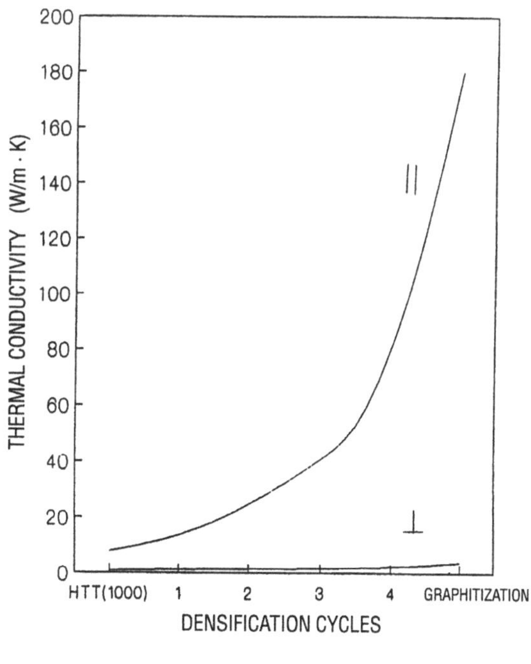

Fig. 7.6. Increase in thermal conductivity of carbon/carbon composites with the number of densification cycles. (1000 – 1000 °C = the heat treatment temperature)

Table 7.1. Thermal conductivity of representative carbon/carbon composites as reported in the literature

Composite description	Fiber type	Matrix precursor	Fiber orientation	Density (g/cc)	Thermal conductivity ($W/m^{-1} K^{-1}$)
Unidirectional	PAN	Phenolic	In fiber direction	1.7	100–140
			Transverse to fibers	NA	10–16
	PAN	Pitch	In fiber direction	NA	280–350
			Transverse to fibers	NA	40–55
	Vapor grown	CVD	In fiber direction	NA	364–564
			Transverse to fibers	NA	11.6–18.9
Experimental 2D	P100	Pitch	In fiber direction	NA	178
			Transverse to fibers	NA	70
	P100	Pitch/CVD	In fiber direction	NA	269
			Transverse to fibers	NA	33
	P120	Pitch	In fiber direction	NA	172
			Transverse to fibers	NA	51
	P120	Pitch/CVD	In fiber direction	NA	184
			Transverse to fibers	NA	51
	P100	Pitch	In fiber direction	NA	182
			Transverse to fibers	NA	19
ACC	PAN	Phenolic	3D	1.79	84
RCC	Rayon	Phenolic	3D	1.58	250
AEROLOR 32	PAN	CVD	3D	1.6–1.7	10
AEROLOR 33	PAN	CVD	3D	1.8–1.9	120

Table 7.1 (continued)

Composite description	Fiber type	Matrix precursor	Fiber orientation	Density (g/cc)	Thermal conductivity ($W/m^{-1} K^{-1}$)
MORTON THIOKOL HIPIC 3D stitched fabric	PAN	Pitch	In cloth plane	1.9	202
			Out of cloth plane	1.9	171
Woven fabric	K1100 T300	Phenolic	In cloth plane	NA	280
MFC	Pitch felt	CVD	In fiber direction	1.85	360–460
			Transverse to fiber	NA	30–50
PCC-1S	PAN		In fiber direction	1.86	200–220
			Transverse to fiber	NA	100–120
CX-2002U	PAN		In fiber direction	1.7	250–280
			Transverse to fiber	NA	100–120

CVD, chemical vapor deposition; NA, data/information not available.

Fig. 7.7. Comparison of thermal conductivity as a function of temperature for similar bidirectional carbon/carbon composites with PAN fibers. The conductivity is significantly higher with pitch-based matrix [18]

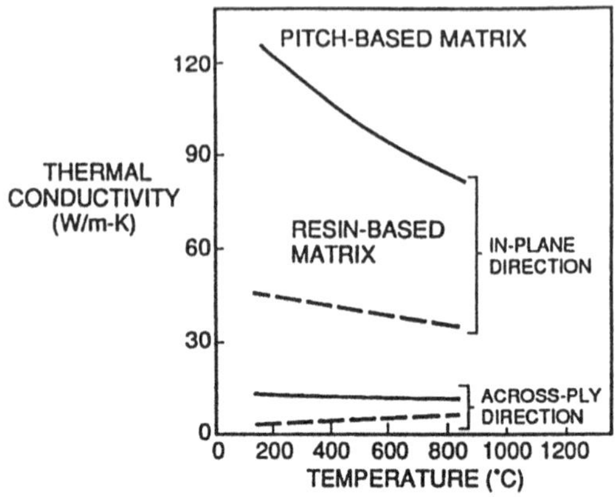

3000 °C [15–17]. Similar effects of the matrix precursor on the thermal conductivity of carbon/carbon composites have been observed in 2D composites (Fig. 7.7) [18] and MD reinforced composites [17–19]. The high thermal conductivity in pitch-derived carbon matrix composites is attributed to the graphitic microstructure of the matrix among the filaments where all of the high conductivity graphitic layers of the matrix lie parallel to the fibers. In such cases the thermal transport properties of the composites are influenced by the properties of the matrix. In pitch-derived carbon

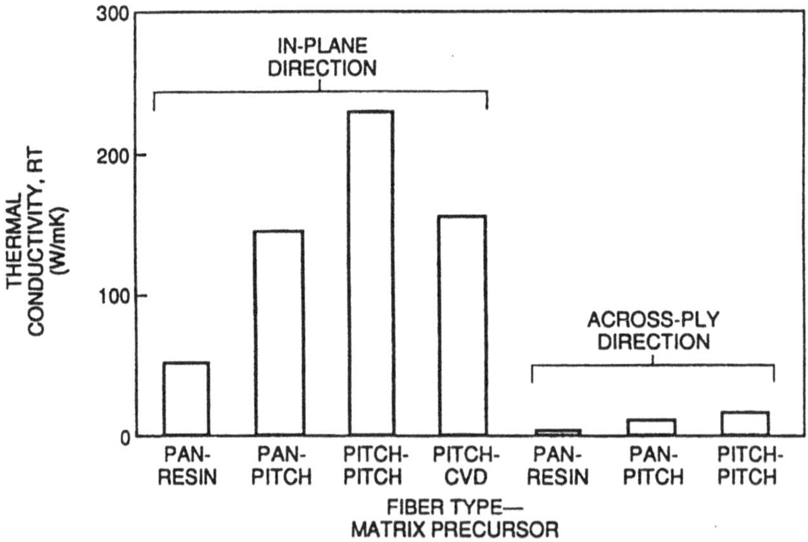

Fig. 7.8. Comparison of thermal conductivity of different fiber/matrix combinations [18]

matrix composites, there is ample scope to improve the matrix microstructure by using highly anisotropic mesophase pitch precursors or through processing routes such as alignment of the mesophase under magnetic field or using HIPIC techniques. These composites with high thermal conductivities can be obtained using pitch matrix precursor. Moreover, when PAN-based carbon fibers are replaced by high-modulus pitch-based carbon fibers in a pitch-derived carbon matrix, the thermal conductivity of the composites is improved by more than 70%. This is evident from Fig. 7.8 [18]. In such cases, both the fibers as well as matrix influence the thermal transport properties of the ultimate composites.

7.4 Composites with Pyrocarbon Matrix

A CVD-derived carbon matrix may exhibit a wide range of microstructures, from isotropic to highly ansiotropic rough laminar structure. Accordingly, the thermal conductivity of carbon/carbon composites with a CVD matrix will also vary with the matrix microstructure and heat treatment temperature (Figs. 7.9 and 7.10) [16, 20]. The thermal conductivity of heat-treated composites are found to be higher than that of composites which have not been heat treated and is highest [20] for rough laminar structure. For most of the carbon/carbon products made from CVD carbon matrix, rough laminar type microstructure is preferred. Whittaker et al. [8, 13, 14] carried out systematic studies on the thermal transport properties of (UD) and bidirectional carbon/carbon composites processed through the CVD route. They found that composite diffusivity parallel to the fiber axis is higher than the diffusivity of the fibers alone by a factor of at least two, indicating that the matrix graphite plays a dominant role in thermal conduction in these composites. The thermal conductivities parallel and perpendicular to the cloth layup of two dimensional carbon/carbon composites have relatively low anisotropy ratios as compared with unidirectional composites. They also studied the matrix microstructure in these composites and found well oriented anisotropic rough laminar carbon matrix around the fibers with a high degree of long range order. This accounts for the high thermal conductivity of the composites in the fiber direction. However, the grain coalescence of the graphite matrix surrounding adjacent fibers forms bridging interfaces which provide a favorable cortuous heat conduction path in a direction transverse to the fiber orientation. This accounts for relatively low anisotropic ratios for thermal conductivities measured parallel and perpendicular to the fiber orientations.

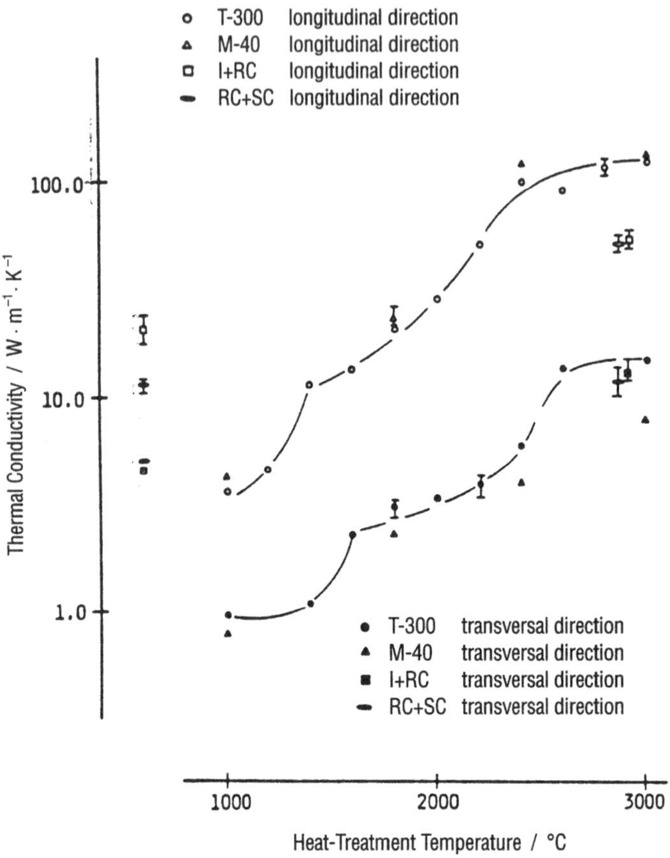

Fig. 7.9. Thermal conductivity of CVD derived carbon/carbon composites [17]

In CVD-derived carbon/carbon composites, irrespective of the matrix precursor, the thermal conductivity is observed to increase with increasing heat treatment temperature and decrease with increasing measurement temperature. This is evident from most of the plots shown in Fig. 7.11 [21]. The decrease in thermal conductivity with increasing test temperature is observed in all carbon/carbon composites pre heat treated to that temperature irrespective of fiber type and architecture (Fig. 7.12) [22]. The extent of variation depends on the matrix precursor and fiber/matrix combinations. The effect is maximum for very high conducting carbon/carbon composites. Based on the thermal conductivity data available for different CVD carbon/carbon composites, a list of thermal properties of the composites has been compiled by Burchell and Oku [22] and by Yamamoto et al.

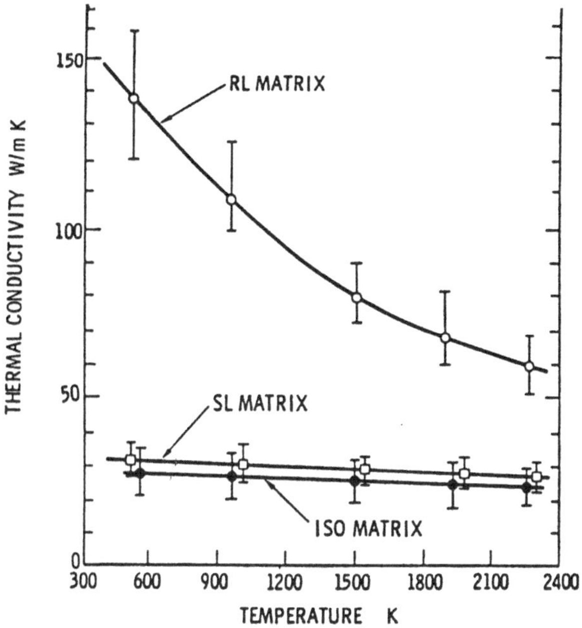

Fig. 7.10. Thermal conductivity of heat-treated CVD rough laminar (RL), smooth laminar (SL), and isotrop (ISO) matrix composites [20]

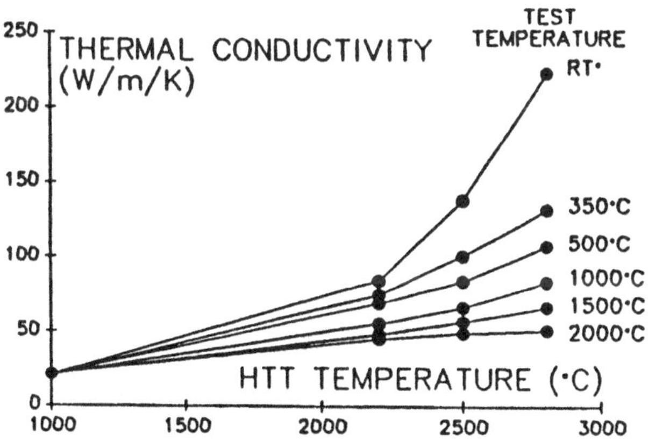

Fig. 7.11. Dependence of thermal conductivity on heat treatment temperature of CVI densified carbon/carbon composites [21]

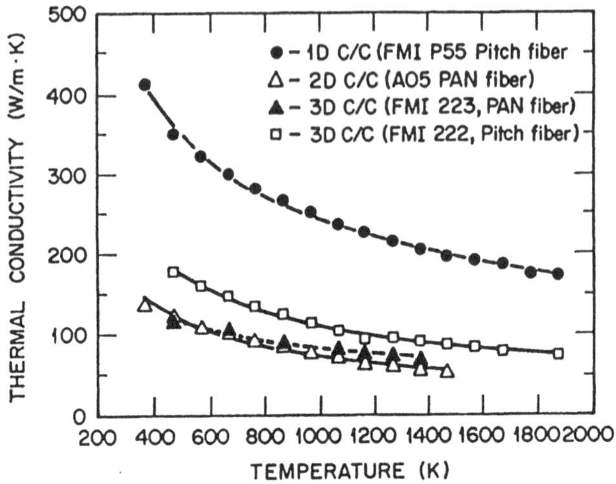

Fig. 7.12. The temperature dependence of thermal conductivity of 1D, 2D, and 3D carbon/carbon composites [22]

[23], both of which give good a overview of the thermal transport properties of carbon/carbon composites.

7.5 High Conductivity Carbon/Carbon Composites

There has been an increasing amount of interest in high thermal conductivity carbon/carbon composites for thermal management applications such as the first wall of fusion reactors, solid rocket exit nozzles, heat sinks, and carbon/carbon brake pads. As discussed above, high thermal conductivity carbon/carbon composites can be produced by choosing the proper combination of reinforcing fibers and matrix precursor. Carbon/carbon composites with very high thermal conductivity in the order of 850 w/mK at room temperature have been developed using the ultrahigh-modulus pitch-based carbon fiber P-130 x and mesophase pitch as matrix precursor [22, 24] (Fig. 7.13). By using ribbon-shaped mesophase pitch carbon fibers in pitch precursor carbon matrix, carbon/carbon composites have been developed with a high thermal conductivity in the directions parallel and perpendicular to the fiber direction (Table 7.2) [5, 25]. The higher thermal conductivity of the composites transverse to the fiber direction suggests a considerable contribution by the matrix to the transverse thermal conductivity of ribbon-shaped carbon-fiber-reinforced carbon matrix composites.

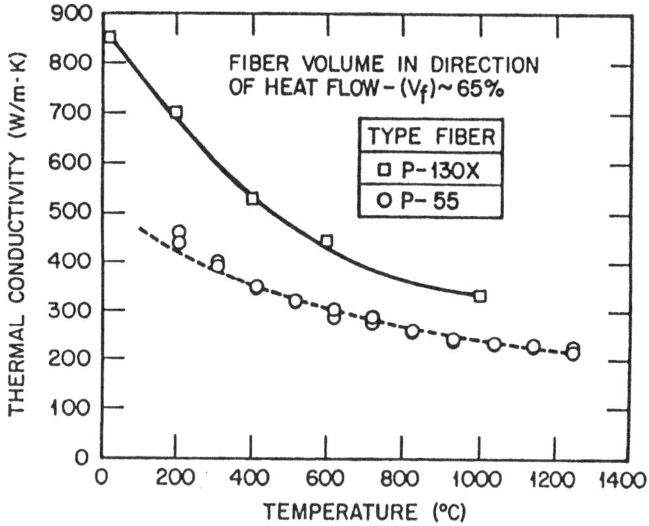

Fig. 7.13. Thermal conductivity of carbon/carbon composites made with different pitch-based carbon fibers [22]

The packing efficiency is better with ribbon shaped carbon fibers than with round-shaped carbon fibers. Because of the packing effect, the matrix aligns itself parallel to the long axis of the fibers with large regions of ordered carbon matrix that were rarely observed in composites reinforced with round fibers [25]. The alignment produces an extremely planar graphite matrix structure. This sheet-like matrix structure may collapse to a smal-

Table 7.2. Thermal properties of carbon/carbon composites formed from mesophase pitch-based towpreg and three different carbon fibers [5]

Composites	Fiber thermal conductivity ($Wm^{-1} K^{-1}$)	Final heat treatment temperature [°C]	Paralle thermal conductivity ($Wm^{-1} K^{-1}$)	Transverse thermal conductivity ($Wm^{-1} K^{-1}$)
T300/AR-120	8.5	1100	4.86 ± 0.21	NA
	76	2400	80.5 ± 2.7	6.86 ± 1.44
P55/AR-120	98	1100	70.8 ± 3.7	2.40 ± 0.40
	196	2400	135.5 ± 4.3	NA
Type I ribbon/ AR-120	236	2400	148.2	213.5

NA, data not available.

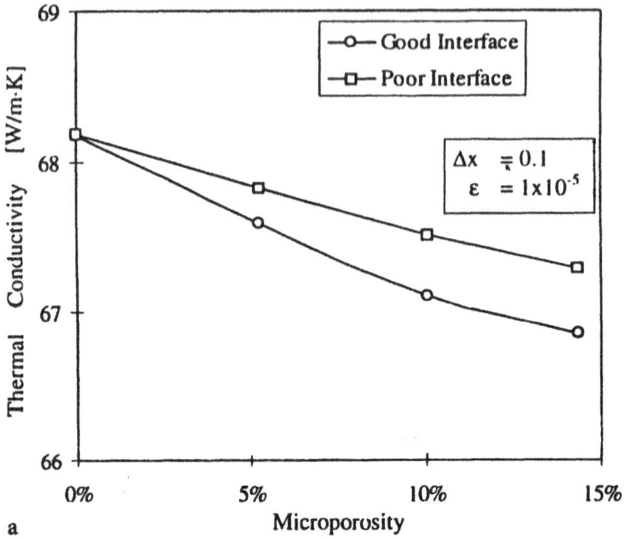

a

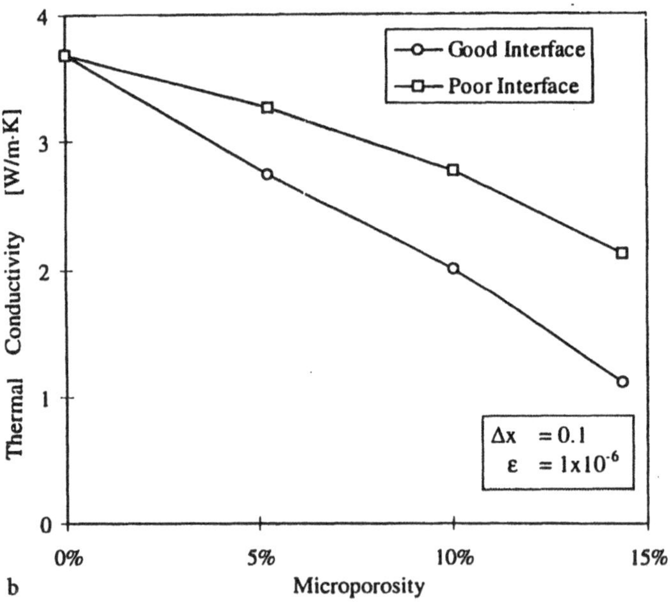

b

Fig. 7.14 a, b. Thermal conductivity of carbon/carbon composites with good and poor fiber/matrix interfaces in the direction parallel (**a**) and perpendicular (**b**) to the fiber direction [5]

ler interplanar spacing upon graphitization, possibly making the matrix of the ribbon fiber composites more heat conductive. In these composites, shrinkage stresses and the thermal expansion mismatch often cause intra-matrix cracks to develop perpendicular to the fiber axis during graphitization, impeding the flow of heat parallel to the fiber axis. Heat flow is uninterrupted parallel to these cracks (parallel to the fiber axis) and, thus, the conductivity of the matrix in the perpendicular direction may be higher than the conductivity of the fiber [25].

Klett [5] has also predicted theoretically the effect of microporosity on the thermal conductivity of carbon/carbon composites for different fiber/matrix interfaces. In the case of a good interface, the defects will be mostly in the matrix region, whereas in case of poor fiber/matrix interface, the bulk of the matrix defects will tend to reduce. Figures 7.14 (a) and (b) show the change in thermal conductivity of carbon/carbon composites along and perpendicular to the fibers, respectively for two types of interfaces. These figures show that a poor fiber/matrix interface will exhibit a higher thermal conductivity both perpendicular and parallel to the fiber axis than a composite with a strong fiber/matrix interface.

Since the ultimate conductivity of carbon/carbon composites is related to the conductivity of carbon fibers, high thermal conductivity composites conveniently can be fabricated using vapor-grown carbon fibers (Fig. 7.15) [22]. Ting and Lake [26] have employed highly graphitic vapor-grown car-

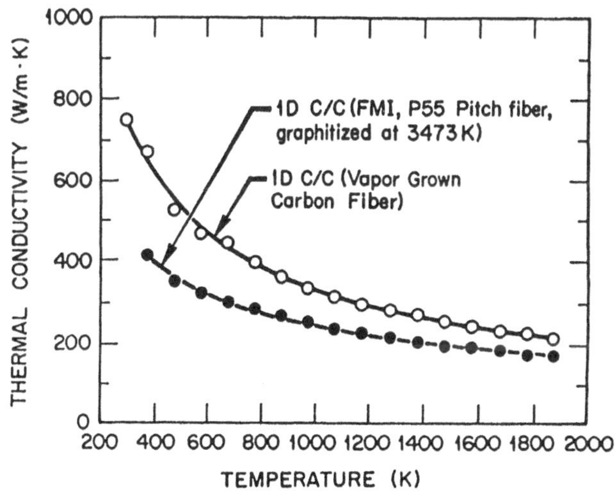

Fig. 7.15. Comparison of thermal conductivities of carbon/carbon composites made with high-performance pitch based and vapor-grown carbon fibers [22]

bon fibers with resin/pitch-based carbon matrix followed by that derived by chemical vapor infiltration. They reported a thermal conductivity of 824 w/mK in the direction of the fibers. According to them, composites made with vapor-grown carbon fibers (VGCF) exhibit a highly graphitic matrix, which is better aligned with the reinforcement than observed in carbon/carbon composites fabricated with less graphitic reinforcement. The thermal conductivity of the composite materials increases with fiber volume content and density. The theoretical estimation of the thermal conductivity of VGCF is reported to be around 1760 w/mK in the fiber direction. This implies that there is ample scope to develop high thermal conductivity carbon/carbon composites with at least 75 % of the thermal conductivity of graphite by using VGCF and suitably developing the carbon matrix around the fibers.

7.6 Thermal Expansion of Composites

The dimensional stability of a material at high temperatures depends on its thermal expansion and coefficient of thermal expansion (CTE). One of the important characteristics of graphite is its low to negative CTE. Figure 7.16 shows a model which explains the mechanism of thermal expansion in two directions in different types of carbon and graphite materials with varying layer structures. As is the case for its mechanical properties, graphite has an anisotropic CTE. It is very low to negative in the planar direction and positive in the C direction. The absolute value depends on the carbon precursor, heat treatment temperature, and processing conditions, i.e., extent of graphitic structure in the matrix. Pyrographite exhibits a thermal expansion anisotropy of about 20. Accordingly, carbon fibers also exhibit a CTE range which depends on the precursor and heat treatment termperature.

In carbon/carbon composites, thermal expansion in the direction of reinforcement is mainly controlled by that of the reinforcing fibers, whereas in the direction perpendicular to it, the type and content of matrix and porosity have a major influence. This is evident from Figs. 7.17 and 7.18, which show the thermal expansion and CTE of UD carbon/carbon composites reinforced with different types of carbon fibers.

Figure 7.19 shows the effect of heat treatment temperature of the composites on the thermal expansion of the composites in two directions. Composites heat treated to graphitization temperatures exhibit a lower thermal expansion in both directions. Fiber volume contents in the com-

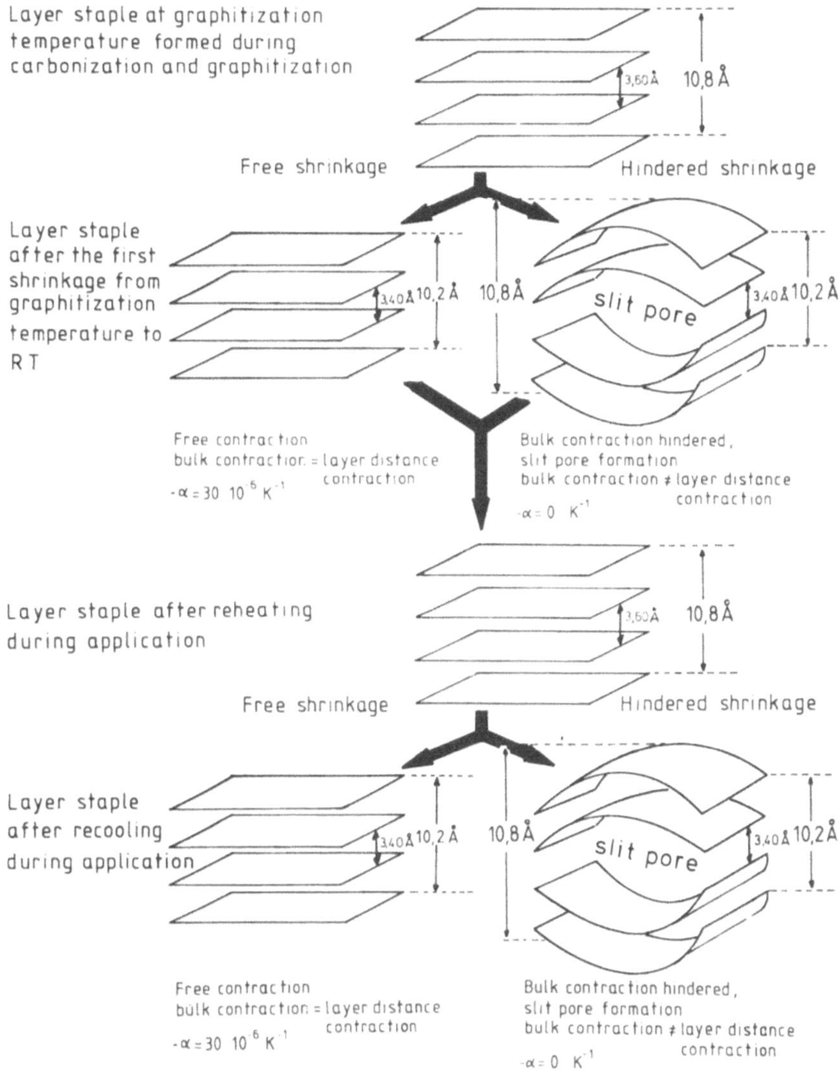

Fig. 7.16. Shrinkage/Expansion behavior of layer staples as a model for anisotropic thermal expansion in bulk graphite

posites control the anisotropy of thermal expansion of the composites. The increase in fiber volume results in an increase in the ratio of the thermal expansion in the two directions. This is is seen in Fig. 7.19. Another factor which controls the expansion in the transverse direction is the porosity in the composite. The network of shrinkage cracks in the matrix also in-

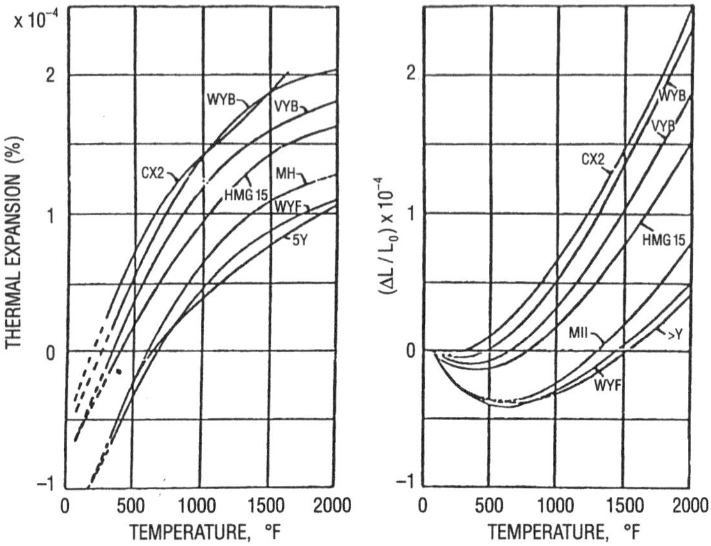

Fig. 7.17. Temperature dependence of thermal expansion (**a**) and coefficient of thermal expansion (**b**) of UD carbon/carbon composites made with different carbon fibers

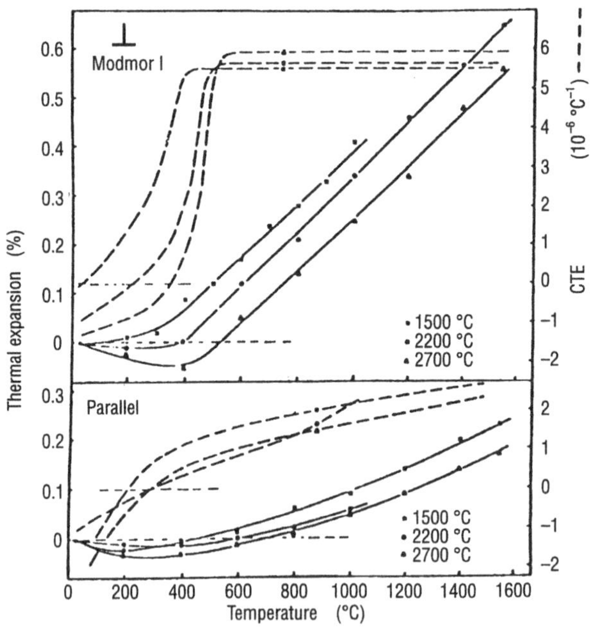

Fig. 7.18. Temperature dependence of thermal expansion (**a**) and coefficient of thermal expansion (**b**) in two directions of high-modulus reinforced thermosetting-resin-derived carbon/carbon composites heated to different temperatures

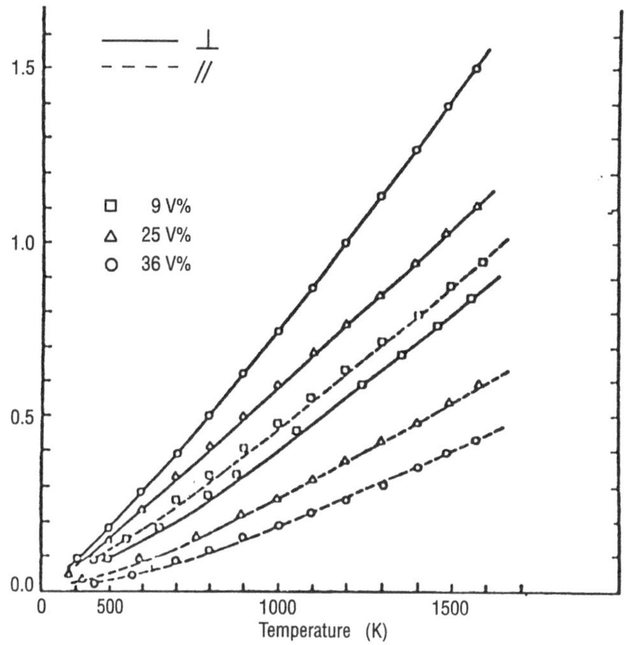

Fig. 7.19. Thermal expansion of CVD/carbon fiber felt composites with different fiber volume contents

fluence the CTE. The expansion in the direction perpendicular to the fibers is compensated by the narrowing of the pores present in the composites. Therefore, composites with a high fiber volume content and low porosity will exhibit a higher thermal expansion in the perpenducular direction than those with a low fiber volume content and a higher porosity. The CTE along and perpendicular to the fiber direction in UD carbon/carbon composites made with advanced pitch and vapor-grown high-performance carbon fibers is shown in Fig. 7.20 [27]. It also shows that the CTE of the composites is dependent on fiber type. The ratio in the two directions increases with the crystalline order of the fibers. Figure 7.21 shows the CTE of 3D carbon/carbon composites in two directions. The CTE increases with test temperature, the extent of the increase being the same in both the directions. The CTE in 3D carbon/carbon composites is lower than in UD or 2D carbon/carbon composites [28].

Meyer et al. [29] carried out a systematic study on the effect of matrix microstructure on the thermal expansion of 3D carbon/carbon composites made with pitch precursor. According to them, the anisotropy in thermal

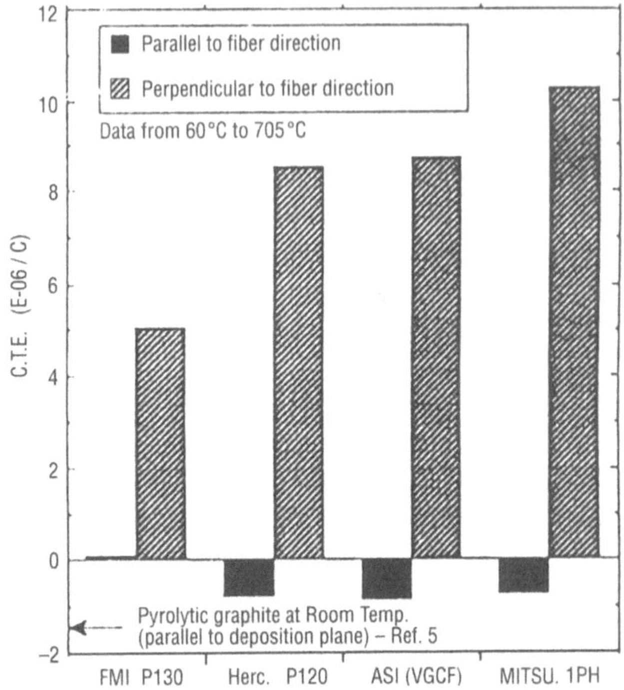

Fig. 7.20. Coefficient of thermal expansion (CTE) of UD carbon/carbon composites made with different types of high-performance carbon fibers [27]

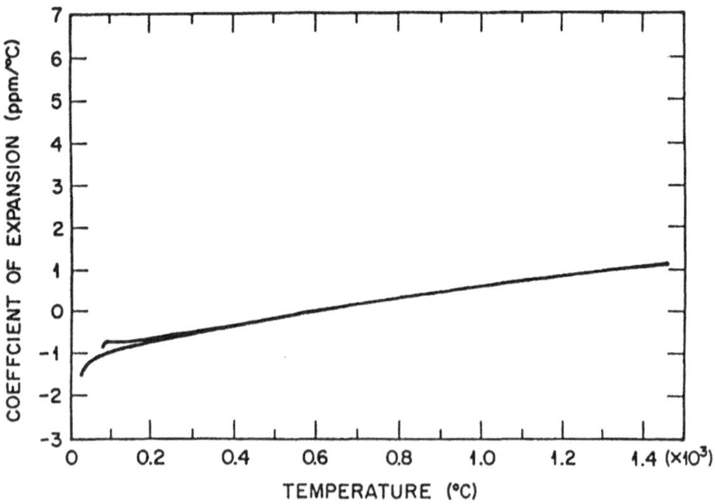

Fig. 7.21. Temperature dependence of the coefficient of thermal expansion of a 3D carbon/carbon composite in X and Y directions [22]

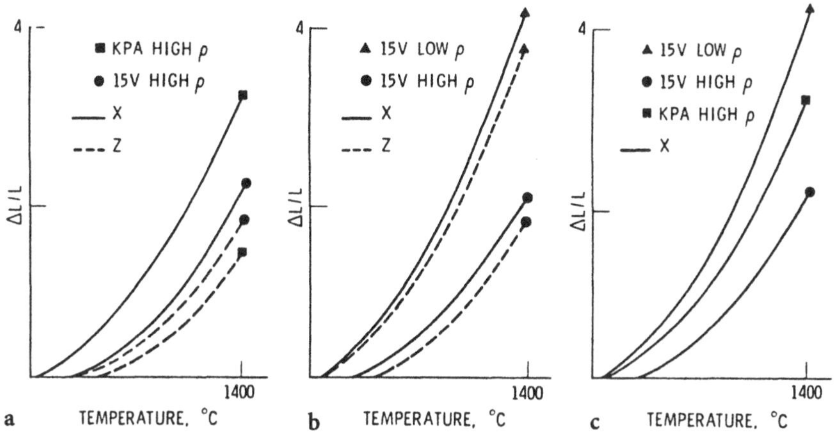

Fig. 7.22 a–c. Relationship of thermal expansion to (a) fiber direction (b) density, and (c) grain size [29]

expansion increases if the matrix microstructure, especially in the weave pockets of the composites, is changed from fine grained to coarse grained. The relationship between thermal expansion of carbon/carbon composites and fiber direction, density, and matrix grain size is shown in Fig. 7.22. The KPA composite interfilament matrix is highly aligned and coarse grained while the 15 V composite contained a predominantly disclinated, randomly oriented fine-grained matrix in between aligned sheaths of matrix around the filaments. Consequently, the former composites 261exhibit a higher anisotropy in thermal expansion than the latter ones.

7.7 Specific Heat of Carbon/Carbon Composites

Like the specific heat of conventional carbon and graphite materials, that of carbon/carbon composites increases with test temperature and heat treatment temperature. Figure 7.23 shows a typical example of the change in specific heat of carbon/carbon composites with temperature. For comparison, the data for graphite is also included in this figure.

7.8 Thermal Shock Resistance of Carbon/Carbon Composites

Thermal shock resistance of a material is an additional required thermal property of material intended for structural applications at high tempera-

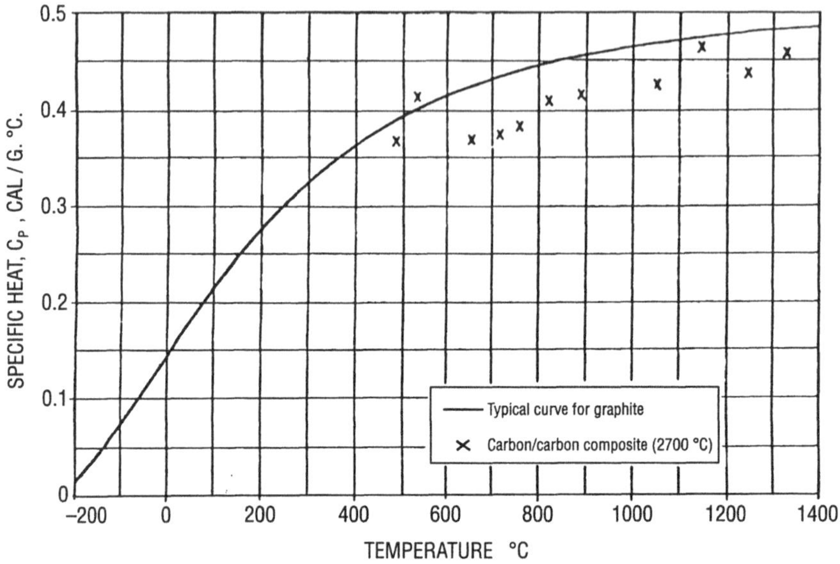

Fig. 7.23. Change in specific heat with temperature of 2D carbon/carbon composite and a pyrographite

ture, specifically under cyclic varying temperature conditions. Thermal shock resistance is theoretically calculated as:

$$\frac{s \cdot k}{E \cdot a}$$

where s is the tensile strength, k is the thermal conductivity, E is Young's modulus and a is the coefficient of thermal expansion. Thermal shock fracture toughness is calculated as :

$$\frac{K_{ic} \cdot k}{E \cdot a}$$

Graphite, having a high thermal conductivity and a low CTE, exhibits very high thermal shock resistance. Carbon/carbon composites, with a much higher tensile strength and thermal conductivity than polycrystalline graphite, exhibit a two to three times higher thermal shock resistance and an order of magnitude higher thermal shock fracture toughness than polycrystalline graphite (Fig. 7.24) [30]. The thermal shock resistance of some of the commercial carbon/carbon composites is given in Table 7.3.

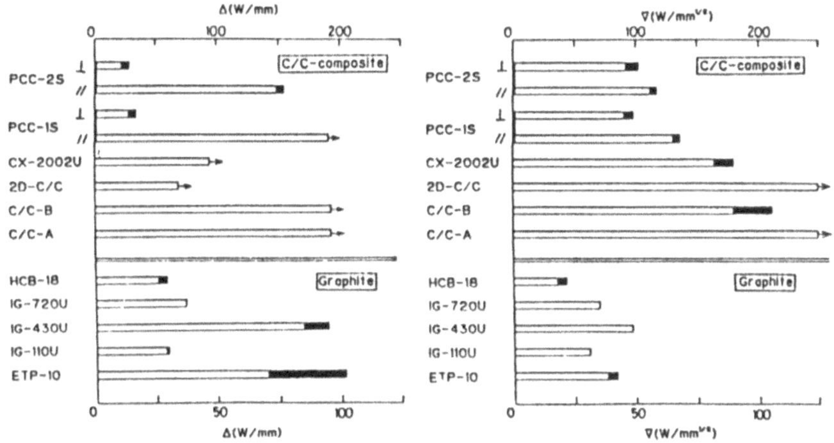

Fig. 7.24. Thermal shock resistance (△) and thermal shock fracture toughness (▽) of graphites and carbon/carbon composites

7.9 References

1. Bokros JC (1965) Physics & Chemistry of Carbon. Vol 5, p 1
2. Gopal ESR (1966) Specific heat at low temperature. Plenum Press NY
3. Kelly BT (1969) Physics & Chemistry of Carbon. Vol 5, p 119
4. Issi JP, Nysten B, Piraux L (1987) Journal of Physics. Vol 3, p 257
5. Klett JW (1994) Heat transfer in C/C Composites materials. PhD Thesis, Clemson University
6. Pierson HO (1993) Handbook of Carbon, Graphite, Diamond and Fullerenes. Noyes Publication
7. Parker WJ, Jenkins RJ, Butler CP, Abbott GL (1961) J App Phys 32, p 1679
8. Whittaker AJ, Taylor R (1990) Proc Roy Soc London A 430, p 199
9. Harris JP, Yates B, Batchelor J, Carrington PJ (1982) J Mat Sci, Vol 17, p 2925
10. Tye RP, Desjarlais AU (1981) 17th Thermal conductivity conference, Washington,
11. Taylor RE, Jortner J, Groot H (1985) Carbon. Vol 23, p 215
12. Brennen J, Bentsen LD, Hesselmann DPH (1985) J Mat Sci, Vol 20, p 2339
13. Whittaker AJ, Taylor R (1981) Proc Roy Soc London 430:167
14. Whittaker AJ, Taylor R (1981) Proc Roy Soc London 430:183
15. Fitzer E, Fritz W, Geigl KH, Vohmann W (1976) High Temperature – High Pressure Vol 8, p 187
16. Kimura S, Yasuda E, Tanade Y (1983) Microstructure and related preperties in C/C Composites. Proc Int Symp on Ceramic Composites for Engines. Japan, p 783
17. Tanamura T, Shioyama H, Ikeda S, Adachi M, Fujii R (1991) Tanso No 149, p 220
18. ZimmerJ (1991) Extended abstracts of the 20th Biennial Carbon Conference, p 390
19. McAllister LE (1983) Multidirection C/C composites. In: Kelly A, Mileiko ST (eds) Fabrication of Composites. PNH Pub

20. Pierson HO, Northrop DA (1975) J Comp Mat. Vol 9, p 118
21. Montaudon M, Gery P, Christin F (1993) 21st Bien Carbon Conf, p 384
22. Burchell TD, Oku T. Materials Property Data for fusion reactor plasma facing carbon/carbon composites. Preprints
23. Yamamosto M et al. (1993) Report Evaluation tests on first wall divertor plate material
24. Baker CF (1993) Proc 21st Biennial Carbon Conf, p 46
25. Klett JW, Edie DD (1995) Carbon. Vol 33, p 1485
26. Ting J, Lake ML (1995) Carbon. Vol 33, p 663
27. Bowers DA, Davis JW, Dinwiddie RB (1994) J Nuclear Materials. 212 p, 1163.
28. Engle GB, Tallon JA, Graves RA (1990) Proc 14 th Conf. on Metal Matrix, Carbon & Ceramic matrix composites, Cocsa Beach Fl NASA Conf. Pub 3097, p 319
29. Meyer RA, Gyatvay SR, Chase AB (1983) 16th Bien carbon Conf, p 505
30. Sato S, Kurumada A, Kawamata K, Ishide R (1990) Int Carbon Symp Tsukuba, p 214

8 Friction and Wear Properties of Carbon/Carbon Composites

Carbon-based materials are widely used in wear related applications, e.g., bearings, seals, and electrical brushes. The development of carbon/carbon composites with superior mechanical properties has opened up a new field of application for carbon materials in the wear domain, i.e., brake applications. The function of the brake is essentially controlled by its ability to absorb energy in the form of heat. Therefore, the thermal characteristics of candidate brake materials are of major importance. In particular, a potential high-performance brake material must have a high specific heat, a high melting point, and adequate mechanical properties at elevated temperature. Conventional monolithic graphite exhibits attractive thermal and physical characteristics. Since carbon materials obtained through different routes differ in the arrangement of hexagonal layers and, hence, in the interlayer cohesive energy, these materials accordingly vary in friction and wear characteristics. The friction and wear characteristics of different types of carbons have been studied in detail [1–4]. Dynamometer test results for a number of carbon materials are presented in Table 8.1. A standard sintered iron-based friction material is also included for comparison.

Table 8.1. Friction and wear characteristics of common friction materials

Material	Density (g/cc)	μ	Weight loss/stop
Monolithic graphite	1.7	0.28	0.7
	1.8	0.33	0.4
Iron based sintered friction material	5.6	0.22	1.7
Carbon/carbon composite (pitch based)	1.6	0.25	0.25
Carbon/carbon composite (CVD based)	1.6	0.19	0.25

μ, friction coefficient.

Vitreous carbon or the so-called glass-like carbon obtained through pyrolysis of the thermosetting resins exhibits very low wear rates in the order of 10^{-7} $mm^3/N^{-1}m^{-1}$. These forms of carbons, because of their low strength and poor thermal conductivity, are not suitable for friction and brake applications, particularly for high speed brakes for aircrafts, etc. These shortcomings have been overcome in carbon/carbon composites which have high structural strength, high thermal conductivity, and thermal shock resistance in addition to excellent friction and wear characteristics.

Aircraft brakes are composed of a number of disks (Fig. 8.1). Half of these (rotors), which are driven by the wheel, are sandwiched between the stators held stationary by the brake structure. The braking action is achieved by pressing the disks together with the use of hydraulic power. Friction is converted to heat, which must be dissipated. Carbon/carbon composites, although initially developed for concord and fighter aircrafts, are considered to be equally suitable for high speed surface transport systems. Dictated by the speed of the moving vehicle, the braking systems for the two categories vary in property requirement. The desired characteristic properties can be obtained by choosing the appropriate engineering design, matrix composition, and fiber layup and processing conditions. Research and development of carbon/carbon composites for aircraft applications has been mainly carried out in commercial establishments and the results are proprieatry of the companies, whereas, research and development in academic institutions concerns medium level energy dissipation brakes.

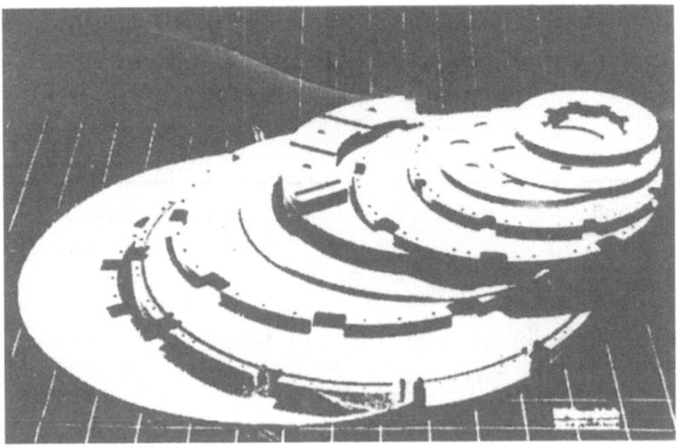

Fig. 8.1. Carbon/carbon brake disks

Most tribo systems reported on in the literature are based on metallic materials, with steel being the most dominant amongst these. However, ceramic materials are also used in high-performance tribo systems, particularly when high temperature performance is important. The crystal bonding of ceramic materials is predominantly ionic and/or covalent and these materials possess a high elastic modulus, and a high chemical and thermal stability. The particular features of ceramic tribo systems are essentially related to their extreme mechanical, chemical, and thermal stability.

The tribo system employed to stop or slow down a moving vehicle basically consists of two components: a rotor driven by the wheel and a stator held stationary by the system. During braking action, the tribo surfaces of the two components are forced to slide against each other.

The key issue in tribology is the mechanical contact between tribo surfaces. In the case of ideal surfaces, which can be atomically smooth, clean and chemically identical to bulk materials, the contact between the two surfaces pressed together by a normal load, L_0, may be considered continuous. If A_0 is the nominal contact area, then the mean or nominal contact pressure P_0 is defined by

$$P_0 = L_0/A_0.$$

However, in practice, the tribo surfaces are seldom perfectly smooth but rather have a typical surface topography which can be described by a two-dimensional waviness exhibiting a random distribution of asperities of different heights and wavelengths. Therefore, in practice, the contact between two engineering surfaces is discontinuous, as schematically illustrated in Fig. 8.2. Under a moderate normal load with no lateral sliding, only a certain number, n, of small surface elements, mainly at the summits of the highest asperities, are in direct or real mechanical contact. Under such conditions each surface element of area Ai carries its corresponding part of the normal load, the load element Li. The mean contact pressure over an element is

$$P_i = L_i/A_i.$$

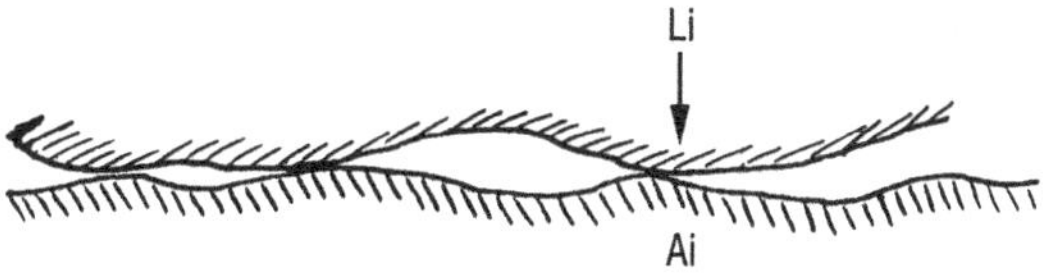

Fig. 8.2. Schematic representation of the discontinuous contact between flat surfaces

The real contact area A is made up of the sum of all contact area elements, according to

$$A = \Sigma A_i = p_i .$$

The mean real contact pressure becomes equal to L_o/A. Under extreme conditions of loading when $p_{i>>} p_o$, p_i corresponds to the definition of mean indentation hardness

$$H = L_o/A .$$

This derivation is based on contact conditions under normal force components. A general tribo system, however, also includes tangential force components which may or may not cause relative sliding between the tribo syrfaces. To maintain sliding at constant velocity the corresponding shear stress

$$t = F/A .$$

Where A is the momentary real contact area and is assumed constant and has to match the shear strength s of the weakest of the materials of the junction and F is the friction force. Under such conditions, the friction force is given by

$$F = s.A \quad \text{or}$$
$$F = s./H.L_o = \mu.Lo$$

where μ is the friction coefficient.

Thus, the coefficient of friction is predicted to vary directly with the strength of the interfacial bond and inversely with the parameter H. Within one and the same class of materials, shear strength and hardness generally vary in a related way so that the ratio remains constant [4]. Theoretically, for metals $H = 3y$ and $s = y/2$ where y is the stress yield under tension.

The friction coefficient $\mu = 1/6$, if only contributions from asperity junctions are considered. In a real system μ is slightly higher, e.g., 0.3 – 1. In addition to the basic asperity junction model outlined above, there are also some other factors which influence the friction phenomena such as junction growth, plowing, influence of temperature, and sliding velocity [2, 3].

The frictional torque accompanying the deceleration of a vehicle results in heat generation in the order of several hundred MJ, which needs to be absorbed by the system while maintaining its structural integrity. Therefore, the braking system is also required to serve as a heat sink and should possess a high thermal capacity and thermal conductivity. In brittle materials, the frictional work cannot be transformed into plastic deformation but is dissipated in tribo surfaces as heat, which for low thermal conductivity systems may result in the formation of a steep temperature gradient.

The surface temperature of a heat sink especially at the asperities, is usually higher (by 100° – 300 °C) than the uniform temperature of the bulk mass since it depends on the rate of energy dissipated per unit area and the rate at which the frictional heat can be conducted in the bulk. Analytical solutions suggest that the maximum surface temperature also depends on the stop time which in turn depends on the coefficient of friction and the applied normal force. High temperatures result from the dissipation of large kinetic energies in short stop times. Very short stop times are achieved by the application of large normal forces and large coefficients of friction. Under such abnormal operating conditions, the conventional brake systems consisting of organic or metallic materials may fail due to physicochemical changes and thermal decomposition of organic materials or to the melting of metals such as copper or steel and the building up of transfer layers, reduction of friction modifiers etc.

Wear processes have been more or less intuitively associated with friction. In many tribo systems improvement of wear resistance is sought through improved lubrication, whereas in others lubricating films may even be unwanted where high friction is desired such as in friction clutches or brakes. A wear mechanism is a micromechanism by which a tribo surface looses materials. This surface phenomenon affects a thin layer. The loss of materials from this layer takes place by the formation of large numbers of fragments on a microscale, gradually accumulating to measurable volumes [5].

For commercial carbon/carbon brake discs for various aircrafts, different types of fibers are used in a number of configurations, including fabrics, continuous fibers, semirandom chopped carbon fibers and carbon fiber mats either alone or in combinations, with or without interlacing. The configuration adopted by a particular manufacturer is their proprietary.

8.1 Design

Carbon fibers along their length are expected to have low wear and good frictional characteristics. Therefore, the early carbon/carbon disks were made with layers of carbon fabric or with random fibers arrayed parallel to the braking surface. Figure 8.3 [6] shows a schematic representation of fi-

Fig. 8.3. Different types of carbon fiber ordering used in carbon/carbon composites for brake pad applications [6]

ber ordering in three conventional types of carbon/carbon brake disks: (a) carbon fabric laminated, (b) chopped carbon fibers, and (c) laminated carbon fiber felts with cross-ply lamination. The matrix for (a) and (c) may be derived from resin or pitch or chemical vapor deposition (CVD) alone or in combination, while for (b) it is essentially initial resin derived carbon matrix densified either by CVD or by pyrolysis of resin. Ever since the development of carbon/carbon brake discs, continuous research and development has focused on improving fiber layup design and processing routes in order to achieve high thermal conductivity and improved mechanical properties and friction characteristics at reduced costs.

Under extreme operating conditions, especially during abortive take offs of heavy and fast-moving aircrafts, the peak temperature at the interface depends on the rate of generation of heat and the rate of its dissipation into the system through conduction normal to the surface of friction. Since the friction efficiency of the system is strongly affected by the temperature, it is desirable to conduct away the heat at faster rates. This necessitates the control of fiber orientation in direction perpendicular to the friction surface. Thermal boundary conditions coupled with structural, mechanical and frictional requirements suggest various fiber orientations in a carbon/carbon brake disk, which are shown schematically in Fig. 8.4. [7]. A three dimensionally woven continuous fiber structure would be ideal, but it may be cost prohibitive. The other engineering solutions suggest:

(a) needled random mat moulded to shape
(b) needled carbon fiber woven cloth with short carbon fibers – interdispersed
(c) continuous carbon fiber strands or woven carbon fiber cloth lying in the plane of the disk with layers rotated relative to one another
(d) combination of the above layup systems in an organised stack.

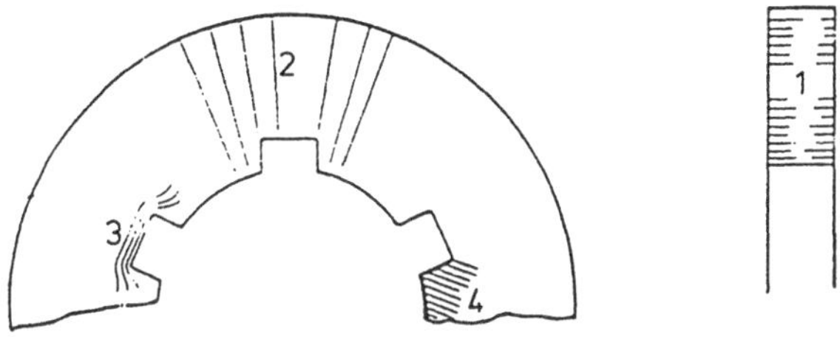

Fig. 8.4. Preferred orientation of carbon fibers in a brake pad for different performance requirements. *1*, normal to disc surface (for heat flow), *2*, radially at surface (for heat flow), *3*, around notches (for strength), *4*, angled at slot (for strength) [7]

The micromechanism of braking involves the formation of wear debris and friction film as shown schematically in Fig. 8.5 [6]. Carbon/carbon composites have a low coefficient of friction and high wear rates during taxiing operations. This results in the formation of ground wear debris containing flake-like turbostratic, less-ordered carbon than the base composite. The friction surfaces are characterized by dull, unpolished wear surfaces. Under landing conditions, the turbostratic flake-type debris is deformed and compacted to a friction film that protects the wear surface. The wear debris produced under these conditions has a higher degree of order than the taxiing debris, and its structure is more closely related to that of the base composite materials. X-ray diffraction (XRD) of RTO debris show a narrow 002 peak, suggesting stress regraphitization of the film (Table 8.2) [8]. Repeated landing operations cause delamination of a part of the friction film resulting in the formation of new wear debris. A major portion of these debris along with turbostratic debris is compacted again, regenerating the friction film. The presence or absence of these wear debris and the film that forms between the sliding surfaces are major determinants of the coeffi-

Table 8.2. X-ray diffraction of RTO debris in carbon/carbon composites

Operation	d_{002} (nm)	L_a (nm)	L_c (nm)
Basic composites	0.338	20	20
Cold taxi debris	0.343	3.4	6.4
Landing wear debris	0.338	NA	20
RTO debris	0.339	15	8.5

d_{002}, in terlayer spacing; L_a, stack width; L_c, stack height; NA, data not available. RTO, rejected take off.

Fig. 8.5. Schematic flow diagram of mechanism of braking and formation of wear debris and film in carbon/carbon brakes [6]

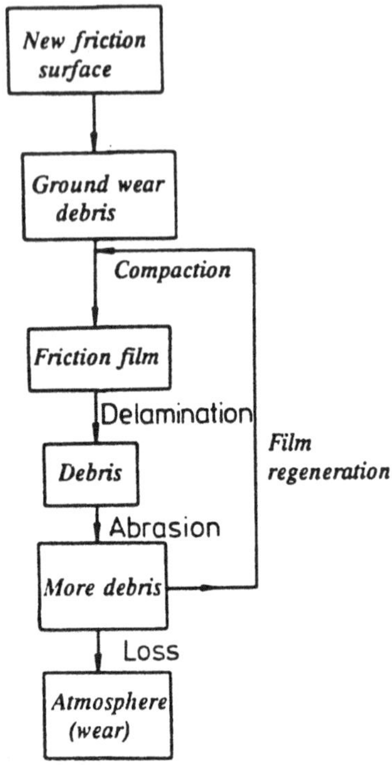

cient of friction and wear ratios of carbon/carbon composites. Therefore, factors which affect the formation and stability of this wear debris are critical to the fundamental understanding of friction and wear of these materials. The particulate debris does not lower the coefficient of friction, whereas film type debris acts as a solid lubricant, lowers the coefficient of friction, and reduces the wear rate [9, 10].

There are two dominant mechanisms of wear, or carbon loss, from the surface of the carbon brakes: losses by gasification resulting from the generation of high frictional temperature in air, and abrasion losses. In the former case, carbon is lost as atoms of carbon in the form of oxides of carbon, carbon monoxide, and carbon dioxide. The gasification processes produce chemisorbed species of oxygen known as surface oxygen complexes. The chemical reactivity of the edges of wear debris, especially those having a turbostratic structure formed during taxiing, is quite high. Therefore, chemisorption of oxygen and water adsorption in carbon/carbon brakes is higher at low speeds, i. e., during taxiing. Carbon is also lost to the atmosphere as wear dust. The type and amount of debris produced is dependent to a great extent on the type of fibers, matrix microstructure, and

Table 8.3. Properties of carbon/carbon composites for aircraft brake discs

Density	1.72 – 1.78
Open porosity	10% – 6%
Extent of graphitization	75% – 90%
Flexural strength	60 – 90 MPa
Compressive strength	120 – 140 MPa
ILSS	7.5 – 11.5 MPa
Specific heat	1.2 – 1.5 J/kg
Thermal conductivity	45 – 110 $Wm^{-1}K^{-1}$

ILSS, interlaminar shear strength.

composite macrostructure. Resin-derived, CVD, and pitch-derived carbon matrixes heated to same high temperature not only possess different graphitic contents and crystalline ordering, leading to variation in thermal conductivity, but also vary in structural characteristics, i.e., fiber/matrix bonding. Accordingly, during a braking operation these composites are heated up to different temperatures, which are highest for resin-derived low conducting carbon and lower for CVD and pitch-derived highly ordered, high thermal conducting carbon matrixes. Therefore, oxidation weight loss for resin-derived composites is higher than for CVD and pitch-derived composites, whereas the formation of debris is greatest for the latter composites. One needs to optimize the contents of different types of structures and hence of carbon materials to be used in order to have an optimum set-up. Table 8.3 gives the characteristic physical properties of carbon/carbon composites required for brake disc applications.

8.2 Low Energy Tribological Behavior of Carbon/Carbon Composites

Asbestos fibers are used as reinforcement in conventional resin matrix composite brakes. Since the recognition that small-diameter asbestos fibers are carcinogenic, the brake manufacturing industries have been searching for asbestos-free friction materials. For efficient performance, the replacement reinforcement should have mechanical and thermal properties which are at least as superior as those of asbestos fibers. Among inorganic fibers, carbon fibers are known to have extraordinary characteristics surpassing even those of asbestos fibers. With the successful application of carbon/carbon composites as brake material in supersonic aircrafts, it is hoped that these materials can also be used for automobile industries with low energy frictional requirements and especially as carbon/carbon brake lining against cast iron discs in conventional automotive brakes. Compared with carbon/carbon composites for aircraft industries, research and development of carbon/carbon composites with automotive brakes has been undertaken in more countries. Basically because these systems require materials with low energy friction properties, composites with less stringent configuration and architecture can be used [11, 12].

Studies of these composites are made with commercially available tribotesters. Figure 8.6 [12] shows a commercial tribotester type RWS 60B made by M/s Krauss Messtechnik, Murr, Germany. The tester is equipped with a commercially available brake system consisting of cast iron discs and a brake saddle with two brake shoes. The friction coefficient is calculated from the friction momentum which is transferred by a lever to a load cell. The temperature is measured with a thermocouple on the circumference of the disc. As for high-performance aircraft brake discs, systematic studies have been prformed on the effect of various parameters, such as type, orientation, amount, and configuration of fibers, heat treatment temperature, processing conditions, and microstructure, on the performance of carbon/carbon brakes under low torque conditions [12–15].

Figure 8.7 [15] shows a schematic representation of carbon fiber orientation of unidirectional (UD) CFRC composites against the disc. The friction coefficient and friction wear obtained with these samples are plotted

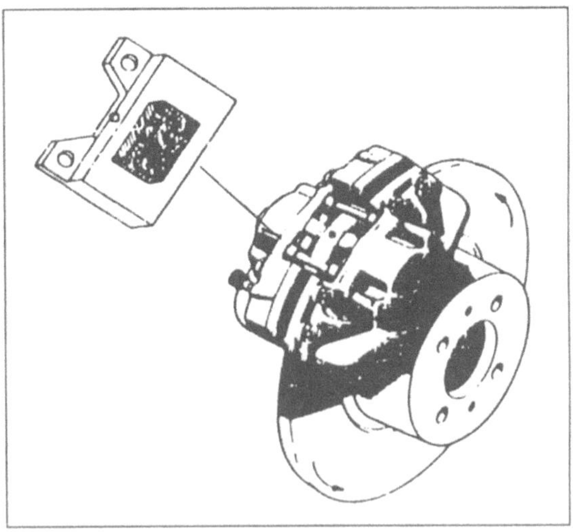

Fig. 8.6. Sample holder in Krauss commercial turbotester [12]

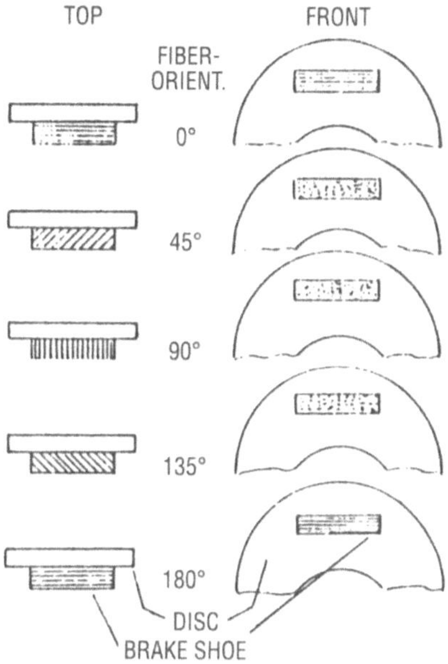

Fig. 8.7. Carbon fiber orientations in the samples against the brake disks [15]

in Fig. 8.8 [12]. The carbon/carbon composites with around 45° fiber orientation have the highest friction coefficient and a high wear rate [12]. Friction characteristics of carbon/carbon composites are also found to be dependent on the fiber volume fraction of the composites and the density of the composites. These are shown in Fig. 8.9 for two-directional (2D) carbon/carbon composites. High-density carbon/carbon composites with low porosity and high integrity exhibit the best performance of all.

Chen and Ju [16] also studied the low energy tribological behavior of various carbon/carbon composites with varying fiber orientations in two directions and matrix microstructure. They observed that like the high energy friction mechanism, under low energy tribological testing also type III debris films are formed, resulting in a decrease of friction after an initial increase. Among various combinations of weave patterns, composites made with chopped, 8H satin weave and pitch-resin CVI matrix configuration exhibited a low friction coefficient, the lowest weight loss, and a stable wear rate. According to Pu Baojion et al. 8HS satin rayon-based carbon/carbon composites, because of their relatively flat surfaces, exhibit better friction patterns than those made with plain fabric (Fig. 8.10) [17]. The tribological characteristics of 2D composites are sensitive to braking pressure, linear velocity, and absorbed energy of carbon disks [18–21]. Figure 8.11 [18] shows the variation of the friction coefficient with different initial

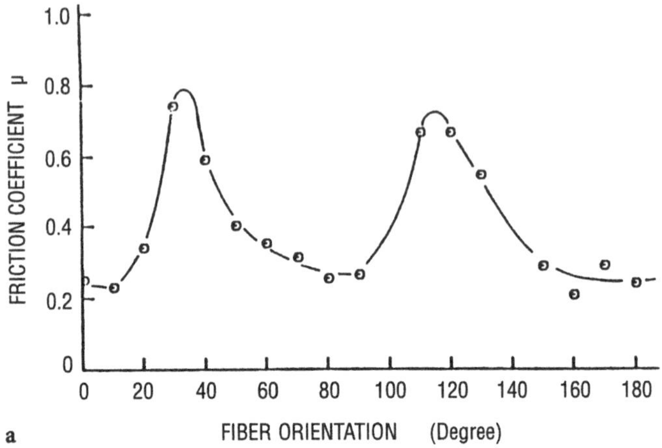

a

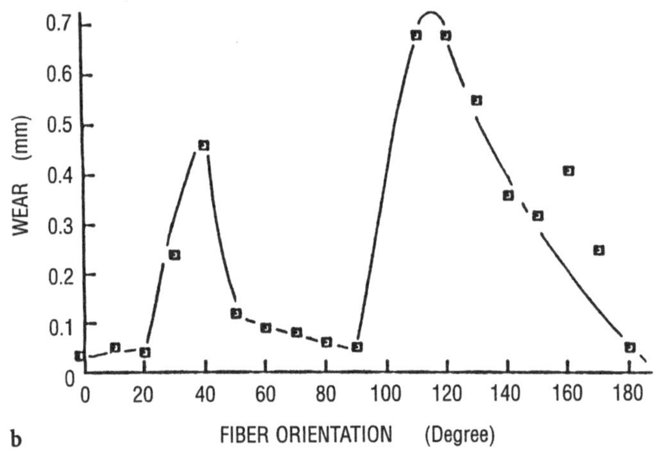

b

Fig. 8.8 a, b. Coefficient of friction (**a**) and friction wear (**b**) as function of fiber orientation in UD composites [12]

velocities. The average friction coefficient is found to increase with decreasing initial velocity. The friction coefficient decreases with increasing pressure.

Detailed research work has shown that also the fiber/matrix bonding has a significant influence on the friction behavior of the composites. The fiber ends do not break uncontrolled, but rather they are grinded corresponding to their angles. Composites with a 3D configuration exhibit not only the highest friction coefficient but also a constant friction over a wide temperature range. Not only the optimum fiber orientation but also the

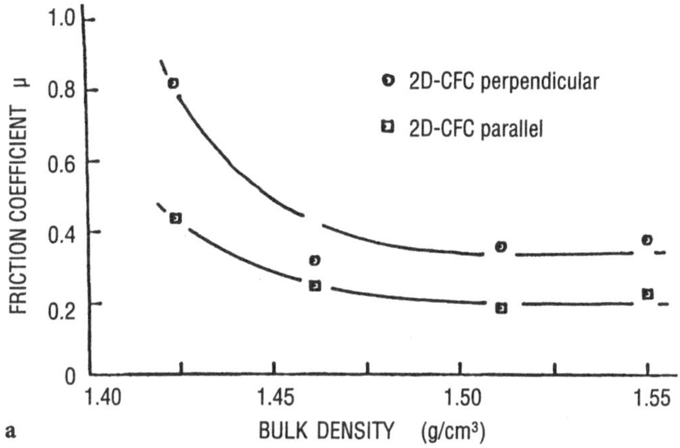

a

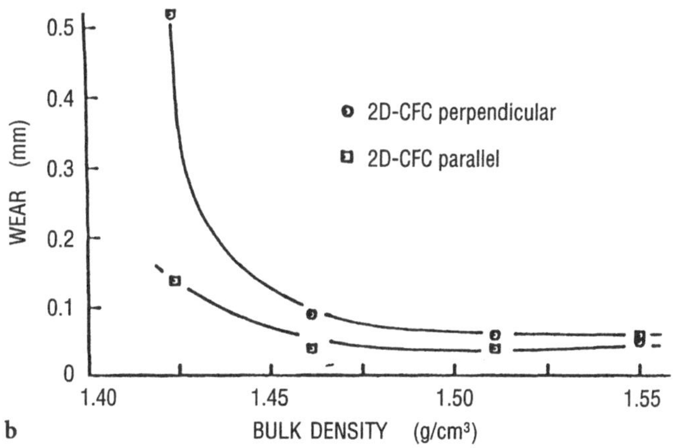

b

Fig. 8.9 a, b. Coefficient of friction (**a**) and friction wear (**b**) of 2D carbon/carbon composites as function of bulk density, i.e., fiber volume fraction [15]

optimum fiber type and volume content must be chosen. According to Kimura et al. [22], the coefficient of friction decreases with increasing Young's modulus of the composites. This implies that the composites should be heat treated at higher temperatures than usual. Figure 8.12 shows the effect of heat treatment temperature on the friction coefficient of UD carbon/carbon composites. Similarly, wear also decreases with increasing heat treatment temperatures of the composites. This has been explained by a reduced abrasion effect due to shearing of graphitic planes in composites heat-treated at high temperatures. This is also evident from the graphitic film formation on the disc surface. A higher heat treatment temperature of

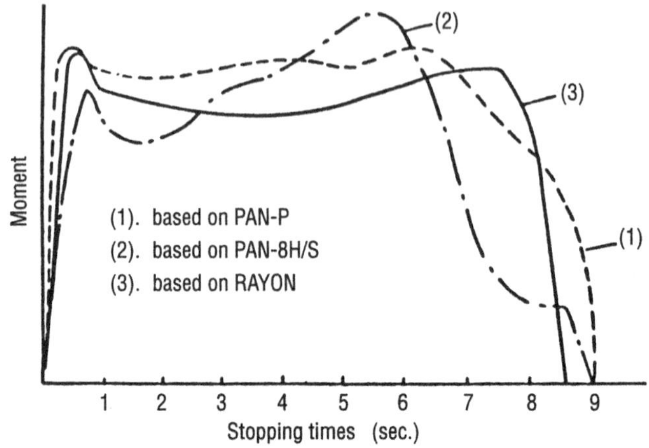

Fig. 8.10. Relative moment curves of three types of carbon/carbon composites [17]

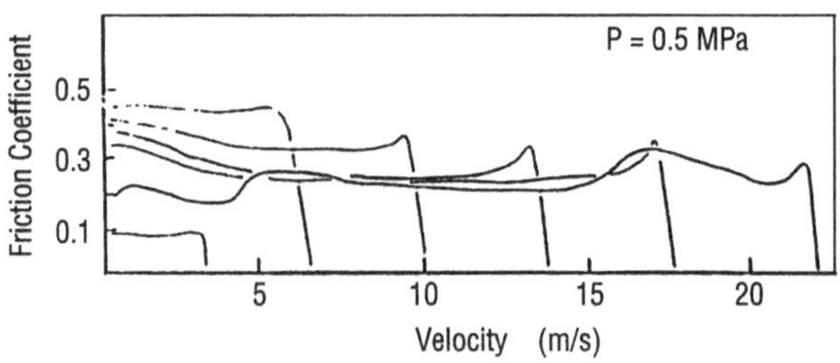

Fig. 8.11. Variation of friction coefficient of carbon/carbon composites with different initial velocities [18]

the composites, i.e., a higher degree of graphitization of the fiber or composites, results in a slow rise in temperature of the friction surface. For example, nongraphitized composites exhibit a friction surface temperature as high as 800 °C, while that of graphitized composites is quite low. The friction coefficient is observed to be higher for composites with higher frictional surface temperatures. Accordingly, composites with a low graphitic content, i.e., with low heat treatment temperature exhibit a high friction surface temperature and hence a high frictional coefficient. A heat treatment temperature for composites of around 1600 °C is found to be optimum.

The frictional characteristics are also found to be influenced by the environment as shown in Fig. 8.13. In a wet environment, the rise in friction surface temperature is low and as is the coefficient of friction. After

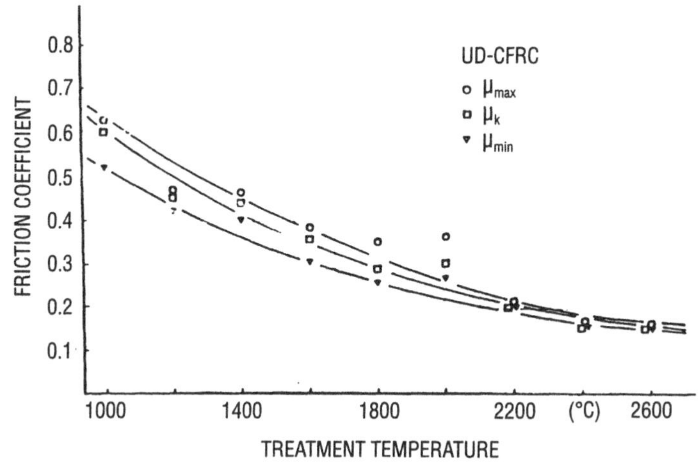

Fig. 8.12. Effect of heat treatment temperature on friction coefficient of UD carbon/carbon composites

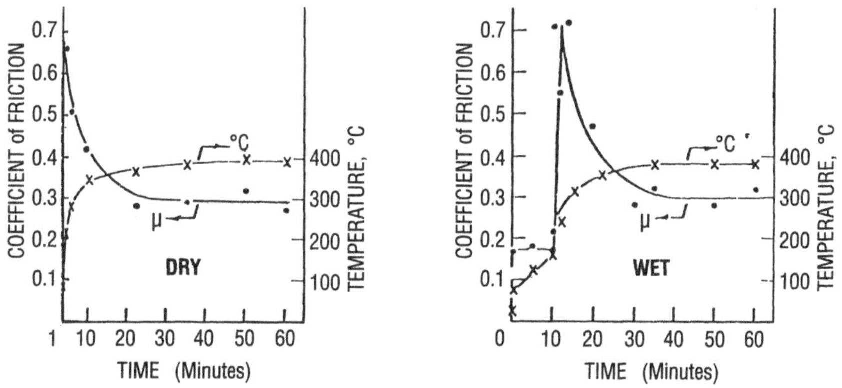

Fig. 8.13. Effect of environment on coefficient of friction and surface temperature as a function of time in a turbotester

about 10 min, the moisture is desorbed, the temperature rises, and the coefficient of friction also increases ultimately behaving as in a dry environment.

Not only the carbon fiber and composite characteristics but also the inorganic additive, if added to the matrix of the composites, influences the friction characteristics of the ultimate composites. TiN or carbides, which are otherwise good abrasive materials, when added to carbon/carbon composites reduce the friction coefficient of the resulting material from 0.8

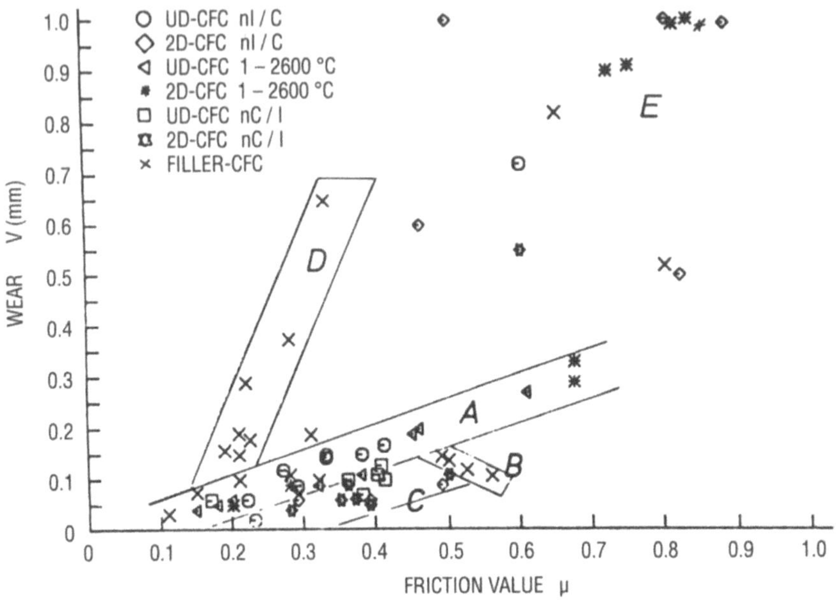

Fig. 8.14. Effect of additives in carbon/carbon composites on coefficient of friction and wear [23]

to 0.5. Figure 8.14 shows the effect of different types of additives on the coefficient of friction and wear of carbon/carbon composites. Region A corresponds to the linear relationship between friction coefficient and wear. Region B corresponds to carbon/carbon composites filled with TiN. Region C corresponds to those additionally impregnated with polymeric materials. D corresponds to inorganic fillers other than TiN. Above a friction coefficient of 0.7, the wear rate becomes very high. This is seen in region E. Likewise, the surface temperature of such composites, which is normally in the range of 400 °C for a friction coefficient of 0.3 to 0.4, rises to as high as 700° – 800 °C for a rise in friction coefficient to 0.6 and above [11]. Similar effects are observed with silicon, boron, or phosphorous-based antioxidation additives and coatings [23[.

In order to use carbon fibers instead of asbestos in friction materials and for their wider applications in industrial brake linings for surface transportations etc, it is essential to produce carbon fibers and carbon/carbon composites in an economically effective way. This may require development of cheap carbon fibers and matrix precursors, alternative composite manufacturing techniques etc. M/s Ashland Petroleum Company and the

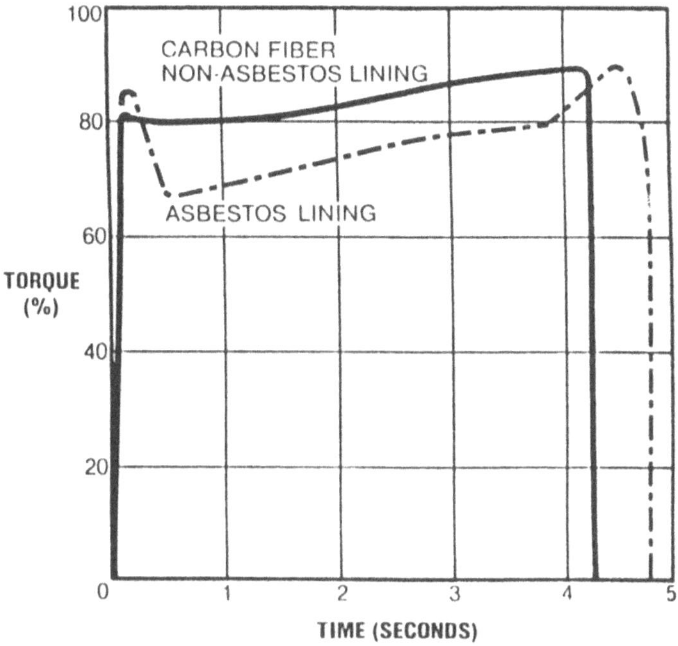

Fig. 8.15. Comparison of carbon fiber composites and asbestos-lined friction surfaces [24]

Industrial Brake Division of Goodyear Aerospace have developed carbon/carbon composite brake and friction materials incorporating inexpensive carbon fibers and matrix precursors. The studies made by them showed that carbon fiber composites exhibit a much better performance than asbestos-lined brake system (Fig. 8.15) [24]. Use of oxidized PAN fibers is increasing in the fabrication of carbon/carbon brake discs. Oxidized PAN fibers are used as continuous fibers, mat, or 2D woven cloth. Besides the low costs involved, the handling of oxidized PAN fibers is much easier than for carbon fibers. These fibers are cut and laid according to a predetermined configuration and heat treated to high temperatures (1600°–2200 °C) before infiltration by the CVI technique. Pu, Wang, and Zhang [25], while studying the tribological behavior of carbon/carbon composties made with different frictional interfaces observed that starting with oxidized PAN fibers, the friction interfaces consisting of cloth-reinforced CVI carbon against mat-reinforced CVI carbon have the highest friction factor and a low wear.

8.3 References

1. Burton RA, Burton RG (1987) Carbon matrix and ceramic matrix tribocomposites. Symp of the Mechanical Failure Preventive Group Cambridge University Press
2. Kennedy F (1987) Thermochemical contact phenomena and wear of sliding composites. Annual Report ONR Contact N00014-81-K0090
3. Wright MA (1995) Brake Materials Research, In: Materially speaking. Vol 10, N 2
4. Lancaster JK (1979) The wear of carbon & graphite. Treatise on materials. Sci & Tech 13, p 141
5. Awasthi S, Wood JL (1988) Advanced Ceramic Materials 3, N 5, p 449
6. Savage G (1993) Carbon/carbon Composites. Chapman & Hall Pub
7. Stimson IL, Fisher R (1980) Trans Roy Soc Lond A 294, pp 583
8. Huton T, McEnaney B (1995) Proc MTC Int Conf on Carbon and Carbonaceous composites materials Malenovice, Czech Republic
9. Murdie N, Ju CP, Don J, Fortuanto FA (1991) Carbon 29:335
10. Ju CP, Lee KJ, Wu HD, Chen CI (1994) Carbon 32, p 971
11. Fitzer E, Gkogkidis A, Heine M (1984) High Temp High Pressure 16, p 363
12. Gkogkidis A (1986) PhD Thesis University of Karlsruhe
13. Fitzer E, Fritz W, Gkogkidis A, Moergenthaler KD (1986) Sprechsaal. Vol 119, p 463
14. Stanek P, Murdie N, Hippo EJ, Howdyshell B (1993) Bien Carbon Conference, p 378
15. Fitzer E et al (1986) 4th Int Conf Carbon, Baden-Baden Carbon 86, p 867
16. Chen JD, Ju CP (1995) Carbon. Vol 33, p 57
17. Baojian P, Huigen X, Xukai Z, Pengzhi O 19th Bien Carbon Conf, p 342
18. Narita N, Kuroski K, Herai T (1990) Friction Mechanism of C/C Composites Int Sym on Carbon Tsukuba Japan, p 386
19. Baojian P, Huang Q (1989) 19th Bien Carbon Conf, p 352
20. Murdie N, Don J, Kocher C, Lie R, Ju CP (1989) 19th Bien Carbon Conf, p 496
21. Li CC, Shahan EE (1981) Friction & wear studies of carbon composites in air and in helium in Wear of Materials ASME, pp 525
22. Kimura S, Yasuda E, Narita N (1983) Friction and Wear of C/C Composites. J Japanese Society of Lubricating Engineers 28, pp 185
23. Mayer BH (1991) PhD Thesis, University of Karlsruhe, Germany
24. Hettinger Jr WP, Newman JW, Krock RP, Boyer DC (1986) Ashland Petroleum Company Information Brochure
25. Baojing P, Wang F, Zhang J (1993) Bien Carbon Conf, p 148

9 Oxidation and Oxidation Protection of Carbon/Carbon Composites

In contrast to the attractive thermal and mechanical properties of carbon/carbon composites discussed in previous chapters, a severe limitation with these composites is their high susceptibility to oxidation in an oxidizing environment at temperatures even below 500 °C. The life span of carbon/carbon composites in a nonoxidizing environment is found to be same at 2500 °C as it is at 350 °C, whereas in the presence of air it has been found to decrease from 400 h to zero at 500 °C [1]. This quite often becomes a major barrier to the use of carbon/carbon composites as a structural material in high-temperature systems operating in oxidizing environments, such as turbine engines and ramjet components that are exposed to hot combustion gases, air breathing propulsion systems, and air frame components experiencing aerodynamic heating. Clearly, for wider applications of carbon/carbon composites as high temperature structural materials, some form of oxidation protection is essential. An understanding of the oxidation mechanism for carbons in general and that for carbon/carbon composites in particular helps in the development of protection systems for these composites.

9.1 Oxidation of Carbon/Carbon Composites

The oxidation process in carbon materials in the presence of oxygen as well as in steam has been studied in great detail over the past 50 years [2, 3]. In graphitized carbon, it is well known that gasification of carbon atoms takes place at exposed edges of the layer planes and also at vacancies and dislocations in the basal planes; however, the rate of reaction in the prismatic planes is much higher than in the basal planes [2]. The difference in reactivity is so high that carbon atoms can be classified into reactive and unreactive carbons. This has given rise to the concept of active surface area (ASA) [4]. Each carbon material posses active surface area to a varying extent; this controls the gasification rate of the material. The active surface area itself changes with different burn off of the host carbon. In disordered carbons the gasification reactivity is influenced by many important factors such as (i) crystalline order controlled by the form of carbon since the thermodynamic activity of nongraphitized (disordered) carbon is higher than that of graphite; (ii) active surface area; (iii) access to catalyst particles, especially those which are strong catalysts of oxidation, e.g., Fe, Co, and Ni; and (iv) surface chemistry related to chemisorption of oxygen at low temperature with subsequent desorption of oxides of carbon.

With the development of carbon/carbon composites, studies on the oxidation behavior of these composites gained importance. Carbon/carbon composites are comprised of multiphase carbons, i.e., carbon fibers with different microstructures and crystalline order and carbon matrix with varying crystalline order and porosity. All of these variations in the constituents will influence the oxidation behavior of carbon/carbon composites. Therefore, a great deal of research work has been carried out to determine rate controlling factors for oxidation of carbon/carbon composites. These include oxidation behavior of constituents such as carbon fibers, chars, and cokes alone as well as in combination [5–11]. An oxidative attack on carbon fibers may take place at the cross section or lateral surfaces depending on the skin core heterogeneity and crystalline order. This is especially the case in pitch-based carbon fibers from wihich fibers with isotropic carbon to highly crystalline microstructure (P120, P130) can be produced. In the case of PAN-based carbon fibers, high-modulus fibers are found to be more oxidation resistant than high-strength fibers, high-modulus fibers do not exhibit any skin core preferential oxidation [8]. PAN-based carbon fibers are found to be more oxidation resistant than pitch-based carbon fibers of the same modulii.

In accordance with general composite theories, the mechanical and physical properties of carbon/carbon composites obey to a great extent the rule of mixture, but the intrinsic rate of oxidation of the composites is not an additive property with respect to the reactivity of the components. Various interfaces of the types fiber/matrix, matrix/matrix, and fiber/fiber discussed in Sect. 5.4 predominantly control the oxidation kinetics of carbon/carbon composites. Based on theoretical analyses, Luthra [12] has proposed different oxidation mechanisms operating at different temperatures. At temperatures below about 500 °C, the oxidation of carbon/carbon composites is controlled by a carbon/oxygen interface reaction. At higher temperatures, the rate limiting step slowly changes to a mixed, gas-phase diffusion-interface reaction. At temperatures above ~700 °C, depending on the gas flow conditions, the reaction is limited by gas-phase diffusion. Chang and Rhee [5] reported similar observations based on their oxidation experiments with thermosetting-resin-derived carbons.

In general the oxidation of carbon/carbon composites yields linear Arrhenius plots for log of gasification rate vs 1/T (Fig. 9.1). The gasification rate increases with decreasing heat treatment temperature. However, activation energies of gasification vary with the composite systems. McKee, on studying the variation of porosity with gasification of a variety of carbon/carbon composites, has proposed that gasification of the composites result in a rapid increase in percent porosity and mean pore diameter [13]. Oxidation first takes place at the interfaces between fibers and matrix. The fiber bundles provide pathways for the rapid diffusion of oxygen into the composite interiors. Oxidation then proceeds either to fibers or to matrix depending on the structural order of the constituents.

Some authors have observed that fibers are more reactive than the matrix especially in the case of composites made with carbon fibers having low modulus and crystallinity [10, 13]. In contrast Yasuda et al. [9] have reported that the matrix has a faster rate of oxidation and in the initial stages the

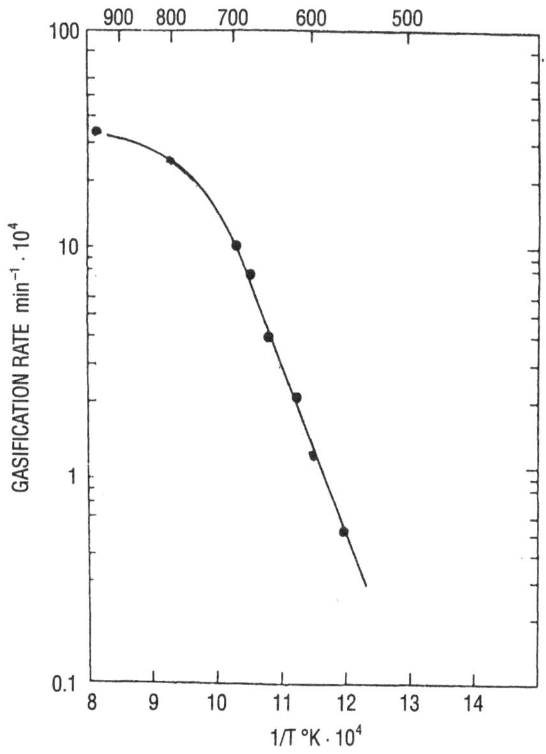

Fig. 9.1. Arrhenius plot (log rate vs. 1/T) for the oxidation of carbon/carbon in air

matrix is oxidized first followed by a slower oxidation of the fibers. Jones et al. [14] have observed that the gasification behavior of carbon/carbon composites is directly linked to the structure of the constituent fibers. Composites made with PAN carbon fibers are found to be more reactive than those made with pitch carbon fibers. This is attributed to the associated binder phase microstructure which is influenced by the type of fibers. Carbon matrix wrapped around pitch carbon fibers in a concentric fashion will obviously possess a lower surface area than a matrix with a rough scale type structure. Although the latter type of matrix microstructure may be advantageous with respect to the mechanical properties of the composites, ingress of oxygen results in a higher gasification and lenticular voids parallel to the fiber surface.

Similar results are obtained with resin-derived carbon/carbon composites made with surface-treated and non-surface-treated carbon fibers [15]. In the former composites, during pyrolysis the carbon matrix shrinks onto the fibers while in the latter composites the matrix shrinks away from the fibers resulting in slit-type porosity. This in turn results in differential gasification of the carbon/carbon composites (Fig. 9.2) [15].

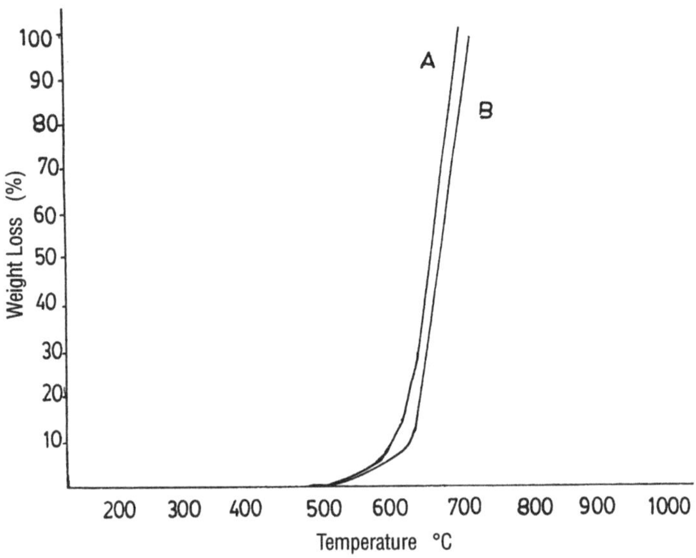

Fig. 9.2. Oxidative weight loss with increase in temperature of carbon/carbon composites (heat treatment temperature of 1000 °C) with (*A*) weak and (*B*) strong fiber/matrix bonding [15]

The matrix microstructure of the composites developed at higher temperatures as a result of differential carbonization shrinkage stresses in the composites is also found to have an effect on the oxidation rate of the two composites. Composites having a kinked graphitic structure and good fiber/matrix bonding exhibit a lower susceptibility to oxidation than those having long range lamellar structure with slit-type porosity (Fig. 9.3) [15]. In such systems, the relative reactivity of fibers and matrix depends, however, on the structure of each phase and the various interfaces present in these composites [15, 16]. The preferential oxidation of fibers or matrix in carbon/carbon composites with different macrostructures and matrix microstructure shown in Fig. 9.4. Under wet oxidation conditions also, composites with an anisotropic matrix exhibit a higher durability than those with isotropic matrixes [17].

Using carbon/carbon composites made with fibers and a carbon matrix heat treated to different temperatures, Fishbach and Uptegrove [11] have observed that the microstructural characteristics of the matrix, such as crystallographic order, preferred orientation, and morphology, have a great influence on the gasification of carbon/carbon composites. The oxidation rate of carbonized composites is about an order of magnitude higher

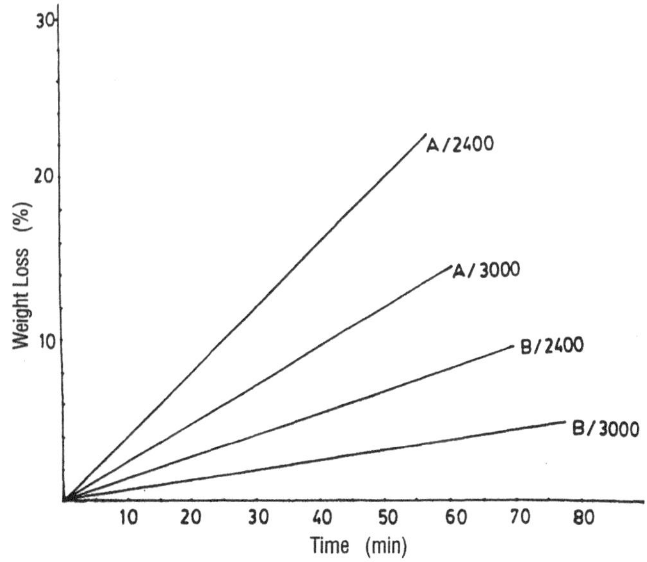

Fig. 9.3. Effect of heat treatment temperature (2400 K and 3000 K) on oxidation behavior of carbon/carbon composites with (*A*) weak and (*B*) strong fiber/matrix bonding [15]

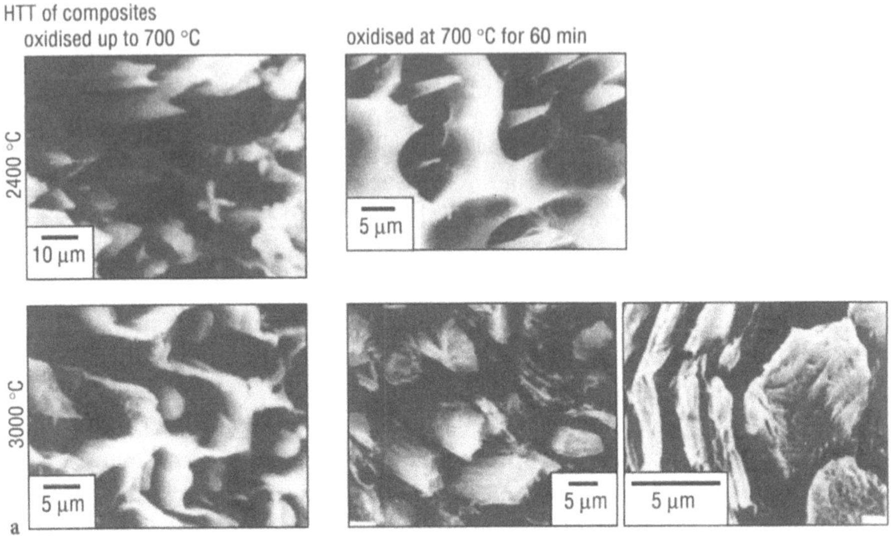

Fig. 9.4 a, b. Scanning electron micrographs of partially oxidized carbon/carbon composites exhibiting preferential oxidation of fibers and matrix in composites processed with different variables; **a** weak fiber/matrix bonding

than that of the graphitized materials. Microscopic examinations of the oxidized composites reveal that oxidation of the graphite matrix composites starts at the lateral surface of the fibers and progresses to the graphitized matrix. However, if the fibers have a thin coating of graphitized matrix adhered to the surface, they are protected and oxidation starts at the matrix/matrix interface. High temperature treatment also lowers the content of low volatile impurities which otherwise act as a catalyst of oxidation, thus reducing the gasification of carbon/carbon composites.

The preferential reactivity of the unorganized carbon matrix or fibers or an interphase layer as discussed above for unidirectional (UD) and cloth-reinforced composites is also found to prevail for three-directional (3D) carbon/carbon composites [18]. It has been established that highly crystalline graphitized carbon exhibits higher oxidation resistance than less-ordered carbon matrix and that in carbon/carbon composites matrix graphitization can be achieved more conveniently by the phenomenon of stress graphitization. Therefore, researchers are trying to produce a graphitic carbon matrix in situ in carbon/carbon composites. The group led by Thrower and Radovic [19] has proposed a new method of improving the carbon structure in carbon/carbon composites and hence the oxidation resistance of the composites. They have observed that improved oxidation re-

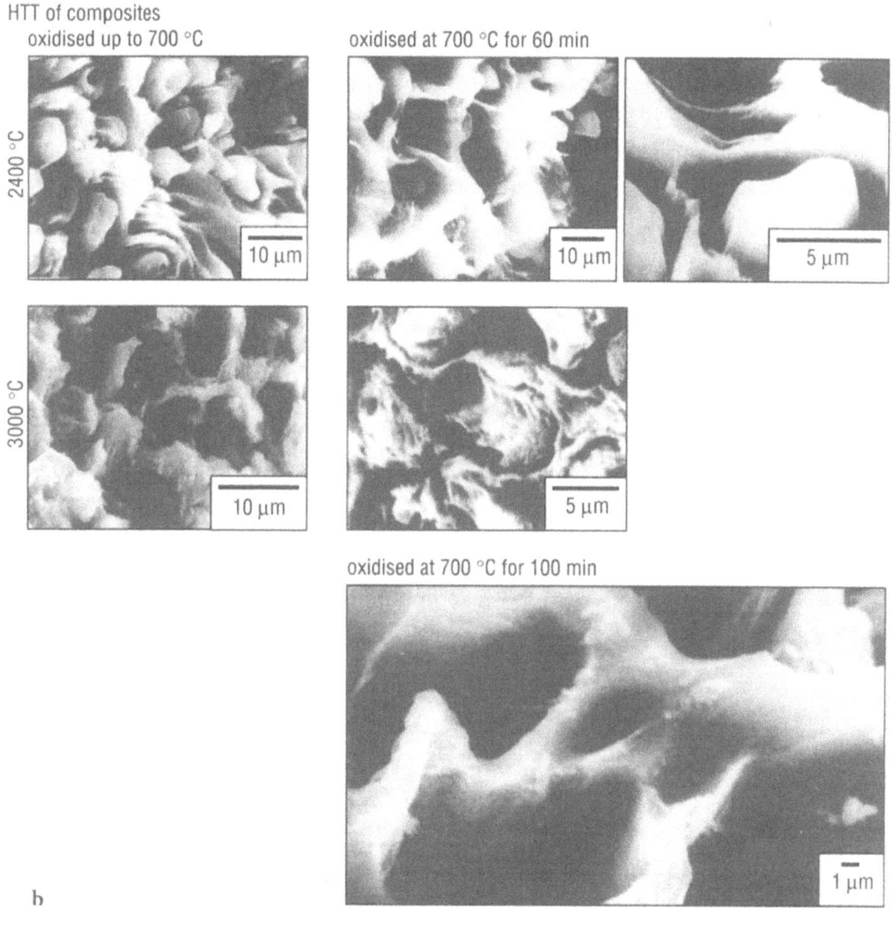

Fig. 9.4 (continued) **b** strong fiber/matrix bonding [15]

sistance is invariably obtained if porosity is developed in the fibers prior to composite preparation. When subsequent carbonization of the matrix takes place within the pores of the fibers, the matrix is graphitized in a way similar to stress graphitization. The resulting carbon at the fiber/matrix interface is more oxidation resistant than that obtained when matrix carbonization occurs in the presence of the fibers that do not possess a developed porous structure.

Chang and Rusnak [6] have reported interesting experimental observations on the effect of gas flow rate. Although the effect of gas flow rate on the gasification of carbon/carbon composites is not significant at 650 °C, it becomes quite prominent at a temperature of 750 °C and higher. The effect of stress on the oxidation rate of carbon has been examined by a few researchers with contradicting results. Thrower and Marx [20] while conducting the experiments under stress did not find any effect on the oxidation rate. Knefeld et al. [21] on the other hand have reported that a tensile stress on graphite increases the oxidation rate significantly at 1000 °C in an environment of argon with a small amount of water vapor. In the case of carbon/carbon composites, at low temperature (625 °C) the oxidation rate is not influenced by tensile stresses, while at high temperatures (~ 1350° – 1400 °C) tensile stresses increase the oxidative weight loss.

Another aspect which is related to oxidation of carbon/carbon composites and their applications as a structural material is the effect of oxidation on the mechanical properties of carbon/carbon composites. Very few researchers have reported on this aspect. Crocker and McEnaney [22] observed that during oxidation of two-directional (2D) carbon/carbon composites, oxygen attacks at the fiber/matrix interfaces within fiber bundles, leading to decoupling of fiber and matrix and a reduction in flexural strength. The mode of fracture also changes depending on the micro- as well as macrostructure of the initial composites. They found a reduction in elastic modulus and flexural strength by 30% and 50%, respectively, after a 10% weight loss at 700 °C. Zao et al. [23], who conducted the experiments at a low oxidation temperature in the order of 600 °C, found that oxidation of carbon/carbon composites has a catastrophic effect on the mechanical properties of carbon/carbon composites. A 20% oxidative weight loss decreases the strength by about 60% and Young's modulus by about 75%. Ahearn and Rand [24] studied the effect of low level oxidation (less than 3.5%) at two temperatures, i.e., 420 °C and 730 °C on the fiber/matrix bonding and mechanical properties of pitch-derived UD carbon/carbon composites. They also observed that the material loss occurs primarily at the fiber/matrix interface. After carbon burn off, composites which were oxidized at 730 °C continued to fail catastrophically at low strength, while those oxidized at 420 °C showed a change in mode of failure from a totally brittle mechanism to a pseudoplastic failure mode. In the latter case, the flexural strength and ultimate failure strain are reported to increase by 85% and 125% respectively, for a weight loss of 1.4%.

9.2 Protection of Carbon/Carbon Composites Against Oxidation

As discussed above, two basic rate-controlling factors responsible for oxidation of carbon/carbon composites are composite texture and structural defects and gaseous diffusion. The relative contribution of the two depends on the application temperature. At lower temperatures (500°-800 °C) the first factor dominates, while the second factor controls the gasification at higher temperatures (800°-2000 °C). Furthermore, with respect to applications, carbon/carbon composites are catagorized as "limited life" and "extended life" application composites [25]. The former involve minimal thermal cycling and short time operations at temperatures above 1650 °C, whereas the latter composites undergo numerous thermal cycling and hundreds of hours of operation at temperatures below 1650 °C. The oxidation rate of carbon can be reduced by interfering with the operating oxidation mechanisms in the temperature range of interest. This will mean different protection systems at different temperatures. Intensive research has been performed over the last 50 years to develop coatings for high-temperature oxidation protection of graphites for various applications. Most of the reports are in patent form and hence very little is discussed in the open literature. McKee [26] has made an excellent comprehensive compilation of the patent literature on oxidation protection of carbon materials. The experiences gained from carbon industries are basically being persued to develop oxidation protection coatings for carbon/carbon composites along with improved techniques such as chemical vapor deposition (CVD) or plasma spray or sol-gel technology.

Various approaches which have been described in th open literature for protection of carbon/carbon composites are summarized in Table 9.1. These can be grouped into two catagories: (i) inhibitors and sealants to block the active sites and slow down the oxidation rate and (ii) diffusion barriers consisting of coatings and overcoats to prevent oxygen ingress and carbon egress. However, most of the current systems of oxidation protection use several of these methods in a compatible or complimentary manner. Needless to say the type of protection system required is dictated by the end application of the carbon/carbon composite. Systems used for coat-

Table 9.1. Methods for oxidation protection of carbon/carbon composites

Inhibitors and sealants	Barrier coatings and overcoats
Halogens	Noble metals
Phosphorus Compounds	Borides
Boron compounds	Carbides/nitrides
Polysiloxanes	Silicides
Silicon	Intermetallic compounds
Borate and silicate glasses	Engel/beewer compounds

ing carbon/carbon composites in cruise missiles may prove to be unsatisfactory for aircraft brake application and vice versa.

9.2.1 Oxidation Protection by Inhibitors

The rate of carbon gasification at low temperature can be effectively retarded by decreasing the porosity and blocking or poisoning the active sites by elements which have an inhibiting effect on oxidation, such as boron, phosphorus, and silicon. In fact, active sites on a carbon surface responsible for gasification of carbon are equally preferred sites for the adsorption of inhibitors such as halogen compounds, boron, and phosphorus compounds. These compounds are generally impregnated into carbon/carbon composites in solution form. The inhibition effect of halogens on the oxidation behavior of carbon appears to be confined to temperatures below about 900 °C; at higher temperatures the halogen atoms are desorbed from the carbon surface, and their protective effect is thereby lost [27]. Phosphorus compounds alone or in combination with halogens are found to have very strong inhibiting effects [28].

Phosphorus in the form of organophosphorus compounds are impregnated into the carbon materials followed by thermal decomposition at around 400 °C. This leaves behind a strongly adsorbed phosphate residue on the carbon surface which poisons the sites most likely to be oxidized [29]. Inorganic derivatives of phosphates can also be impregnated into the carbon/carbon composites either as premix or as a solution [1, 28]. These are found to have a marked effect on the oxidation inhibition of carbon/carbon composites at moderate temperatures. Figure 9.5 shows the oxidation behavior of carbon/carbon composites before and after impregnation with $Zn_2P_2O_7$ [30].

The effectiveness of phosphorus compounds as inhibitors of oxidation is also limited to a temperature of 850° – 900 °C. Phosphates in combination with other inhibitors, such as silica, silicon carbide, and boron oxide, have been found to reduce the oxidation of carbon at 1500 °C to one tenth.

Borates are generally more effective than phosphates in reducing the oxidation rate of carbon. Therefore, extensive studies have been made on the effect of boron and boron compounds on the inhibition of carbon–oxygen reactions. These compounds have been widely used for oxidation protection of fibers, carbon matrixes, as well as the carbon/carbon composites. Carbon fibers impregnated with boron have been found to exhibit an increased thermal stability in an oxidizing environment as a result of active

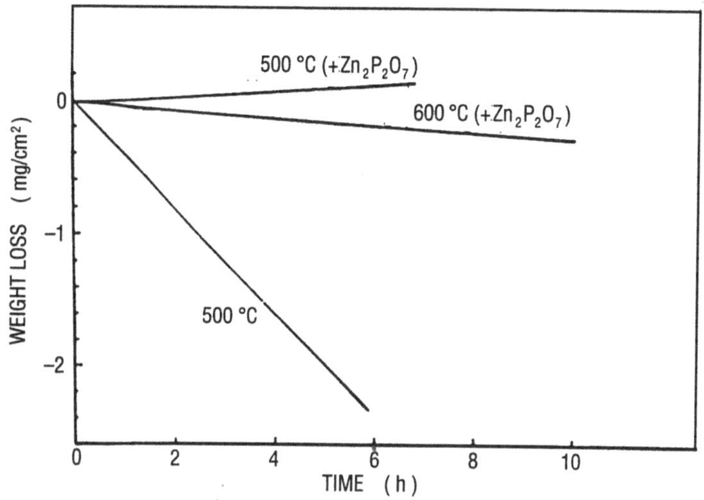

Fig. 9.5. Effect of $Zn_2P_2O_7$ inhibitor on isothermal oxidation behavior of carbon/carbon composites

site poisoning [31]. Boron oxide is found to inhibit oxidation in two ways: (i) by poisoning the active sites; (ii) through sealants by making a glassy barrier film. Ehrburger et al. [31] reported that even with a low boron oxide content (up to 3.5%) there is a significant inhibiting effect which does not markedly change with the level of burn off. They observed that the inhibition effect of boron oxide increases with concentration up to a loading of 5% or so. Above this, it remains almost constant. However, if the composites impregnated with boron oxide to more than 10% are annealed at 700 °C for 60 h, the inhibition efficiency increases. Although the rates of carbon gasification are reduced appreciably by treatment with boron, the apparent activation energy is not affected by the presence of boron oxide. This indicates a lowering of the number of oxidation sites in the treated material with no change in oxidation mechanism. With increased amounts of boron oxide, a continuous layer can be formed which fills and seals the pores and contributes to a further decrease in the carbon oxidation rate due to the oxygen diffusion process.

McKee et al. [32] proposed that carbon impregnated with organoborates and subsequently heat treated are much less susceptible to oxidation. According to them, the polymeric borate groups $(BO_3)_n$ are preferentially absorbed on zigzag sites of the graphitic crystallites as shown in Fig. 9.6 [27] and inhibit oxidation. Addition of B_2O_3 to 10% by weight results in a

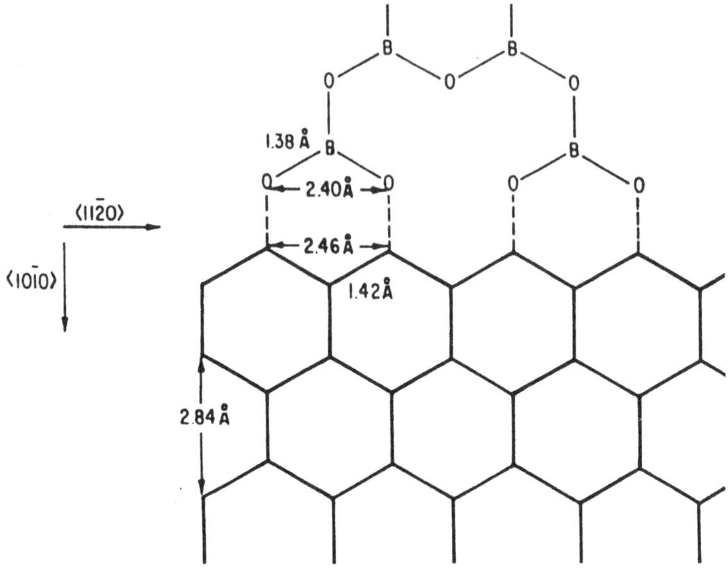

Fig. 9.6. Schematic representation of bonding of $(BO_3)_n$ polymer to {1010} face of graphite lattice [27]

decrease in the oxidation rate of carbon by an order of magnitude. Jones and Thrower [33] have found similar oxidation inhibiting effects of boron in high-modulus pitch-based carbon fibers doped with boron. Halogenated glasses, particularly those containing 5–10 mol % MgF_2 have been found to be more effective oxidation inhibitors at temperature above 1250 °C than B_2O_3 glass inhibitors [34]. Based on theoretical considerations, Luthra has estimated the maximum use temperature for long-life applications of boron-inhibited carbon/carbon composites to be around 700 °C.

Mixtures of metallic borides such as those of hafnium, zirconium, elemental boron, silicon carbide, and other glass-forming particulates are also often added as inhibitors in the matrixes during the formation of carbon/carbon composites. On exposure to oxygen, these additives form mobile borates or borosilicate glasses and protect the carbon/carbon composites from oxidation.

Of all the materials used as oxidation inhibitors for carbon/carbon composites, boron in one form or another, i. e., either elemental boron or boron compounds, has been found to be the most appropriate material for oxidation protection at temperatures less than 1500 °C. This is because of the appropriate viscosity of borate glasses formed during the initial reaction with oxygen, coating the constituents of the composites, and sealing the matrix inner and surface cracks [35]. Investigations on the oxidation behavior of oxidation inhibitor modified matrix carbon/carbon composites have shown that the initial gasification of such composites is very high but decreases with time as the glassy inhibitor phase spreads through the composites [36)]. Scanning electron micrographs of the partially oxidized composites reveal a glassy phase between the fiber bundles (Fig. 9.7) [36].

Bromine also inhibits oxidation of carbon by accepting electrons from carbon, thereby lowering its Fermi level [37]. Ho and Chung [38] studied the effect of bromination of carbon fibers and carbon/carbon composites on their oxidation behavior by exposing them to bromine vapors at room temperature. They found that bromination reduces the oxidation rate of carbon/carbon composites by up to 43% at 650 °C and that this is effective only for composites heat treated to medium temperature only. The inhibitor-modified matrix technique or the so called admix technique is to be utilized with caution. A nonuniform distribution of additives, which may be most common, leads to quality control problems. Addition of these inorganic materials in the precursor matrix affects the microstructure of the ultimate carbon matrix and, hence, the attractive physical and thermal properties of carbon/carbon composites. In certain cases, the mechanical properties of the matrix-modified carbon/carbon composites have been found to be 20% – 30% lower than that of the unmodified composites. Borate glas-

Fig. 9.7. Section through "matrix-inhibited" carbon/carbon composites after 30% in weight burnoff in air at 800 °C [36]

ses, in particular, pose problems of moisture sensitivity, long-term compatibility with the substrate, and long-term high temperature stability. Borate-containing carbon/carbon composites may require external protection coatings for an efficient performance at high temperatures [39].

9.2.2 Protection by Coatings

For high temperature oxidation protection of carbon/carbon composites, certain coatings and overcoats have to be applied to the composites. The choice of materials for these coatings is restricted by certain physical and chemical criteria listed in Table 9.2.

In practice, it is difficult for a single material to satisfy all of these criteria. Materials with thermal properties which match those of the substrates and low oxygen permeability are not stable at high temperatures. On the other hand, in the case of materials which are stable at very high temperatures, their thermal properties differ appreciably, resulting in the development of tensile stresses in the coating during rapid thermal cycling which in turn leads to cracks in the coating. Many studies have been carried out on certain metallic and nonmetallic materials having compromising properties in the search for a suitable material for oxidation protection coating for carbon/carbon composites.

Table 9.2. Important characteristics of barrier coatings for carbon/carbon composites

Physical compatibility
The barrier coating should be well adhered to the substrate and encapsulate it uniformly
Should be free of cracks and connected pores and remain so over complete application temperature
Should have coefficient of thermal expansion very close to that of substrate
Chemical properties
Should be stable up to very high temperatures and be nonvolatile in the presence or absence of oxygen
Should have low oxygen permeability and low carbon egression
Should not chemically interact with carbon substrate and with the overcoats etc.

9.2.2.1 Noble Metal Coatings

Noble metals such as platinum and iridium are reported to have been tested as high temperature protective coatings for carbons [40–42]. It is believed that hese high temperature metals inhibit cracking arising due to thermal mismatch by virtue of their ductile nature. Although coating of platinum by the CVD technique yields nonporous layers, its suitability as a high temperature oxidation protection coating is found to be limited since it exhibits pinhole oxidation at high temperature. Iridium is of particular interest as a high temperature protective barrier coating [39, 43]. Iridium has a high melting point (2400 °C) and is an excellent oxygen and carbon barrier [43]. It does not react with carbon to form carbides. In the early days of the US space program, considerable interest in iridium coating was shown especially for the protection of graphite nose tips. Iridium coatings have been successfully applied by means of slurry dipping, clading, and electroplating. However, at very high temperatures, iridium is susceptible to vaporization by the formation of volatile oxides such as Ir_2O_3 and IrO_3 [44, 45]. The coefficient of thermal expansion of iridium is also much higher than that of carbon. These specific inherent problems along with the prohibitive cost of iridium are the factors which limit the application of this metal as an oxidation protection coating for carbon/carbon composites.

9.2.2.2 Silicon Coatings

Silicon-based ceramic coatings, due to their suitable thermophysical properties, have been quite attractive for carbon/carbon oxidation protection at temperatures exceeding 1000 °C for extended periods. As is evident from Figs. 9.8 and 9.9 [39], silicon-based ceramics have the best thermal expansion compatibility with carbon/carbon composites and exhibit lowest oxidation rate amongst high temperature ceramics. Moreover, they adhere well to carbon/carbon composites. Fused silica has a very low oxygen permeability, especially at high temperatures. Dense, high purity silicon carbide and silicon nitride coatings have the ability to afford protection at temperatures of 1700°–1800 °C under highly oxidizing conditions. These coatings are applied to the carbon/carbon substrate by pack cementation, CVD, vapor phase deposition, plasma spray, electrodeposition, etc. Most of the work reported in the literature is based on pack cementation and the CVD process since these processes produce materials in the purest form.

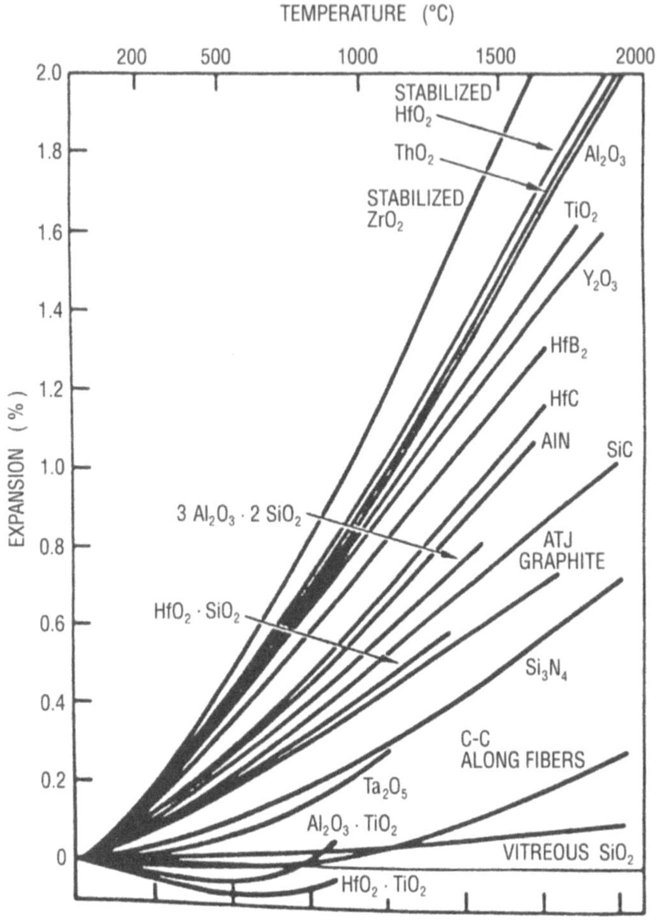

Fig. 9.8. Thermal expansion characteristics of ceramics [39]

There is, however, an additional way to obtain a well-bonded silicon carbide layer on carbon substrate. This is by chemical reaction of the carbon substrate with molten silicon to form silicon carbide surface layers [46–50]. Figure 9.10 demonstrates the application of this liquid-impregnation-reaction technique for a fine grained graphite bearing shell. This technique has also been applied to carbon/carbon composites as shown in Fig. 9.11.

However, care must be taken to avoid the preferred reaction of silicon melt with the carbon fibers rather than with the carbon matrix. This problem can be solved by the controlled formation of the highly reactive carbon matrix. This process is in commercial use for oxidation protection of car-

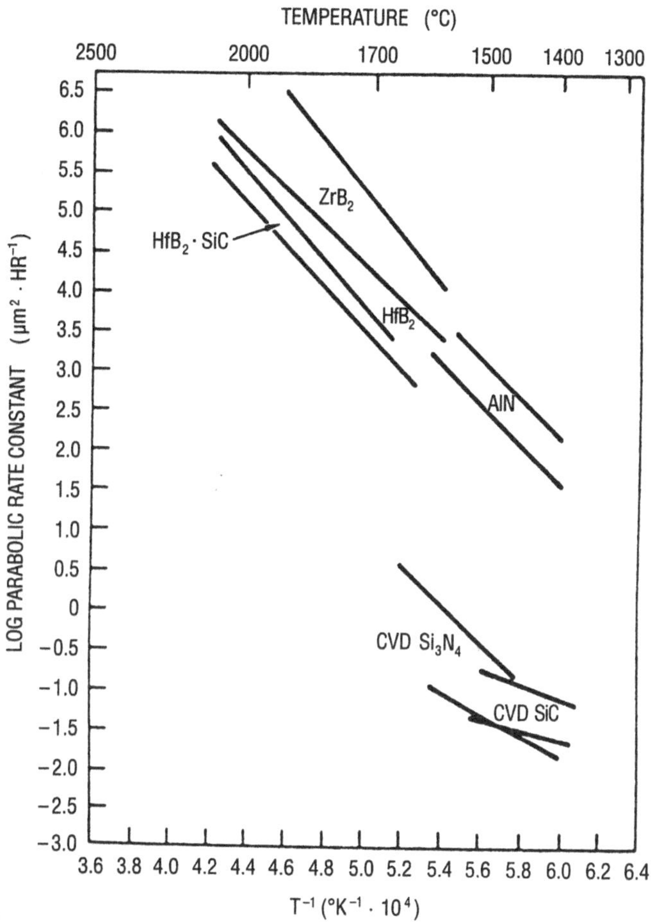

Fig. 9.9. Oxidation characteristics of high temperature ceramics [39]

bon/carbon composites in various temperature ranges above 1600 °C. Sometimes, double layers of SiC (the first one formed by siliconizing with a CVD overcoat) are found more advantageous. A recent development that is particularly useful for filling the pores of carbon/carbon composites is chemical vapor infiltration (CVI). This is applied for infiltration of both silicon carbide and silicon nitride. Silicon nitride has a lower thermal expansion coefficient than silicon carbide and should be a more effective coating than silicon carbide.

One of the principal limiting factors for the success of SiC as an oxidation protection coating on carbon/carbon composites is the formation of

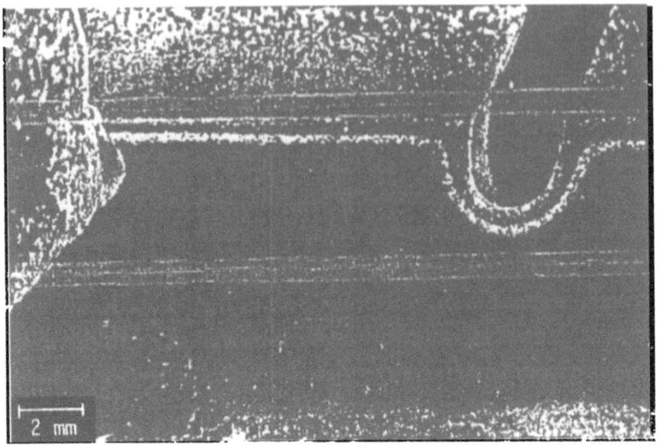

Fig. 9.10. Application of liquid impregnation reaction technique to fine grained graphite bearing shell. A layer of silicon (*top*) well adhered to graphite block (*below*)

Fig. 9.11. Total siliconized surface layer in carbon/carbon composite

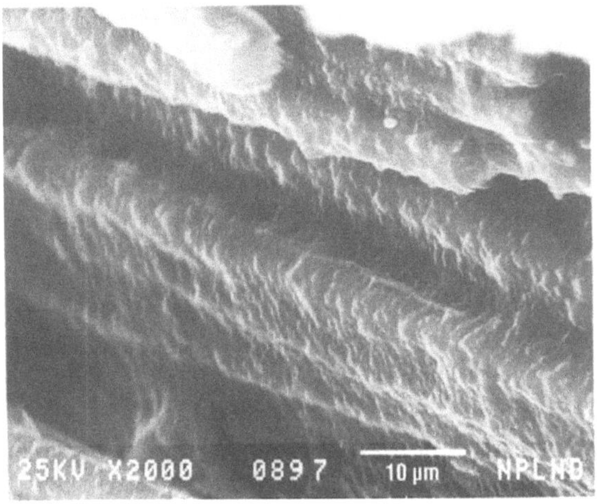

crack patterns in the SiC coating. Mismatch in coefficient of thermal expansion of the coating and substrate induces tensile strains in the coatings and hence cracking during cooling from high temperature processing and during high temperature thermal cycling. These cracks, however, may give way to ingress of oxygen and oxidation of carbon/carbon composites underneath. For low temperature oxidation protection, borate glasses are used as sealants for these cracks.

Figure 9.12 shows the oxidation behavior of a carbon/carbon composite coated first with silicon carbide and subsequently with molten boron oxide

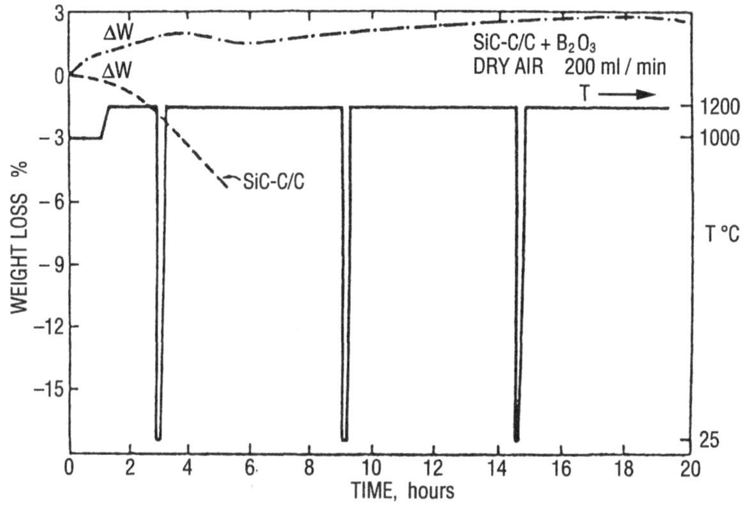

Fig. 9.12. Oxidation behavior of SiC-C/C specimens, before and after coating with borate, in dry flowing air at 1200 °C (with intermittent cooling to 25 °C). Weight changes vs time [7]

during thermal cycling [7]. As is evident from this figure, composites with a silicon carbide coating alone suffer a rapid oxidative loss possibly through the cracks while for those having an overcoat of borate glass, no appreciable weight loss is detected for 20 h at 1200 °C. Although oxidation-protected carbon/carbon composites containing borate glasses within the matrix as well as surface crack healers have shown encouraging results for extended life protection up to 1200 °C, the moisture senstivity of borate glasses and the relatively high oxygen permeability of B_2O_3 are quite sensitive issues with respect to their reliability and safety factors. Coatings containing borate glasses may develop defects and spall due to hydrolysis and expansion. This is especially true if the glass is present beneath the coating as well as in the coating cracks. Therefore, for the development of coatings for protection above 1200°–1400 °C, especially for extended life applications, alternative material and coating configurations need to be adopted. In the temperature range 1200°–1400 °C, the outer surface of silicon carbide or silicon nitride is oxidized via the following reactions depending on the pressure and temperature to give a thin silica surface film:

$$SiC + 3O_2 \rightarrow SiO_2 + CO$$
$$SiC + 2O_2 \rightarrow SiO_2 + CO_2$$

$$SiC + O_2 \rightarrow SiO(g) + CO$$
$$SiC + 3/2O_2 \rightarrow SiO(g) + CO_2$$
$$Si_3N_4 + 3O_2 \rightarrow 3SiO_2 + 2N_2$$

The reaction products can also combine between themselves as

$$SiO(g) + 1/2O_2 \rightarrow SiO_2$$
$$SiO(g) + CO \rightarrow CO_2 + Si$$
$$SiC + 2CO \rightarrow SiO_2 + 3C$$
$$Si\,O_2 + CO \rightarrow CO_2 + SiO\,(g)$$

Figure 9.13 [51] shows the standard free energy of the reactions giving SiO and SiO_2. Silica acts as a self-healing and protective oxygen diffusion barrier film. Once formed, it protects the underneath silicon carbide and silicon nitride coatings. Considering the advantages of silica films as oxygen diffusion barriers, tetraethyl orthosilicate (TEOS) impregnation techniques and sol-gel routes have been recently adopted to deposit silica films either on carbon/carbon composites or on the SiC coated carbon/carbon composites [52–54]. Figure 9.14 shows a comparison of the loss in mass between carbon/carbon composites with and without TEOS overcoats. Silicon in the form of silicides ($MoSi_2$) has also been used for oxidation pro-

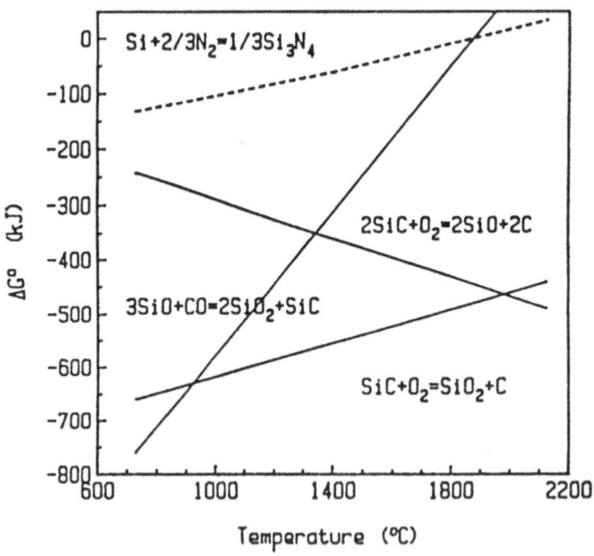

Fig. 9.13. Free standard energy of reactions as a function of temperature [51]

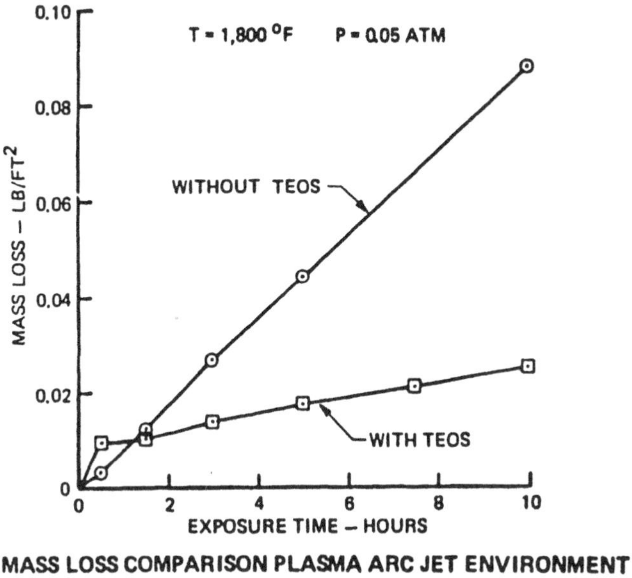

Fig. 9.14. Comparison of mass loss of carbon/carbon composites with and without tetraethyl orthosilicate (TEOS) coating

tection of carbons [48, 53]. However, one has to be very cautious since molybdenum oxide acts as catalyst for oxidation of carbon materials.

Silica coatings, although they are good oxygen barriers, exhibit structural problems at high temperatures. Silica coatings heated at temperatures exceeding 1200 °C for long periods of time tend to devitrify to cristobalite or tridymite structures, resulting in volume contraction and cracking. Chemical reaction between silica and the material underneath may also cause problems. The following reactions may occur between SiO_2 and silicon coatings [39].

$$SiC + 2SiO_2 \rightarrow 3SiO + CO$$
$$SiC + SiO_2 \rightarrow 2SiO + C$$
$$2SiC + SiO_2 \rightarrow 3Si(s) + 2CO$$
$$Si_3N_4 + 3SiO_2 \rightarrow 6SiO + 2N_2$$

According to Ehrburger [51] the oxidation protection of carbon/carbon composites by silicon based ceramic coatings through formation of a silica layer can be divided into two temperature regimes:

(i) below transition temperature (Tc), the oxidation of SiC and Si_3N_4 give essentially solid SiO_2 (passive oxidation)
(ii) above Tc, gaseous SiO formation dominates (active oxidation).

The first reaction is accompanied by a slight increase in weight, while in second case, weight loss occurs as a result of the evolution of SiO. The temperature of transition will depend on the pressure of oxygen and reactants. If the pressure of oxygen is greater than the sum partial pressure of SiO and CO or SiO and N_2, then oxidation will be passive. If the opposite is the case, oxidation will be active. The thermodynamic calculations and experimental verifications have shown that at temperatures around 1700°–1800 °C, the total gas pressure at the interface may exceed 1 atm causing disruption of the protective silica layer. This limits the application temperature of SiC and Si_3N_4 coatings. This drawback can be overcome to some extent by having free silicon in the system. The presence of free silicon increases the temperature of bubble formation and lowers oxygen, carbon, and nitrogen pressures, thus raising the application temperature limit of SiC and Si_3N_4 coatings [39, 42, 55, 56]. Because of this, the pack cementation process, whereby the desired amount of free silicon can be introduced during SiC formation, is most often used. In this process, carbon/carbon material is placed in a pack containing SiC with Al_2O_3, Si, MgO etc. When heated, this mixture generates a protective layer on the composite surface. Impurities or additives used for lowering the viscosity of silica also cause blistering of the coating and must therefore be avoided.

New innovative approaches are being tried to overcome the problems of cracking in SiC and Si_3N_4 coatings and their oxidation [57–59]. These are based upon: (a) functionally gradient CVD of C/SiC/Si_3N_4 by continuously varying the coating composition from a carbon-rich silicon carbide to pure silicon carbide and then a depositing a Si_3N_4 coating on top of the SiC coating [58]; (b) functionally gradient chemical vapor reaction (CVR) C/SiC/Si/Si_3N_4 coating, wherein functionally gradient C/SiC coating is formed via SiO reaction with the carbon/carbon substrate, Si is formed via SiO chemical vapor reaction deposition and the top Si layer is converted to Si_3N_4 layer via nitridation. These coatings exhibit excellent adherence to the carbon/carbon substrate combined with a crack-free appearance. This has been attributed to the following factors: (i) functionally gradient CVD coatings redistribute the stresses over the entire coating thickness and form a carbon-rich silicon carbide interfacial layer; (ii) codeposition of carbon with SiC allows a gradual change in the coating's thermal expansion and Young's modulus. This in addition to a very fine microporosity results

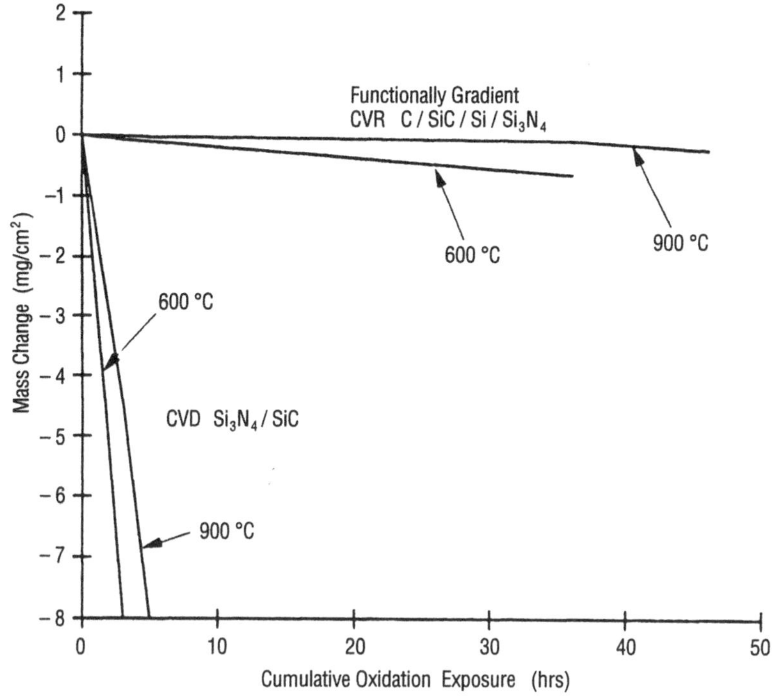

Fig. 9.15. Mass change during isothermal oxidation of CVD Si_3N_4/SiC- and CVR C/SiC/Si/Si_3N_4-coated carbon/carbon composites [58]

in improved thermal shock resistance. Figure 9.15 compares the oxidation behavior of CVD- and CVR-coated carbon/carbon composites [58]. These coatings are stable even up to a temperature of 1650 °C and exhibit negligible oxidation weight loss at 1300 °C under isothermal oxidation or thermal cycling in air. Although all of these modifications could improve the performance of SiC and Si_3N_4 at higher temperatures, the ultimate utility of these materials is most likely limited to under 2000 °C.

9.2.2.3 Coatings Based on Refractory Metals

A number of coatings based on refractory metals have been studied as potential oxidation protection systems for carbon/carbon composites. These include borides, carbides, nitrides, and oxides of zirconium, chromium, aluminium, hafnium etc. [25, 26, 55]. These coatings can be formed through CVD, PVD, pack cementation or decomposition of organometallic compounds with subsequent reaction with the substrate. These coatings have been found to be quite effective at low temperatures up to 900 °C. At higher temperatures, their advantageous properties are outweighed by their disadvantages since carbides and borides of zirconium and hafnium have least carbon egress. At the same time, they are susceptible to oxidation in the temperature range 1200° – 1400 °C, resulting in the formation of oxides. Although these oxides of zirconium and hafnium are highly erosion resistant, they have a high oxygen permeability [26, 39]. HfO_2 and ZrO_2 exhibit permeabilities in the order of 10^9 g/cm. Therefore these as such, cannot be used for oxidation protection of carbon/carbon composites at high temperatures. However, they can be used as overcoats as erosion-resistant coatings or as undercoats as carbon-egress-resistant coatings in multilayer coating systems [26]. Moreover, a mixed combination of several metallic and nonmetallic materials may be formed by using different forming routes such as reaction forming, diffusion forming, or a combination of these to produce oxidation-resistant coatings with desired characteristics.

As an example, an oxidation protection coating consisting of zirconium–boron–silicon can be formed by coating the composites with boron carbide either by CVD or by pack cementation followed by an overcoat of zirconium and silicon using the slurry method. The complete mass is further pack cemented in the presence of silicon carbide powder to give a coating consisting of zirconium-boron-silicon. Such coatings prepared as per predetermined configurations possess the required characteristics of good adhesion, sealants, and oxygen barriers. Zircon-based (zirconium silicate) coatings are also found to be very effective in increasing oxidation resistance of carbon/carbon composites [60]. Moreover, SiC concentration gradient coating on carbon/carbon composites with an overcoat of a thin film of zircon (1.5 µm) exhibits no weight loss at 1000 °C and a weight loss of about 1% of that of the original composite at 1400 °C (Fig. 9.16) [61].

9.2.2.4 Multilayer Coatings

High temperature oxidation protection of carbon/carbon composites invariably requires external coating systems. The most essential requirements of these coatings are that they must possess durability for cycling to high temperatures and have the inherent ability to restrict the ingress of oxygen and egress of carbon. Since no single coating can meet all of the

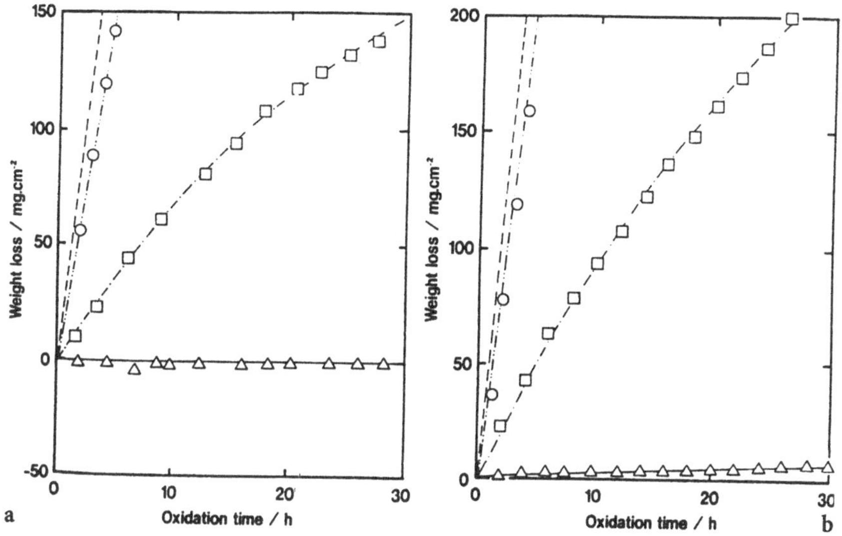

Fig. 9.16 a, b. Changes in weight loss with time of oxidation at (a) 1000 °C and (b) 1400 °C for the as-received C-C composites (---), the C/C composite with SiC concentration gradient (o) and the SiC gradient C/C composite coated with a zircon film of 0.5 μm thick (□) and 1.5 μm thick (△) [61]

necessary requirements, attention is currently being focussed on multilayer coatings [26, 39, 55–67]. Numerous patent reports describe multilayer coatings consisting of Mo, Zr, Cr, Al_2O_3, silicon, boron etc. as such or in the form of their inorganic compounds. Morimoto et al. have proposed lanthanoid (+ yttrium) silicate to be the best candidate for an outer layer coating because of its low coefficient of thermal expansion, low evaporation rate, and low oxygen permeation at high temperatures [64].

A four-layer coating system for high temperature oxidation protection of carbon/carbon composites has been proposed by Strife and Sheehan [39] consisting of refractory oxide/silica glass/refractory oxide and refractory carbide. In principle, the outer layer of refractory oxide provides protection against erosion by hot flowing gases and the inner layer of silica glass acts as an oxygen barrier and sealant. The second layer of refractory oxide acts as a carbon diffusion barrier, and the carbide layer, which has matching thermal expansion, and is in intimate contact with the carbon/carbon composite, facilitates a good adhesion to the substrate.

The concept of such a multilayer coating is an ideal one. However, the choice of processing routes to form these coatings with perfection and the chemical inertness of the layers with the possible problem of interlayer diffusion during processing as well as during service are of considerable concern. Mechanical compatibility and matching thermal expansion coefficients of various layers are also important factors which determine the effectiveness of such oxidation protection systems in terms of limited temperature protection or extended time protection under temperature cycling.

Efforts are being made to develop techniques for the protection of carbon/carbon composites against oxidation at elevated temperatures concurrently with the development of composites. However, no general solution has yet been proposed for protection at all temperatures. The key issues to be tackled in order to develop effective protection systems have been identified, the most important of wich include:

(i) Thermal stability of the material,
(ii) Adhesion with the carbon/carbon substrate,
(iii) Chemically compatible interfaces throughout the coating architecture,
(iv) Oxygen permeation and carbon egression
(v) Thermal mismatch stresses causing spallation and crack formation
(vi) Erosion in hot flowing gases.

On the basis of intensive studies on high temperature properties of various materials and their compounds under oxidizing environments, the range of candidate materials has been narrowed down. Boron, silicon, zirconium, hafnium, and iridium in one form or another are the most suitable materials for oxidation protection coatings.

The most common compounds of the elements used and their major roles in the protection of carbon/carbon composites are given in Table 9.3. However, no single element or its compound can fulfill all of the abovementioned issues. Moreover, the carbon/carbon composite materials are to be used in a diverse range of applications under different working temperature conditions, such as extended life low (< 1000 °C) and medium temperature (1000° – 1600 °C) applications or limited life high temperature (> 2000 °C) applications. Therefore, the specific methods for achieving effective oxidation protection are related to particular operating conditions and performance requirements. Regardless of application, a multilayer external coating system is essential for high temperature oxidation protection of carbon/carbon composites.

Figure 9.17 shows a general scheme of such a multilayer coating system. A permutation/combination of the elements and their compounds mentioned in Table 9.3 can be selected. The exact configuration and the materials for the layers will depend on the users, their requirements w.r.t. the com-

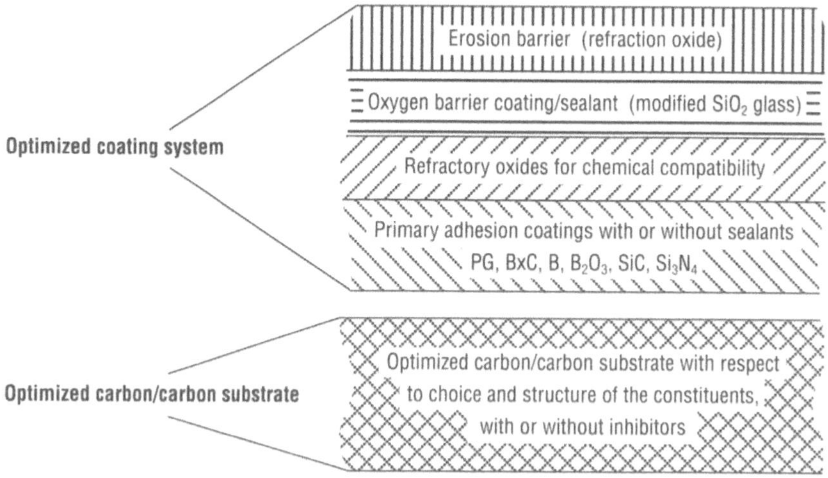

Fig. 9.17. A general scheme of multilayer coatings

ponent/structure, the fabrication/processing facilities, and all of the operating conditions and performance requirements. A complete, long-term protection system is not yet available. The questions of what extent of inhibition against burn off and what duration and temperature of operation under well-defined conditions is possible remain to be answered.

Table 9.3. Most commonly used elements and their compounds

Element	Compound	Major function/protection role
Carbon	Pyrolytic graphite	Substrate adhesion/protection from chemical attack
Phosphorus	Phosphates	Low temperature oxidation protection
Boron	Alone	Substrate adhesion
	Boron carbide	Substrate adhesion
	Boron nitride	Oxidation inhibiting matrix addition
	B_2O_3	Sealants
	Borates	Oxidation inhibition
Silicon	Silicon carbide	Substrate adhesion
	Silicon nitride	Oxidation protection
	SiO_2	Oxygen barrier
	Silicides	Protective layer
Iridium	Alone	Carbon egression barrier

9.3 References

1. Fitzer E (1977) Carbon. Vol 16, p 3
2. Thomas JM (1965) In: Walker PL (ed) Chemistry and Physics of Carbon. Vol 1, Marcel Dekker New York Pub, p 1
3. McKee DW (1981) In: Walker PL, Thrower PA (eds) Chemistry and Physics of Carbon. Vol 1, Marcel Dekker New York, Pub, p 1
4. Hoffman WP (991) Carbon. Vol 29, p 769
5. Chang HW, Rhee SK (1978) Carbon. Vol 16, p 17
6. Chang HW, Rusnak RM (1979) Carbon. Vol 17, p 407
7. McKee DW (1987) Carbon. Vol 25, 551
8. Bahl OP, Dhami TL (1987) High Tem.-High Pressure. Vol 19, p 211
9. Yasuda E, Kimura S, Shilsua Y (1980) Trans JCSM. Vol 6, p 14
10. Goto KS, Han KH, St Pierre GR (1986) Trans iron & Steel Inst Japan. Vol 26, p 597
11. Fishbach DB, Uptegrove DR (1977) Proc 13th Bien Conf Carbon
12. Luthra KL (1988) Carbon. Vol 26, p 217
13. McKee DW (1988) Oxidation protection of C/C composites: 5th An Conf MTC, SIU, p 81
14. Jones LE, Thrower PA, Walker jr PL (1986) Carbon. Vol 24, p 51
15. Manocha LM (1994) Carbon. Vol 32, p 213
16. Dhami TL, Bahl OP, Manocha LM (1993) Carbon. Vol 31, p 751
17. Iwashita N, Sawada Y, Shimizu K, Shinke S, Shiyoyama H (1995) Carbon. Vol 33, p 405
18. Han JC, He XD, Du SY (1995) Carbon. Vol 33, p 473
19. Rodriguez-Mirasol JR, Thrower PA, Radovic LR (1995) Carbon. Vol 33, p 545
20. Thrower PA, Marx DR (1977) Proc 13th Bien Carbon Conf
21. Knefeld R, Linkenheil G et al. (1972) Proc Carbon 72, Baden-Baden
22. Crocker P, McEnaney B (1991) Carbon. Vol 29, p 881
23. Zao JX, Bradt RC Walker PL jr (1985) Carbon. Vol 23, p 9
24. Ahearn C, Rand R (1996) Carbon. Vol 34, p 239
25. Sheehan JE (1989) Carbon. Vol 27, p 709
26. McKee DW (1991) Oxidation Protection of Carbon Materials. In: Chemistry and Physics of Carbon. Vol 23, p 173
27. McKee DW, Spiro CL (1985) Carbon. Vol 23, p 437
28. Hawtin P, Gibson JA (1970) 3rd Conf Ind Carbon & Graphite, London, p 309
29. McKee DW, Spiro CL, Lamby EJ (1984) Carbon. Vol 22, p 285
30. Gkogkidis A (1986) PhD Thesis University of Karlsruhe
31. Ehrburger P, Baranne P, Lahaye J (1986) Carbon. Vol 24, p 495
32. Mckee DW, Spiro CL, Lamby EJ (1984) Carbon. Vol 22, p 507
33. Jones LE, Thrower PA (1987) Ext Abs 18th Bien Carbon Conf, p 468
34. LoBiondo NE, Jones LE, Clare AG (1995) Carbon. Vol 33, p 499
35. Martin GR (1976) US Patent 3936574
36. McKee DW (1988) Carbon. Vol 26, p 659
37. Ho CT, Chung DDL (1990) Carbon. Vol 28, p 521
38. Ho CT, Chung DDL (1990) Carbon. Vol 28, p 815
39. Strife JR, Sheehan JE (1988) Ceramic Bulletin. Vol 67, p 369

40. Harding JT, Fry VR (1986) Proc 10th Int Precious Metal Conf, p 431
41. Wright TR, Weyand JD, Kizer DE, Simmons WC (1967) Proc 4th Space Cong
42. St Pierre GR (1991) Proc Seventh Annual Conf MTC, SIU, p 181
43. High Temperature Oxidation Resistant Coatings. National Acad Sc Pub No ISBM 0-309-01769-6 (1970)
44. Rogers DC, Shuford DM, Mueller JI (1975) Seventh National SAMPE Tech Conf, Vol 7, p 319
45. Rogers DC, Scott RO, Shuford DM (1976) 8th National Sampe Tech Conf, p 308
46. Huettner W, Weiss R, Dietrich G, Meistring R (1990) Proc ESA symp ESTEC Noordwijk (NL) March ESA SP-303-1990), p 91
47. Naslain R, Hagenmuller P, Christine F, Herault L, Choury JJ (1980) Advances in Composite materials. Vol 2, p 1084
48. Fitzer E, Herbst H, Schilichting J (1974) Proc 4th Carbon & Graphite Conf Soc Chem Ind London, p 401
49. Evans CC, Parmee AC, Reinbird RW (1974) 4th Carbon & Graphite Conf Soc Chem Ind London, paper, 113
50. Gaddow R (1984) PhD Thesis, University of Karlsruhe
51. Ehrburger P (1990) Protective layers for special types of composites. In: Figueiredo JL et al. (ed) Carbon fibers, filaments and composites, Kluwar Acad pub, p 327
52. Curry DM, Scott HC, Webster CN (1979) 24th Natl SAMPE Symp, p 1524
53. Fitzer E, Heym M (1978) High Temp High Press. Vol 10, p 29
54. de Castro LD, McEnaney B (1992) Carbon 92, Essen, p 738
55. Strife JR (1990) 6th Annual Conf MTC, SIU, p 166
56. Dickinson RC, Baker DM (1988) Proc 5th Annual Conf MTC, SIU, p 151
57. Hirai T (1993) Ceramic Transaction. Vol 34, p 11
58. Kowbel W, Withers JC, Ransone PO (1995) Carbon. Vol 33, p 4156
59. Tsou H, Kowbel W (1995) Carbon. Vol 33, p 1279
60. Manocha LM, Manocha SM (1995) Carbon. Vol 33, p 441
61. Yamamoto O, Sasamoto T, Inagaki M (1995) Carbon. Vol 33, p 359
62. Honig ES (1988) Proc 5th Annual Conf MTC, SIU, p 163
63. Murdie N, Hippo EJ (1988), ibid p 137
64. Morimoto T, Ogura Y, Kondo M, Ueda T (1995) Carbon. Vol 33, p 351
65. Buchanan FJ, Little JA (1995) Carbon. Vol 33, p 491
66. Lamouroux F, Bourrat X, Naslain R (1993) Carbon. Vol 31, p 1273
67. Walker TB, BookerLA (1995) 22nd Bien Carbon Conf, p 148

10 Applications of Carbon/Carbon Composites

As is evident from the previous chapters, carbon/carbon composites of the desired shape with properties required for particular application can be produced by meticulously choosing the type, architecture, and amount of carbon fiber and matrix precursor, and the processing conditions. Therefore, carbon/carbon composites are considered to be a class of materials with a wide spectrum of properties and applications. Moreover, the ongoing development of high-performance carbon fibers, high char yielding synthetic resins, and pitches continuously add to the spectrum of carbon/carbon properties and products. It is with these milestones, carbon/carbon composites, which, were developed about 30 years ago to meet the anticipated needs of the emerging space program, are regarded nowadays as logical addition to the members of fine-grained carbons and graphites. Carbon/carbon composites offer a large potential as high-performance engineering material. Therefore, in addition to the special defense, aircraft, and spacecraft applications, a steady interest is also growing in civil market segments. Based on worldwide consumption, a general trend of carbon/

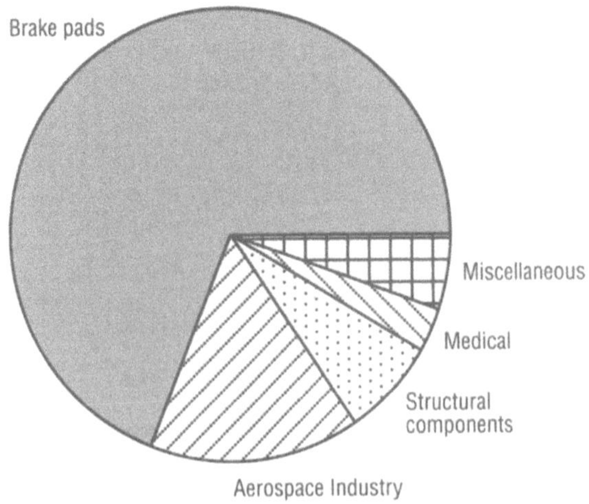

Fig. 10.1. Worldwide consumption of carbon/carbon composites

carbon consumption can be discerned as shown in Fig. 10.1. In terms of mass consumption and money, the main applications of carbon/carbon composites today are still in military, space, and aircraft industry while those in the field of general mechanical engineering are picking up steadily. In the following sections general applications of carbon/carbon composites are arranged according to the various specific properties of the composites. There is great potential for the wider application of these materials.

10.1 Carbon/Carbon Composites as Brake Discs

The development of high-speed large-capacity aircrafts not only required improved aeroengines but also necessitated the development of aircraft braking systems made of light weight materials with a smooth frictional behavior capable of withstanding high temperature. This was particularly important for aborted takeoffs of large inertia planes. During aborted takeoffs the temperature of the frictional surface may rise to more than 2000 °C in about 20 s. The conventional steel disk brakes, if successful in stopping

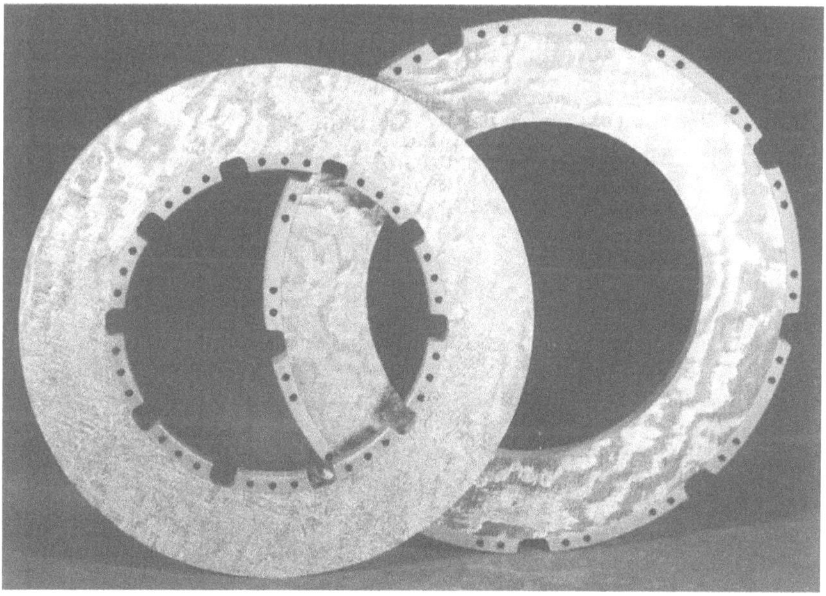

Fig. 10.2. Sepcarb® Carbon/carbon brake discs of Airbus Aircraft (Courtesy of Sepcarb, France)

the aircraft in an emergency are likely to get destroyed by warping or melting due to the intense heat generated during the stop [1]. Carbon/carbon composites which are stable at high temperatures and have excellent friction and wear characteristics were the obvious solution to this problem. Concerted research and development efforts in the design and manufacture of carbon/carbon brake systems led to the widespread use of commercial carbon/carbon composite aircraft wheel brakes in 1973. Dunlop was the first company to manufacture and fit carbon/carbon brakes on a VC 10 aircraft followed by their trials on concorde [2]. Since then there has been continuous research and development going on to further reduce the weight and improve performance. As a result, at present almost all military aircrafts such as the US F.14, F.15, F.16, F.18, and French Mirage 2000, as well as commercial aircrafts such as Boeing 747, Airbus, Concorde, Canadair challenger, and Gulfstream III have employed carbon/carbon composites as brake disk materials [2–4]. Figure 10.2 shows carbon/carbon brake disks of the Airbus. The superiority of C/C over other metals as friction materials is seen in Fig. 10.3 [5].

Carbon/carbon brake materials have a density of about one-fourth of steel-cermet brakes and more than twice the heat capacity. These physical characteristics along with the frictional and thermal superiority of carbon/carbon composites provides them with nearly four times the stopping power of copper or steel brakes. Airbus A320 can operate 2500 landings per overhaul with carbon/carbon brakes as compared with 1500 landings per overhaul with metal brakes.

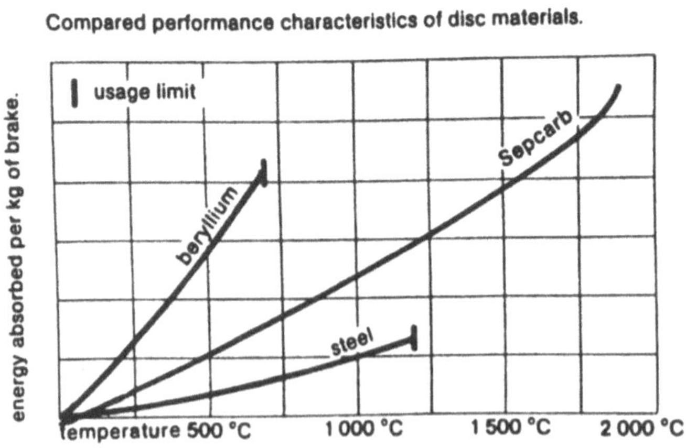

Fig. 10.3. Performance charateristics of conventional brake materials [5]

Fig. 10.4. Carbon/carbon brake assembly for automotives [10]

The continuous research and development efforts and advances in carbon/carbon technology have not only improved the performance of these composites but have also reduced their cost. When carbon/carbon brakes were introduced, the cost of this material was around US$ 1000 per kilogram; at present it is about $ 250.

Although with the increase in air traffic there is an enormous market potential for carbon/carbon brakes, the market is being supplied by only a few companies. These include Hitco, Aircraft Braking Systems, and Allied Signals in the USA, SEP in France, and Dunlop in the UK.

The potential for significant growth in carbon/carbon brakes is further underlined by the interest shown by manufacturers of racing cars, heavy duty surface transport systems and passenger cars in utilizing carbon/carbon in their advanced automotive braking systems [7–9]. Figure 10.4 [10] shows a photograph of a carbon/carbon disk for an automotive brake. For surface transport systems, the temperature required for brake operation is quite low (400° – 600 °C), and therefore relatively cheap designs and normal processing methods of 2D carbon/carbon composites can be utilized for such applications.

10.2 Carbon/Carbon Materials for the Aerospace Industry

The aerospace field continues to be one of the primary areas of application of carbon/carbon composites. The classic examples are the thermal protection system of space shuttles, rocket nozzles, exit cones, and nose tips of re-entry vehicles etc. Figure 10.5 shows a typical rocket nozzle design which includes of carbon/carbon components [11]. All of the refractory, thermal, thermophysical, mechanical, thermomechanical and structural properties of carbon/carbon composites are utilized in these applications.

The utility of carbon/carbon composites was first demonstrated in space shuttle program as wing leading edges and nose caps, which during reentry operations may encounter temperatures close to 2000 °C; this is well beyond the operation temperature of metals [12, 13]. Each orbiter wing contains 22 leading edge carbon/carbon airfoil panels and 22 sealing strips of carbon/carbon. The nose cap is around 140 cm in diameter [14]. Figure 10.6 shows a prototype nose tip. These components provide thermal protection to the instrumentation from the searing heat of reentry and main-

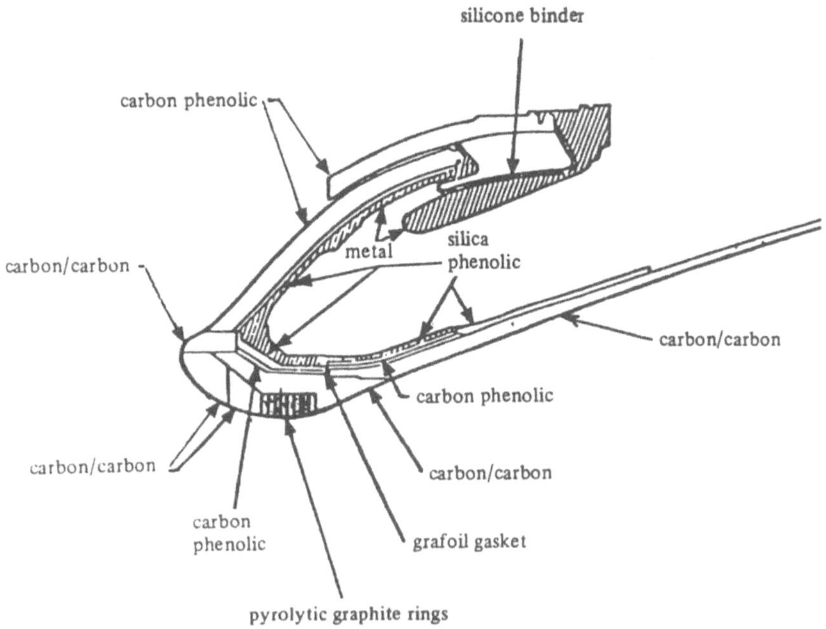

Fig. 10.5. Schematic of Rocket nozzle – design concept

Fig. 10.6. Carbon/carbon prototype nose tip after test [15]

tain structural integrity over multiple missions, especially under conditions of subzero temperatures in outer space to close to 1700 °C during entry into the atmosphere. In addition to the space shuttles, all American and European strategic nuclear missiles employ carbon/carbon nose cones and heat shields. Carbon/carbon composites for such applications are required to possess high thermal conductivity and high specific heat so that the component operates as a heat sink, absorbing the heat flux without any problem. The porosity should be low and uniform so as to have a uniform low ablation/erosion. The superior ablation performance of carbon/carbon composites is demonstrated in Fig. 10.7 [15]. These stringent requirements in carbon/carbon composites are best achieved through three-directional (3D) fibrous reinforcement and the HIPIC densification route. However, the leading edges are formed from two-directional (2D)

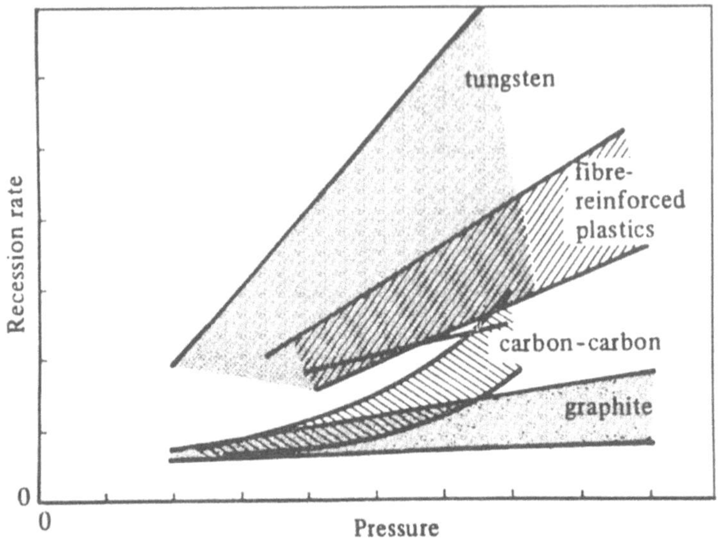

Fig. 10.7. Relative ablation performance of carbon/carbon composites [15]

carbon fabric with a carbon matrix derived from phenolic resin/pitches. Furthermore, these are generally protected against oxidation through siliconization/heat treatment or tetraethyl orthosilicate (TEOS) sealant.

Advance carbon/carbon composites (ACC) with silicon carbide and ceramic multilayer coatings for oxidation protection are utilized in the European spaceplane Hermes and National Aerospace Plane (NASP) with anticipated temperatures of 2750 °C on the nose cap and approximately 1950 °C on the wing and tail leading edges [16]. Other components used for aircraft industries are gas rudders and thrust deflectors for military aircrafts (Fig. 10.8) and expansion nozzle of a hypersonic propulsion unit (Fig. 10.9) [17].

In rocket motors with a solid propellant system, the throat and exit cone are made of carbon/carbon composites. The exhaust gases from the propellant chamber pass out through the throat and then finally out through the exit cone. Figure 10.10 shows a schematic representation of the structural parts of a rocket motor assembly and Figs. 10.11 and 10.12 show the carbon/carbon components used in rocket motors [18].

Although these materials are only for short time in operation, the property requirements are extremely stringent. Stress analyses conducted in the throat region has shown these stresses to be isotropic in nature, thus necessitating the choice of high-density 3D carbon/carbon materials.

Fig. 10.8. Carbon/carbon gas rudder for military aircraft. (Courtesy of Sigri Great Lakes Carbon, Germany)

Fig. 10.9. Carbon/carbon expansion nozzle of hypersonic propulsion unit. (Courtesy of Sigri Great Lakes Carbon, Germany)

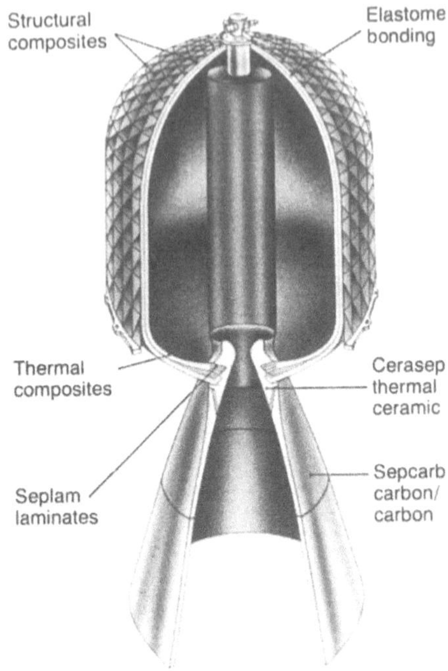

Fig. 10.10. Schematic representation of structural parts of a rocket motor. (Courtesy of Sepcarb, France)

Fig. 10.11. Carbon/carbon nozzle throat of Ariana V booster. (Courtesy of Sepcarb, France)

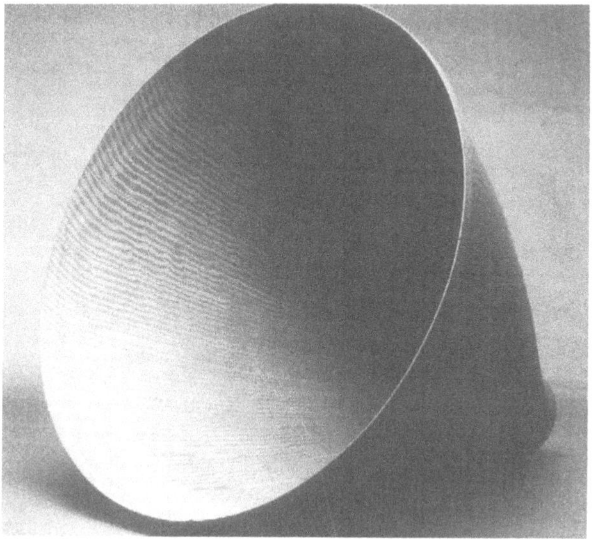

Fig. 10.12. Carbon/carbon exit cone for MAGE 2 geostationary rocket motor. (Courtesy of Sepcarb, France)

10.3 Aeroengine and Turbine Components

The efficiency of heat engines is greatly improved by operating them at high temperatures, permitting high combustions. The choice of carbon/carbon composites in jet engine rotors and stators offers the possibility of operating at temperatures well in excess of 500°–600 °C higher than those used in conventional engines with high temperature alloys. Furthermore, the use of carbon/carbon composites results in a significant reduction in weight and a reduced engine size and fuel consumption. Concerted efforts are being made in the USA and Europe to develop turbine components using carbon/carbon composites. The basis of this development is the utilization of the high tensile strength of the fiber materials, which allows the compensation of the centrifugal stress on the blades. This in turn, subjects the blades to additional compressive prestresses. The LTV Missiles and Electronics group in the USA has developed various carbon/carbon jet engine components which include turbine wheels, combustion chambers, and exhaust nozzles [14]. Figure 10.13 shows a carbon/carbon turbine rotor. A 365-mm-diameter carbon/carbon turbine wheel has been demonstrated to successfully spin at 28 000 rpm at a temperature of 1760 °C. ACC

Fig. 10.13. Carbon/carbon turbine rotor

composites are being used in products such as the nozzle in the F-100 jet engine afterburner, with turbine wheels operating at >40000 rpm [19].

10.4 Diesel Engine Components

Piston-driven engines such as diesel engines could also operate with increased efficiency and reduced weight if cooling requirements were minimized through the use of materials which function efficiently at high temperatures. Carbon/carbon is currently being evaluated for many such applications. The components for which carbon/carbon materials may potentially be used include piston crowns, valves, cylinder liners, and connecting rods [14, 15, 17, 18, 20, 21]. Carbon/carbon pistons have a lower density than the conventional aluminium light weight piston. Moreover, with the latter pistons, because of the advantageous low coefficient of thermal expansion (CTE) of carbon/carbon composites, the piston to cylinder wall clearance can be kept very low so that piston rings and skirts are not necessary. At the same time the carbon/carbon piston would perform in the same way as any other piston in a reciprocating internal combustion engine, while reducing the weight and increasing the mechanical and thermal efficiency of the engine. Figure 10.14 [15] shows a carbon/carbon piston

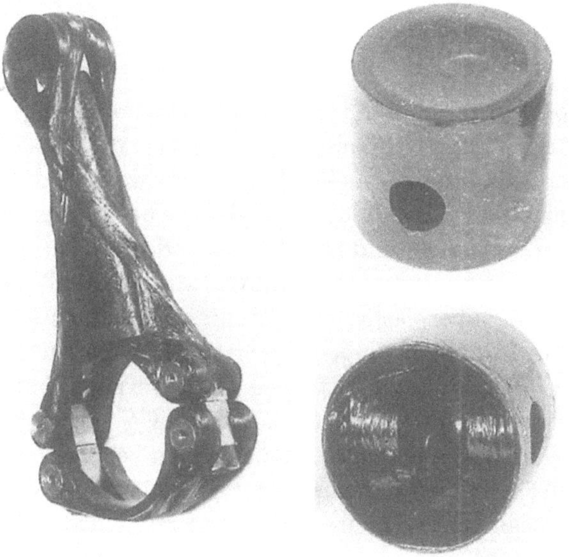

Fig. 10.14. Carbon/carbon piston and connecting rod for diesel engine [15]

and connecting rod used in demonstration diesel engine. For such a piston, which has been tested successfully at NASA Langley Research Center [14, 21], carbon fiber reinforcements of four-dimensional angle interlocked and a knitted multilayer warp sock with resin carbon, and the chemical vapor deposition process for carbon densification have been used. Carbon/carbon pistons for experimental motor cycle engines and for rotary valves in two- and four-stroke small engines are being developed by Hitco [14]. The efficiency of an adiabatic diesel engine can be enhanced to approximately 48% thermal efficiency by increasing the operating temperature to about 1100 °C from about 300 °C for a basic diesel engine. This is feasible with carbon/carbon composites only.

10.5 Carbon/Carbon Composites as Refractory Materials

It is well known that polygranular carbon and graphite materials are refractory materials par excellence in a nonoxidizing environment. With carbon fiber reinforcement, the mechanical properties of these carbon refractories can be further improved, thereby opening up new fields of industrial applications.

10.5.1 Hot Press Dies

The traditional material used in die manufacturing for high-temperature high-pressure sintering of ceramics and metals is a thick-walled fine-grained polycrystalline graphite. As an alternative, thin-walled carbon/carbon composite dies with a high hoop strength have been tested for cold molding in powder metallurgy and have been found to withstand repeated applications. One interesting and innovative use is the metallurgical hot forging process in the superplastic stage. The air inlet channels of the European JF 90 are made from high temperature titanium alloys having superplasticity at about 1000 °C. The hot forming is performed by a simple blowing up of tubes that are closed at one end as shown in Fig. 10.15 [15].

The mold is made of two-dimensional carbon/carbon composites which may be up to 5 m in length (Fig. 10.16). Figures 10.17–10.19 show carbon/carbon tool segments, pressure plates, and resistance elements used as reusable molds for cobalt-based products and hot sintering applications [17, 22, 23]. The use of carbon/carbon cylindrical heating elements for hot isostatic presses provides a constant resistance; hence, fewer power adjustments are required during operations and operation costs are reduced.

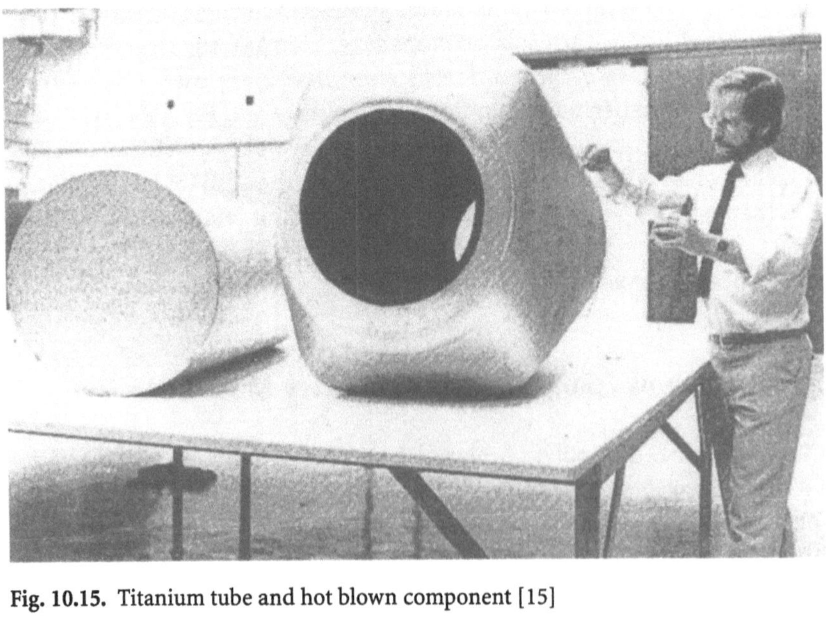

Fig. 10.15. Titanium tube and hot blown component [15]

Fig. 10.16. Carbon/carbon mould for blowing titanium

Fig. 10.17. Carbon/carbon hot press die components [23]

Fig. 10.18. Carbon/carbon resistance heating elements [17, 23]

a

b

Fig. 10.19 a, b. Carbon/carbon protection tubes (a) and grids (b) [23]

This not only allows a higher application temperature but also increases the life span of the mold by a factor of 10 compared with a graphite mold.

10.5.2 High Temperature Corrossion Resistant Fasteners

Carbon/carbon bolts, screws, nuts, and washers (Fig. 10.20) are used where high temperatures and severe chemical conditions are present. The strength and stiffness at high temperatures guarantees high fastening stability. The systems made of carbon/carbon composites are self-fastening at high temperatures due to the anisotropic CTE. These parts are used in the semiconductor industry, in furnace construction, and in other high temperature apparatus and equipment such as liners, plates, crucibles sleeves, and other auxiliary aids (Figs. 10.21 – 10.23). Carbon/carbon composites are also used as heating elements, especially if extreme conditions of

Fig. 10.20. Carbon/carbon fasteners [17]

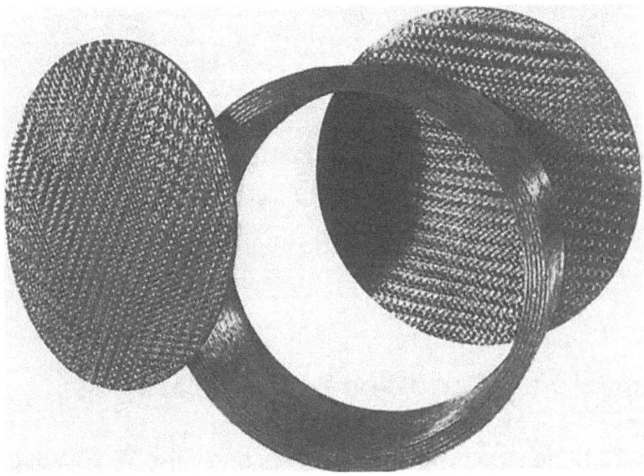

Fig. 10.21. Carbon/carbon liners and plates [23]

Fig. 10.22. Carbon/carbon crucible [23]

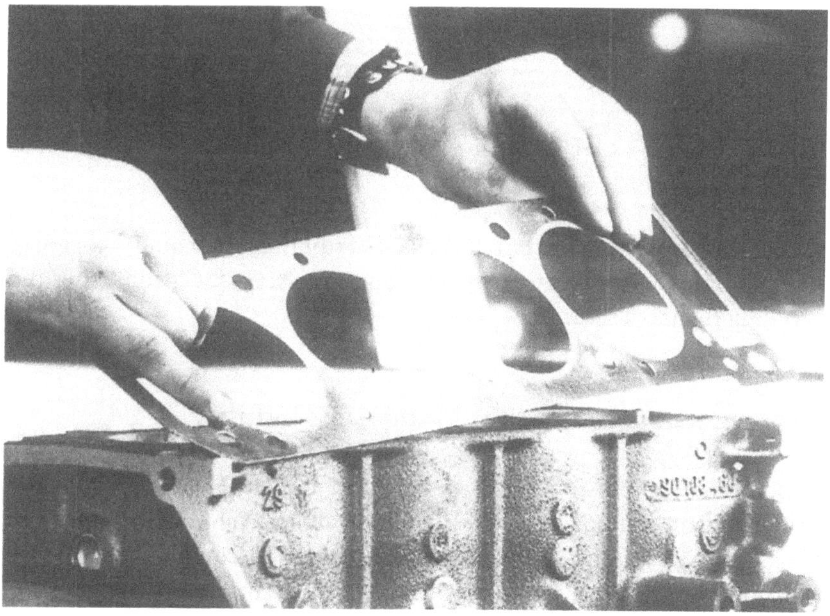

Fig. 10.23. Carbon/carbon tool segments

Fig. 10.24. Sepcarb® Carbon/carbon sintering tray for carbonizing and carbiding furnaces [18]

mechanical properties and temperature distribution are required together. Other commercial applications include sintering trays for carbonizing and carbiding furnaces (Fig. 10.24) [18].

10.5.3 Carbon/Carbon Composites in the Glass Industry

Carbon and graphite is not wetted by molten glass. Additionally, carbon/carbon composites possess a higher thermal shock resistance and impact toughness, and their porosity and thermal conductivity may be controlled; hence, there is no impairment of glass surfaces. Therefore, carbon/carbon materials are used in various parts of glass-container-forming machines [14–17, 20–24]. They are used in channeling systems to carry the gobbets of molten glass, as molds for crystal glass products, and as an asbestos replacement for hot-end glass contact elements for moving hot glassware articles (Figs. 10.25, 10.26) [17]. A carbon/carbon channeling system does not require an inner cooling system like for metal scoops, nor does it need to be treated with paint finishes or an oil spray coating. The

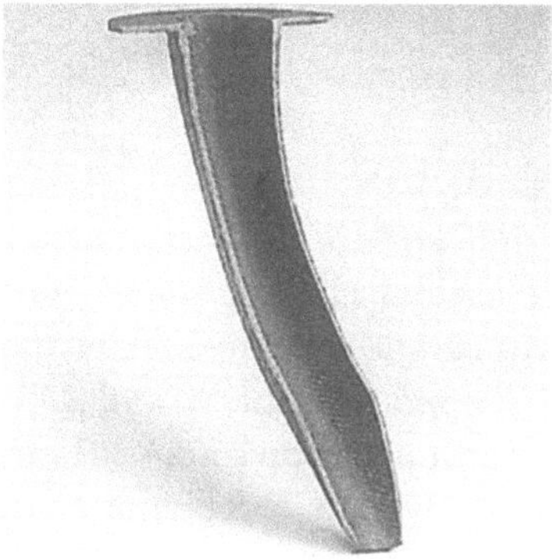
a

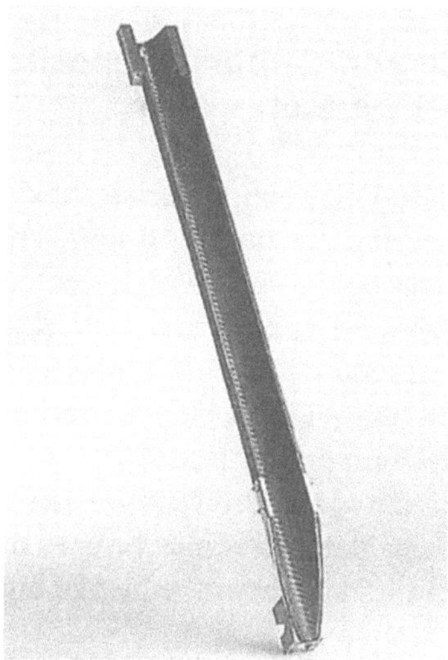
b

Fig. 10.25 a, b. Carbon/carbon channeling systems to carry gobbets of molten glass (17)

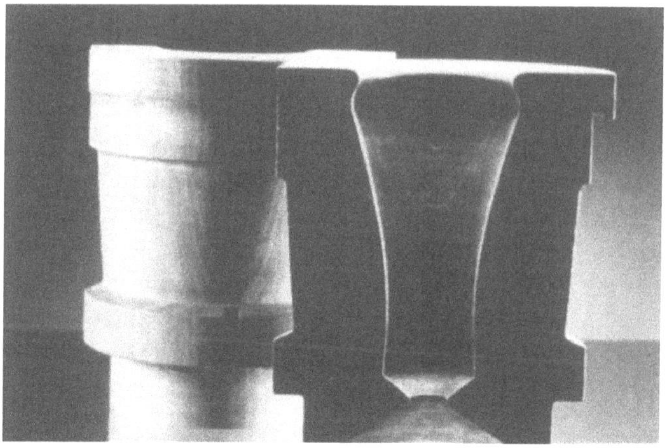

Fig. 10.26. Carbon/carbon molds for crystal and lead crystal drinking glasses [17]

low weight of a fast-moving scoop reduces the moment of inertia and lowers the stresses imposed on the gobbet distribution mechanism. The mechanical and thermal stability leads to plastic deformation of the scoop at the gobbet impact point and ensures a long service life as well as low glass contamination. The controlled low thermal conductivity of the contact elements results in a presents rapid heat distribution and thus avoids cooling cracks.

10.6 Carbon/Carbon Composites for Use at Intermediate Temperatures

Carbon/carbon materials, having high thermal and electrical conductivities and resisitance to most chemicals, are used as electrodes and other structural components in fuel cells. One such example is the fuel cell system based on a phosphoric acid electrolyte which operates at 200 °C and generates 200 kW to 11 MW of electrical energy. Carbon/carbon composites are the materials of choice even for vanes for rotary vane compressors and vacuum pumps [9, 14].

Carbon/carbon composites may be preferred for nuclear waste storage containers. They may be used on laser shields to protect space-based satellite systems from the heat of high-powered laser beams.

10.7 Carbon/Carbon Components for Chemical Process Technology

Elemental carbon, because of its exceptional purity and its complete inertness in acid, alkaline, and salt solutions and in organic solvents, is one of the most corrosion-resistant materials. It is not resistant to a media with powerful oxidizing action, e.g., nitric acid, chlorine bleaching solution, and oleum, especially at elevated temperatures. Synthetic granular carbon is used in many chemical industries, in particular, as heat exchangers [15]. Carbon/carbon composites are finding application in new areas of the chemical industry. Carbon/carbon packings are highly effective for separation columns used to separate liquid mixtures in distillation rectification plants. They are also used as stirrers, feed pipes, support grids, filter plates, and thin-walled heat exchangers. Figures 10.27–10.29 show some of the components for chemical industries marketted by Sigri Great Lakes Carbon, Germany [17].

10.8 Carbon/Carbon Composites as Biocompatible Materials

Elemental carbon is considered to be the most biocompatible of all known materials. It is compatible with blood, soft tissues, and bones, as is evidenced by the successful replacement of ligaments with carbon fibers in human knees and ankles [15]. The biomedical applications of carbon fibers in

Fig. 10.27. Structured packings for separation columns [17]

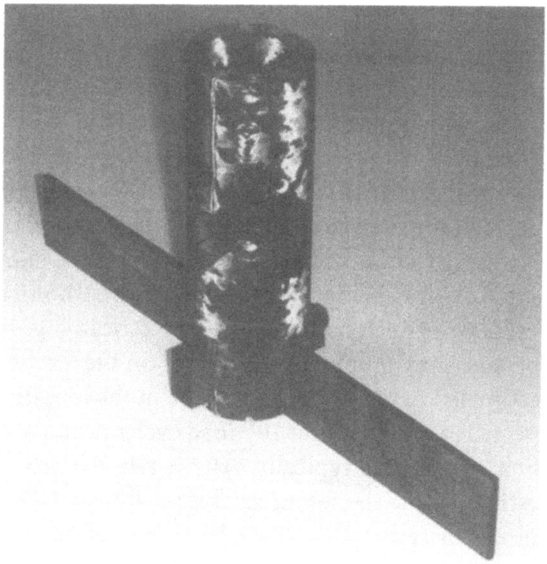

Fig. 10.28. Carbon/carbon stirrer [17]

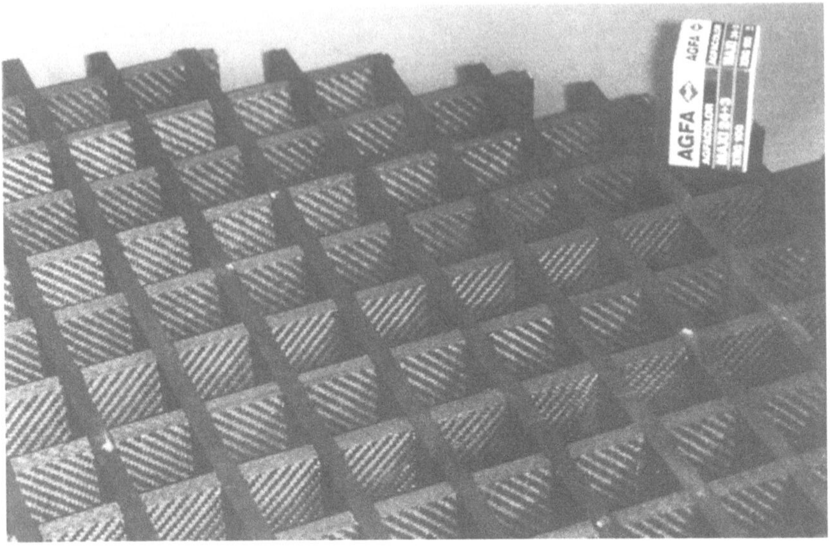

Fig. 10.29. Carbon/carbon support grids and filter plates used in chemical industries [17]

combination with carbon matrix were demonstrated by us for bone plates in osteosynthesis and as endoprostheses to replace conventional steel plates which have a modulus an order of magnitude higher than that of bone [25]. However, the tougher carbon-fiber-reinforced polymer matrix composites, which were subsequently developed, especially with epoxy and polysulfone matrixes, are considered competitors of carbon/carbon composites. Extensive projects were undertaken in Germany with Government support for hip-joint replacement with carbon/carbon composites. The advantages of such materials are depicted in Fig. 10.30. For a normal healthy person, maximum tensile stresses on the femur during walking have been estimated to be about 50 – 60 MPa in the longitudinal direction and are expected to undergo over 10^{12} load cycles per year. General-grade carbon/carbon composites typically exhibit two to three times the tensile strength estimated for the femur neck and almost a negligible loss in mechanical properties even after 10^7 cycles at 90 % of the original stress. The major advantage of such implants is the cementless fixation of the stem within the femur. Figure 10.31 shows a model of a carbon/carbon hip prosthesis with a carbon neck [18]. The high impact properties of carbon/carbon composites together with their biocompatibility renders these materials suitable for use as heart components. Figure 10.32 shows a model of carbon/carbon

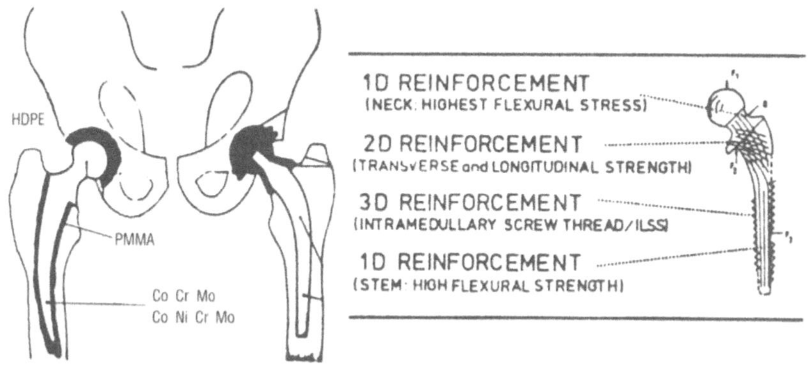

Fig. 10.30. Comparison of conventional and carbon/carbon hip prosthesis. Possibility to imitate the femur structure

Fig. 10.31. Carbon/carbon total hip prosthesis [18]

Fig. 10.32. Model of carbon/ carbon artificial heart for animals [15]

artificial heart components for animals [15]. Although the biocompatibility of carbon/carbon composites has been demonstrated beyond doubt and many potential applications have been demonstrated, their commercial exploitation has been limited due to the lengthy regulation formalities required to obtain licenses for application in human medicine.

10.9 Emerging Applications

The latest applications of carbon/carbon composite structural materials are in advanced aircrafts and thermal management systems. These application are possible on account of the development and use of ultrahigh-modulus, high thermal conducting, pitch-based, and vapor-grown carbon fibers. Carbon/carbon composites, which have highly flexible properties and can be formed into complex shapes, are the candidate materials of choice for newly evolving applications such as parabolic radio frequency (RF) antennae, reflectors for satellite communications, RF limiters, plasma-facing surfaces and diverter plates of the Tokamak Fusion reactor [14, 26–28], and cladding elements for the Joint European Torus nuclear fusion reactor [17]. Figure 10.33 shows the carbon/carbon macrobloc system for a nuclear fusion reactor [18].

The applications and market for carbon/carbon composites are continually developing for different sectors. These materials have a clear potential for mechanical engineering applications in hot-load-bearing structures and corrosive surroundings.

Growth and application in this sector has been slow due to cost, lack of open literature and data bases concerning the properties, performance, and design, and the classified nature of the materials in some countries. However, the demand for carbon/carbon materials is steadily increasing on account of (a) the continuous improvements in properties which are being

Fig. 10.33. Sepcarb® Carbon/carbon macrobloc system for nuclear fusion reactors [18]

made, especially with respect to the modulus and thermal conductivity of carbon fibers; (b) the development of new matrix precursors and new fabrication methods with a view to controll the carbon matrix microstructure in the composites; (c) the development of oxidation-resistance coatings which can withstand much higher temperatures; (d) the development of micromechanical analysis techniques for these composites. Furthermore, the ever-increasing demand of new high-performance aerosystems such as hypersonic planes, mechanical components of high-power nuclear reactors, and materials for these is also responsible for the renewed generation of interest in carbon/carbon matrials. The enormous corrosion problems in waste-water treatment and in the washing out of polluted gases represent another broad field of new applications.

10.10 References

1. Bauer DW, Kotlensky WV (1971) 10th Biennial Conf. on Carbon Bethlehem, Paper FC-37
2. Stimson IL, Fisher R (1980) Trans Royal Soc London, A 294, p 583
3. Ruppe JP (1987) Can Aeronaut Space. J Vol 27, p 212
4. Thomas CR (ed) (1993) Essentials of Carbon/carbon composites. Royal Society of Chemistry (Pub), p 33
5. Fisher R, Weaver JW (1974) 4th London Carbon conf, p 708
6. West P (1988) Adv Mater Vol 10 No 20
7. Fitzer E, Fritz W, Gkogkidis A, Morgenthaler KD (1986) Proc Carbon 86, Baden Baden
8. Klein AJ (1986) Materials and Processes 130 (5), p 64
9. Savage G (1992) Carbon/carbon Composites. Chapman and Hall (Pub)
10. Gokogkidis A (1986) PhD Thesis, University of Karlsruhe
11. Kotlensky WC, Flock DM, Rogers DM, Davies HO, Ellis RA (1977) SAMPE (Quarterly), p 1–9 & 20–37 (July)
12. Becker PR (1981) Am Ceramic Soc Bull 60 (11), p 1210
13. Curry DM, Latchem JW, Whisenhunt GB (1983) Space shuttle orbiter leading edge structural subsystem Development AIAA 83-0483, Jan
14. Robin L (1992) Applications of carbon/carbon. In. Buckely JD, Edie DD (eds) carbon/carbon materials and composites; NASA Reference Publication 1254
15. Fitzer E (1987) Carbon 25, 163
16. Grenie Y (1987) Looking Ahead for Materials and Processes. In: Bossu JD, Briena G, Lissac P (eds) Elsevier Science Publisher, p 377
17. Carbon/carbon Composites; Company Brochure, Courtesy: Sigri Great Lakes carbon, Germany
18. Courtesy: SEPCARB, France
19. Marsh G (1989) Aeros. Compos & Mater. Vol 1, No 4, p 28
20. Taylor AH (1985) NASA Tech Brief Winter, p 156
21. Buckley JD (1988) Ceramic Bulletin. Vol 67, p 364

22. Huettner W (1990) Potential of Carbon/carbon composites as structural materials. In: Figueirado JL et al. (ed) Carbon fibers, filaments and composites, Kluwar Academic Press
23. Carbon Fiber Reinforced Composites, Product Brochure, Schunk Kohlenstoff, Germany
24. Gellhorn E von, Greiber H (1986) Proc Carbon 86, Baden Baden
25. Fitzer E, Huettener W, Manocha LM (1978) 5th London Carbon conference
26. Ando T et al. (1992) J Nuclear Mat 191–194, p 1423
27. Langley RA (1990) MRS Bulletin. July, p 42
28. Ando T et al. (1992) Tanso, No 152, p 120

Subject Index

SPRINGER NATURE

GPSR Compliance

The European Union's (EU) General Product Safety Regulation (GPSR) is a set of rules that requires consumer products to be safe and our obligations to ensure this.

If you have any concerns about our products, you can contact us on ProductSafety@springernature.com

In case Publisher is established outside the EU, the EU authorized representative is:

Springer Nature Customer Service Center GmbH
Europaplatz 3
69115 Heidelberg, Germany

Kolibri 360 GmbH
Ferdinand-Jühlke-Straße 7
99095 Erfurt, Deutschland
produktsicherheit@kolibri360.de